T0215812

HADRONS AND QUARK–GLUON PLASMA

Before matter as we know it emerged, the universe was filled with the primordial state of hadronic matter called quark–gluon plasma. This hot soup of quarks and gluons is effectively an inescapable consequence of our current knowledge about the fundamental hadronic interactions: quantum chromodynamics. This book covers the ongoing search to verify the prediction experimentally and discusses the physical properties of this novel form of matter. It provides an accessible introduction to the recent developments in this interdisciplinary field, covering the basics as well as more advanced material. It begins with an overview of the subject, followed by discussion of experimental methods and results. The second half of the book covers hadronic matter in confined and deconfined form, and strangeness as a signature of the quark–gluon phase. A firm background in quantum mechanics, special relativity, and statistical physics is assumed, as well as some familiarity with particle and nuclear physics. However, the essential introductory elements from these fields are presented as needed.

This text is suitable as an introduction for graduate students, as well as providing a valuable reference for researchers already working in this and related fields.

This title, first published in 2002, has been reissued as an Open Access publication on Cambridge Core.

JEAN LETESSIER has been CNRS researcher at the University of Paris since 1978. Prior to that he worked at the Institut de Physique Nucleaire, Orsay, where he wrote his thesis on hyperon–nucleon interaction in 1970, under the direction of Professor R. Vinh Mau, and was a teaching assistant at the University of Bordeaux. Since 1972 he has made numerous contributions to the research area of thermal production of particles and applied mathematics.

JOHANN RAFELSKI obtained his Ph.D. in 1973 and has been Professor of Physics at the University of Arizona since 1987. He teaches theoretical subatomic physics both at undergraduate and at graduate level. Prior to this, he held a Chair in Theoretical Physics at the University of Cape Town, and has been Associate Professor of Theoretical Physics at the University of Frankfurt. He has worked at the CERN laboratory in Geneva in various functions since 1977, and has been instrumental in the development of the relativistic-heavy-ion experimental program. Professor Rafelski has been a Visiting Professor at the University of Paris and other European and US institutions. He is a co-author of several hundred scientific publications, has edited and/or written more than ten books, and has contributed numerous popular scientific articles.

CAMBRIDGE MONOGRAPHS ON
PARTICLE PHYSICS
NUCLEAR PHYSICS AND COSMOLOGY
18

General Editors: T. Ericson, P. V. Landshoff

1. K. Winter (ed.): *Neutrino Physics*
2. J. F. Donoghue, E. Golowich and B. R. Holstein: *Dynamics of the Standard Model*
3. E. Leader and E. Predazzi: *An Introduction to Gauge Theories and Modern Particle Physics, Volume 1: Electroweak Interactions, the 'New Particles' and the Parton Model*
4. E. Leader and E. Predazzi: *An Introduction to Gauge Theories and Modern Particle Physics, Volume 2: CP-Violation, QCD and Hard Processes*
5. C. Grupen: *Particle Detectors*
6. H. Grosse and A. Martin: *Particle Physics and the Schrödinger Equation*
7. B. Andersson: *The Lund Model*
8. R. K. Ellis, W. J. Stirling and B. R. Webber: *QCD and Collider Physics*
9. I. I. Bigi and A. I. Sanda: *CP Violation*
10. A. V. Manohar and M. B. Wise: *Heavy Quark Physics*
11. R. K. Bock, H. Grote, R. Frühwirth and M. Regler: *Data Analysis Techniques for High-Energy Physics, Second edition*
12. D. Green: *The Physics of Particle Detectors*
13. V. N. Gribov and J. Nyiri: *Quantum Electrodynamics*
14. K. Winter (ed.): *Neutrino Physics, Second edition*
15. E. Leader: *Spin in Particle Physics*
16. J. D. Walecka: *Electron Scattering for Nuclear and Nucleon Structure*
17. S. Narison: *QCD as a Theory of Hadrons*
18. J. F. Letessier and J. Rafelski: *Hadrons and Quark–Gluon Plasma*

In memory of
Helga E. Rafelski

HADRONS
AND
QUARK–GLUON PLASMA

JEAN LETESSIER

Université Paris 7

JOHANN RAFELSKI

University of Arizona

CAMBRIDGE
UNIVERSITY PRESS

Shaftesbury Road, Cambridge CB2 8EA, United Kingdom

One Liberty Plaza, 20th Floor, New York, NY 10006, USA

477 Williamstown Road, Port Melbourne, VIC 3207, Australia

314–321, 3rd Floor, Plot 3, Splendor Forum, Jasola District Centre, New Delhi – 110025, India

103 Penang Road, #05–06/07, Visioncrest Commercial, Singapore 238467

Cambridge University Press is part of Cambridge University Press & Assessment, a department of the University of Cambridge.

We share the University's mission to contribute to society through the pursuit of education, learning and research at the highest international levels of excellence.

www.cambridge.org
Information on this title: www.cambridge.org/9781009290708

DOI: 10.1017/9781009290753

First published 2002
Reissued as OA 2022

A catalogue record for this publication is available from the British Library.

ISBN 978-1-009-29070-8 Hardback
ISBN 978-1-009-29073-9 Paperback

Cambridge University Press & Assessment has no responsibility for the persistence or accuracy of URLs for external or third-party internet websites referred to in this publication and does not guarantee that any content on such websites is, or will remain, accurate or appropriate.

Contents

Preamble   **xi**

I   A new phase of matter?   **1**
1   Micro-bang and big-bang   1
       1.1   Energy and time scales   1
       1.2   Quarks and gluons   6
       1.3   The hadronic phase transition in the early Universe   8
       1.4   Entropy-conserving (isentropic) expansion   10
       1.5   The dynamic Universe   11
       1.6   Looking for quark–gluon plasma: strangeness   14
       1.7   Other probes of quark–gluon plasma   20
2   Hadrons   24
       2.1   Baryons and mesons   24
       2.2   Strange hadrons   27
       2.3   Charm and bottom in hadrons   36
3   The vacuum as a physical medium   37
       3.1   Confining vacuum in strong interactions   37
       3.2   Ferromagnetic vacuum   40
       3.3   Chiral symmetry   43
       3.4   Phases of strongly interacting matter   46
       3.5   The expanding fireball and phase transformation   50
       3.6   QGP and confined hadronic-gas phases   52
4   Statistical properties of hadronic matter   54
       4.1   Equidistribution of energy   54
       4.2   The grand-canonical ensemble   57
       4.3   Independent quantum (quasi)particles   58
       4.4   The Fermi and Bose quantum gases   61
       4.5   Hadron gas   64
       4.6   A first look at quark–gluon plasma   68

II	**Experiments and analysis tools**	**72**
5	Nuclei in collision	72
	5.1 Heavy-ion research programs	72
	5.2 Reaction energy and collision geometry	78
	5.3 Rapidity	81
	5.4 Pseudorapidity and quasirapidity	85
	5.5 Stages of evolution of dense matter	90
	5.6 Approach to local kinetic equilibrium	95
	5.7 The approach to chemical equilibrium	97
6	Understanding collision dynamics	100
	6.1 Cascades of particles	100
	6.2 Relativistic hydrodynamics	104
	6.3 The evolution of matter and temperature	107
	6.4 Longitudinal flow of matter	108
7	Entropy and its relevance in heavy-ion collisions	112
	7.1 Entropy and the approach to chemical equilibrium	112
	7.2 Entropy in a glue-ball	116
	7.3 Measurement of entropy in heavy-ion collisions	120
	7.4 The entropy content in $200A$-GeV S–Pb interactions	122
	7.5 Supersaturated pion gas	124
	7.6 Entropy in a longitudinally scaling solution	128
III	**Particle production**	**130**
8	Particle spectra	130
	8.1 A thermal particle source: a fireball at rest	130
	8.2 A dynamic fireball	137
	8.3 Incomplete stopping	144
	8.4 Transverse-mass fireball spectra	148
	8.5 Centrality dependence of $m_\perp$-spectra	155
9	Highlights of hadron production	159
	9.1 The production of strangeness	159
	9.2 Hadron abundances	165
	9.3 Measurement of the size of a dense-matter fireball	171
	9.4 Production of transverse energy	176
	9.5 RHIC results	178
IV	**Hot hadronic matter**	**187**
10	Relativistic gas	187
	10.1 Relation of statistical and thermodynamic quantities	187
	10.2 Statistical ensembles and fireballs of hadronic matter	191
	10.3 The ideal gas revisited	193
	10.4 The relativistic phase-space integral	195
	10.5 Quark and gluon quantum gases	199

	10.6	Entropy of classical and quantum gases	204
11		Hadronic gas	207
	11.1	Pressure and energy density in a hadronic resonance gas	207
	11.2	Counting hadronic particles	211
	11.3	Distortion by the Coulomb force	215
	11.4	Strangeness in hadronic gas	217
	11.5	The grand-canonical conservation of strangeness	219
	11.6	Exact conservation of flavor quantum numbers	223
	11.7	Canonical suppression of strangeness and charm	228
12		Hagedorn gas	235
	12.1	The experimental hadronic mass spectrum	235
	12.2	The hadronic bootstrap	241
	12.3	Hadrons of finite size	247
	12.4	Bootstrap with hadrons of finite size and baryon number	251
	12.5	The phase boundary in the SBM model	254
V		**QCD, hadronic structure and high temperature**	**258**
13		Hadronic structure and quantum chromodynamics	258
	13.1	Confined quarks in a cavity	258
	13.2	Confined quark quantum states	262
	13.3	Nonabelian gauge invariance	267
	13.4	Gluons	271
	13.5	The Lagrangian of quarks and gluons	273
14		Perturbative QCD	274
	14.1	Feynman rules	274
	14.2	The running coupling constant	277
	14.3	The renormalization group	280
	14.4	Running parameters of QCD	281
15		Lattice quantum chromodynamics	287
	15.1	The numerical approach	287
	15.2	Gluon fields on the lattice	289
	15.3	Quarks on the lattice	290
	15.4	From action to results	293
	15.5	A survey of selected lattice results	298
16		Perturbative quark–gluon plasma	303
	16.1	An interacting quark–gluon gas	303
	16.2	The quark–gluon liquid	306
	16.3	Finite baryon density	309
	16.4	Properties of a quark–gluon liquid	311

VI   Strangeness   **316**
17   Thermal production of flavor in a deconfined phase   316
       17.1   The kinetic theory of chemical equilibration   316
       17.2   Evolution toward chemical equilibrium in QGP   322
       17.3   Production cross sections for strangeness and charm   326
       17.4   Thermal production of flavor   330
       17.5   Equilibration of strangeness at the RHIC and SPS   337
18   The strangeness background   340
       18.1   The suppression of strange hadrons   340
       18.2   Thermal hadronic strangeness production   343
       18.3   The evolution of strangeness in hadronic phase   349
19   Hadron-freeze-out analysis   352
       19.1   Chemical nonequilibrium in hadronization   352
       19.2   Phase space and parameters   355
       19.3   SPS hadron yields   357
       19.4   Strangeness as a signature of deconfinement   361

       References   371

       Index   389

Preamble

Most physicists and astrophysicists believe that space, time, and all the matter and radiation in the Universe were formed during the big-bang some 15 billion years ago. A key challenge is to understand how the Universe we live in today evolved from the cosmic fireball created in the big-bang. As our understanding of the laws of physics improves, we are able to look further back in time, and unravel the structure of the early Universe and its subsequent evolution.

It is widely believed that almost equal amounts of matter and antimatter were created in the big-bang, and that most of the antimatter, if not all of it, annihilated on matter after the Universe had cooled and expanded. This annihilation, which started about 20 μs after the big-bang, occurred after most of the matter we see in the Universe today was already in the form of neutrons, protons, and other hadrons made of quarks. Before the Universe hadronized, it existed in a phase of quarks and gluons in which the matter–antimatter asymmetry which makes the Universe around us today had been a small and insignificant aberration. We are attempting to recreate this phase today, and to study it in the laboratory.

This primordial state of hadronic matter called quark–gluon plasma (QGP) for all purposes an inescapable consequence of our current knowledge about the fundamental hadronic interactions, which is qualitatively rooted in the $SU(3)$-gauge theory, quantum chromodynamics (QCD). We are seeking to verify this prediction and to understand this novel form of matter. To accomplish this, we 'squeeze' the normal nuclear matter in relativistic nuclear collisions at sufficiently high energy. The individual nucleons dissolve, and we hope and expect that their constituents will form the sought-after state, the (color-charged) plasma of freely moving deconfined quarks and gluons.

Pertinent experiments are being carried out today at the European Laboratory for Particle Physics, CERN, located on the French–Swiss border 20 km north of the lake and city of Geneva, and in the USA at the Brookhaven National Laboratory, BNL, on Long Island, some 100 km east of New York City. The most violent central encounters, in which large chunks of projectile–target matter participate, are of particular interest. Therefore, beams of lead and gold ions are made to collide with each other. The available energy in the center-of-momentum (CM) frame exceeds by far the rest energy of each participating nucleon. In a press release, in February 2000, the CERN laboratory has formally announced that it views the collective evidence obtained from seven relativistic nuclear collision experiments as being conclusive proof that some new form of matter has been formed:

> *A common assessment of the collected data leads us to conclude that we now have compelling evidence that a new state of matter has indeed been created, at energy densities which had never been reached over appreciable volumes in laboratory experiments before and which exceed by more than a factor 20 that of normal nuclear matter. The new state of matter found in heavy-ion collisions at the SPS features many of the characteristics of the theoretically predicted quark–gluon plasma.*

The study of highly excited and dense hadronic matter by means of ultra-relativistic nuclear collisions has been and remains a multidisciplinary area of research, which is subject to a rapid experimental and theoretical evolution. This research field is closely related both to nuclear and to particle physics, and, accordingly, this book encompasses aspects of these two wide research areas. It employs extensively methods of statistical physics and kinetic theory. Looking back at the early days, it was primarily the theoretical work on multiparticle production by E. Fermi [121] in the USA, and L. Landau [173, 175] in the USSR, which paved the way to the development in the early sixties [137, 140] of the statistical bootstrap model description of hadron production by R. Hagedorn. This approach was refined as the understanding of hadronic structure advanced, and ultimately it has been modified to allow for the possibility that individual, confined hadron-gas particles dissolve into a liquid of quarks and gluons, which we refer to as the QGP.

The multiparticle-production work was primarily the domain of particle physicists. However, since the early seventies interest in nuclear 'heavy-ion' (not fully stripped heavy atoms) collision experiments at relativistic energies had been growing within the nuclear-physics community. The initial experimental program was launched at the Lawrence Berkeley Laboratory, LBL, at Berkeley, USA, and at the Joint Institute for Nuclear Research, JINR, in Dubna, USSR.

At the LBL, a transport line was built to carry heavy ions from the heavy-ion accelerator HILAC to the BEVATRON which was made famous by the discovery of antiprotons in the early fifties. This BEVALAC facility permitted the acceleration of nuclear projectiles to about* $1A$ GeV/c. Lighter projectiles,

---

* We follow the convention of presenting the beam energy or momentum per nucleon in the nucleus thus: $200A$ GeV implies a projectile with the total energy $200 \times A$ GeV, or momentum $200 \times A$ GeV/c, where A is the number of nucleons in the projectile. We rarely differentiate between the units of mass [GeV/c^2], of momentum [GeV/c], and of energy [GeV], in the relativistic domain of interest to us in this book. This corresponds to the commonly used convention which sets the units of time such that $c = 1$.

which could be completely ionized and had more favorable charge over mass ratios, were accelerated to above $2A$ GeV/c. At the JINR in Dubna, a similar program of research with an acceleration capability restricted to lighter ions has been developed. More recently, another heavy-ion accelerator complex, the SIS (SchwerIonenSynchrotron), of comparable energy to BEVELAC, has been erected at the Gesellschaft für Schwerionenforschung laboratory, GSI, in Darmstadt, Germany. About the time the more modern SIS started up, the BEVELAC closed down in 1993. The energy scale $\mathcal{O}(\infty)$ GeV per nucleon yields compressed nuclear matter at few times normal nuclear density, and yields final-state particle (spectral) 'temperatures' at or below 100 MeV, conditions which are generally considered inadequate for elementary quarks and gluons to begin to roam freely in the reaction volume.

The success of the initial heavy-ion experimental program, specifically the demonstration of the possibility of studying the properties of compressed and excited nuclear matter, gave birth to the research programs at the BNL and CERN. Much of this interest has been driven by the hope and expectation that, within the reach of existing elementary-particle-accelerator facilities, one may find the point of transition from the hadronic gas (HG) phase of locally confined nucleons and mesons to the new QGP phase in which color-charged quarks and gluons could propagate.

The first oxygen beam at $60A$ GeV was extracted from the Super Proton Synchrotron (SPS) accelerator at CERN and met the target in the late autumn of 1986, about the same time as the BNL started its experimental program at the Alternate Gradient Synchrotron (AGS) accelerator with a $15A$-GeV silicon-ion beam. Very soon thereafter, the energy of the SPS beam could be increased to $200A$ GeV and a sulphur-ion source was added. In order to study the relatively large volumes and longer lifetimes expected in dense matter formed in collisions of the heaviest nuclei, an upgrade of the SPS injector system was approved, which, as of 1994, allowed one to accelerate lead (^{208}Pb) ions to $158A$ GeV. At the BNL, a gold (^{197}Au)-ion beam with energy up to $11A$ GeV became available at that time. The smaller beam energy per nucleon of the heavier Pb ions compared with that for sulphur reflects their smaller ratio of particle charge to particle mass, given a fixed magnetic field strength used to bend the beam into a circular orbit in an accelerator.

Today, we are redirecting our efforts toward new experimental facilities. At the BNL, the Relativistic Heavy Ion Collider (RHIC), completed in 1999 with colliding nuclear beams at up to $100A$ GeV, will dominate the experimental landscape for the foreseeable future. It is allowing the exploration of an entirely new domain of energy, ten times greater than that of CERN-SPS. The Large Hadron Collider (LHC) project set in the 27-km CERN-LEP tunnel comprises an important heavy-ion program at energies about a factor of 30 greater than those of the RHIC. As this book goes to press, the expectation is that the experimental data from the LHC will become available in 2007.

In this book, our objective is to offer both an introduction and a perspective on the recent accomplishments and near-term aims of this rapidly developing field. The material derives from our research work, including several reviews, summer courses, and graduate lecture series that we have presented during the past 20 years. The selection of material and emphasis represents our personal experience in this rather wide interdisciplinary field of research, that today cannot, in its entirety, fit into a single volume.

We assume that the reader is familiar with quantum mechanics, special relativity, and statistical physics, and has been introduced both to nuclear and to particle physics. However, we recapitulate briefly as needed the essential introductory elements from these fields. We begin with a 70-page overview, followed by more extensive treatment of the core of our personal research experience, and mention other domains of research as appropriate.

No book is complete and this book is no exception. We will not address in depth many interesting areas of active current research. We treat the two particle intensity interferometry measurements superficially, and have not discussed the elliptical flow measurements which point to early thermalization. We do not explore the theoretical models which interpret suppression of charmonium in terms of QGP, and only key experimental results from this wide research area are shown. We do not discuss the production of photons and dileptons, since this goes beyond the scope of this book, and also in consideration of the inherent difficulties in isolating experimentally these QGP signatures. Instead, we have put a lot of effort into a detailed introductory presentation of hadron physics, as the title of this book announces.

We are hoping that our text can serve both as a reference text for those working in the field and a class text adaptable for a graduate course. One of us (J. R.) has tried out this presentation in the Spring 2001 semester at the University of Arizona. This experience further refined our presentation. Doubtless, later editions will build upon practical experience of how to handle this very diverse material in a classroom. Rather than conventional homework exercises, we leave in the text topics for further research, '*We will not discuss further in this book* ...,' which students can address in class presentations.

We have updated the contents by incorporating advances made up to October 2001, including a selection of run 2000 RHIC results. Most of the material we present has not yet been covered in any other monograph. Complementary books and reports that we found useful are the following.

1. *The QCD Vacuum, Hadrons and the Superdense Matter.* World Scientific, Singapore (1988). E. V. Shuryak presents an early view of the structure of the QCD vacuum.

2. *Finite-Temperature Field Theory.* Cambridge University Press (1989). J. Kapusta offers a lucid introduction to the theoretical aspects of QCD and hot quark–gluon matter, an area that has since developed rapidly.

3. *Thermal Field Theory.* Cambridge University Press (1996). M. Lebellac presents a modern introduction which complements and updates the text of J. Kapusta cited above.

4. *Vacuum Structure and QCD Sum Rules.* North Holland, Amsterdam (1992). M. A. Shifman develops a more comprehensive view of the vacuum structure and presents applications of the sum-rule method.

5. *Particle Production in Highly Excited Matter.* Plenum Press, New York (1993), editors H. H. Gutbrod and J. Rafelski. This volume comprises several comprehensive introductory and survey articles pertinent to interpretation of data.

6. *Introduction to Relativistic Heavy Ion Collisions.* J. Wiley and Sons, New York (1994). In this text, L. P. Csernai emphasizes the transport phenomena in the process of collision and presents applications of matter flow models, including an analysis of the LBL, GSI and Dubna energy ranges, these subjects are not covered in depth in this book.

7. *Introduction to High Energy Heavy-Ion Collisions.* World Scientific, Singapore (1994). C.-Y. Wong emphasizes the role of the parton structure in the collision, considers model dynamics of color strings and the associated pair production mechanisms, and addresses among physical observables more comprehensively the electromagnetic probes of dense hadronic matter.

8. *Bose–Einstein Correlations and Subatomic Interferometry.* John Wiley, Chichester (2000). R. Weiner presents a detailed and technical discussion of the HBT particle-correlation method used to study the space–time geometry in heavy-ion collisions and related topics. Our book offers a short introduction to this monograph.

9. *Quark–Gluon Plasma.* World Scientific, Singapore, Volumes I and II (1990 and 1995) editor R. Hwa. Useful collections of articles on a variety of topics, contributed by hand-picked authors.

10. *Hot Hadronic Matter: Theory and Experiment.* Plenum Press, New York (1995), editors J. Letessier, H. H. Gutbrod, and J. Rafelski. This volume, dedicated to Rolf Hagedorn, comprises in particular a comprehensive survey of the bootstrap model of confined hadronic matter.

11. Proceedings of *Quark Matter* meetings held about every 18 months have in recent years been published in *Nuclear Physics* A. These proceedings present regular comprehensive updates of the experimental results, speckled with a mostly random assortment of theoretical contributions.

12. Proceedings of *Strangeness in Hadronic Matter* have in recent years been published in *Journal of Physics* G. These volumes comprise a comprehensive survey of the strongly interacting heavy flavor probes of phases of hadronic matter.

13. A very useful reference is the bi-annual reissue of the *Review of Particle Physics*, published as separate issues of *Physical Review* D, alternating with

the *European Physical Journal* and accessible online.

A closely related area of research is the study of the properties of quantum chromodynamics by numerical methods within the lattice-gauge-theory approach. We can barely touch this huge research field in this book. Some standard texts are the following.

14. *Quarks, Strings, and Lattices.* Cambridge University Press (1983), by M. Creutz.
15. *Quantum Fields on the Computer.* World Scientific, Singapore (1992), by M. Creutz.
16. *Quantum Fields on a Lattice.* Cambridge University Press (1994), by I. Montvay and G. Münster.
17. Proceedings of *Lattice* meetings, published in *Nuclear Physics*, are the best places to find the most recent results.

The publisher has used its best endeavors to ensure that the URLs for external websites referred to in this book are correct and active at the time of going to press. However, the publisher has no responsibility for the websites and can not guarantee that a site will remain live or that the content is or will remain appropriate.

We would like to thank our friends and colleagues who over the years helped us reach a better understanding of the material addressed in this book: we thank in particular Drs Mike Danos (Chicago and Washington, deceased), Hans Gutbrod (GSI), Rolf Hagedorn (CERN), Berndt Müller (Duke University), and Emanuele Quercigh (CERN and Padua).

This volume is dedicated to Helga Rafelski. Helga has been a companion from day one in the field of relativistic heavy-ion collisions; her presence at the finale will be sorely missed.

Jean Letessier and Johann Rafelski, Paris and Tucson, November 2001.

I

A new phase of matter?

1 Micro-bang and big-bang

1.1 Energy and time scales

When atomic nuclei, generally called heavy-ions, collide at very high energies, such that the kinetic energy exceeds significantly the rest energy, dense hadronic* matter is produced. We refer to these reactions as (ultra)relativistic heavy-ion, or nuclear, collisions. The energy density of hadronic matter with which we are concerned has a benchmark value of

$$\epsilon = 1 \text{ GeV fm}^{-3} = 1.8 \times 10^{15} \text{ g cm}^{-3}. \tag{1.1}$$

The corresponding relativistic matter pressure is

$$P \simeq \tfrac{1}{3}\epsilon = 0.52 \times 10^{30} \text{ bar}. \tag{1.2}$$

Dense matter with these properties must have existed in the early Universe about 10 µs after the big-bang. It might have been recreated extremely rarely in interactions of very-high-energy cosmic-ray particles. Some astrophysical objects may reach these extreme conditions. It had been speculated that a catastrophic change in the Universe could ensue when these conditions are recreated in laboratory experiments, but these fears have been refuted [85].

Experimental study of the physics of the early Universe requires in principle a large, practically infinite, volume of matter. For this reason, it is necessary to study high-energy collisions of the heaviest nuclei, rather than the more elementary and simpler-to-handle interactions of protons or leptons. However, we cannot study in the laboratory physical systems

---

* In Greek, *barys* means strong and heavy; *leptos* is weak, light; *mesos* is intermediate, and *hadros* is strong. Hadronic (strong) interactions involve baryons and mesons (heavy and semi-heavy particles) but not leptons, the light and relatively weakly interacting electrons, muons, the heavy tau, and nearly massless neutrinos.

larger in volume than lead (Pb) or gold (Au). Hence, it would seem that we will not be able to explore experimentally the properties of the phase transition involving the dissolution of hadronic particles, since it is known that genuine phase transitions cannot develop in finite physical systems. However, only for *non-relativistic finite systems* it is impossible to observe experimentally the discontinuous phase properties. In our case, the ability to produce particles from energy and the presence of virtual fluctuation effects greatly enhance the number of physical states accessible. We therefore hope to identify in collisions of relativistic heavy-ions a (nearly) singular manifestation of a phase transition from the nuclear, hadronic phase to a matter phase consisting of quarks and gluons.

We use units in which the Boltzmann constant $k = 1$. In consequence, the temperature T is discussed in units of energy, which, in this book, are either MeV $\simeq 2m_e c^2$ (m_e is the electron mass) or GeV$= 1000$ MeV $\simeq m_N c^2$ (m_N is the mass of a nucleon). The conversion scale of typical temperature involves ten additional zeros:

$$100 \text{ MeV} \equiv 116 \times 10^{10} \text{ K}. \tag{1.3}$$

To appreciate the magnitude of this temperature, let us recall that the center of the Sun is believed to be at $T = 11 \times 10^6$ K, and the scale of temperature of interest to us is in fact $100\,000$ times greater.

In general, the units in this book are chosen such that the numerical values $\hbar = c = 1$, e.g., the mass of particles will also be measured in units of energy and the energy density can appear as the fourth power of an energy unit. With the conversion factor $\hbar c = 0.197$ GeV fm, the reference energy density in normal nuclei is

$$\frac{m_N}{V_N} = 0.17 m_N \text{ fm}^{-3} \simeq 0.16 \text{ GeV fm}^{-3} = 1.27 \times 10^{-4} \text{ GeV}^4. \tag{1.4}$$

Experimental results have shown that ultra-relativistic heavy-ion collisions lead to the formation of a dense hadronic fireball, well localized in space, with an energy density exceeding 1 GeV fm^{-3}. Such a spatially localized drop of highly excited, hot, and dense *elementary matter* will be rapidly evolving, indeed exploding, driven by the high internal pressure. The fireball has a short life span characterized by the size of the system $\tau \simeq 2R/c$.

In relativistic heavy-ion reactions, the collision energy is shared among numerous newly produced hadronic particles. Therefore, in the final state we observe many soft (low-energy) newly produced hadronic particles, rather than a few particles of high-energy as is the case in hard, elementary interactions. An important objective of our research is the understanding of the processes that lead to the conversion of kinetic collision energy into high particle multiplicity. Because of the large numbers of

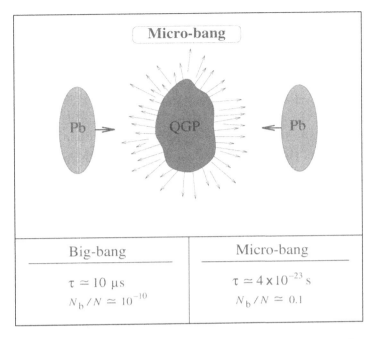

Fig. 1.1. Top: Lorentz-contracted nuclei collide in the center-of-momentum frame and form a region of dense matter, which evolves into a final state of hadrons. Bottom: two key differences involving baryon number N_b and total particle number N between the micro-bang and the cosmological big-bang.

particles produced, many thousands in recent experiments, we believe that this can be accomplished using statistical mechanics. This method has the advantage that it does not require a complete description of the microscopic production and dynamics of particles. It will be introduced in great detail in this book.

A qualitative image of the high-energy nuclear-collision 'micro-bang' is depicted in Fig. 1.1: two nuclei are shown, Lorentz-contracted in the direction of motion, approaching from two sides and colliding in the center-of-momentum (CM) laboratory frame, forming a region of dense matter (dark-shaded), the fireball. Subsequently, the collective expansion flow of fireball matter develops, and evolves in the final state into free streams of individual particles, indicated by individual arrows.

The temporal evolution of a fireball into a final state comprising a multitude of different hadronic particles is similar to, though much faster than, the corresponding stage in the evolution of the early Universe. Relativistic heavy-ion collision leads to a rapidly evolving fireball of quark–gluon plasma (QGP), in which the short time scale involved is probed by the equilibration of abundance of quark flavors. We can not hope to be able to recreate the 'slow big-bang' of the Universe in the laboratory in the last detail. Our objective is to obtain precise information about the physical processes and parameters which govern the rapidly changing hadronic phase. Within a theoretical framework, we can hope to unravel what happened when the Universe hadronized.

The bottom portion of Fig. 1.1 reminds us of the two important differences between the two 'bangs', the big-bang of the Universe and the micro-bangs generated in the nuclear-collision experiments.

1. The time scale of the expansion of the Universe is determined by the interplay of the gravitational forces and the radiative and Fermi pressure of the hot matter, whereas in the micro-bangs there is no gravitation to slow the expansion, which lasts at most about 10^{-22} s. The time scale of the heavy-ion collision, indicated in Fig. 1.1, suggests that the size and the (local) properties of the exploding nuclear fireball must change rapidly even on the scale of hadronic interactions, contrary to the situation in the early Universe. It is convenient to represent the expansion time constant τ_U of the Universe in terms of the Newtonian gravitational constant G and the vacuum energy $\mathcal{B}$:

$$\tau_\mathrm{U} = \sqrt{\frac{3c^2}{32\pi G \mathcal{B}}} = 36\sqrt{\frac{\mathcal{B}_0}{\mathcal{B}}}\ \mu\mathrm{s}, \quad \mathcal{B}_0 = 0.19\,\mathrm{GeV\ fm^{-3}} = (195\,\mathrm{MeV})^4.$$

(1.5)

The range of values of the 'bag' constant $\mathcal{B}$ found in the literature, $145\,\mathrm{MeV} < \mathcal{B}^{1/4} < 235\,\mathrm{MeV}$, leads to $66\,\mu\mathrm{s} > \tau_\mathrm{U} > 25\,\mu\mathrm{s}$.

2. The early radiative Universe was practically baryonless, whereas in the laboratory we create a fireball of dense matter with a considerable baryon number N_b per total final particle multiplicity N. Thus, unlike in the early Universe, we expect in a laboratory micro-bang a significant matter–antimatter asymmetry in particle abundance. The matter–antimatter symmetry of particle spectra is in turn an important indicator suggesting that the matter–antimatter symmetry has been restored in other aspects.

The matter–antimatter-abundance asymmetry is easily overcome theoretically, since it implies a relatively minor extrapolation of the baryo-chemical potential μ_b introduced to fix the baryon density. In fact, RHIC experiments at CM-energy $130A$ GeV per pair of nucleons ($\sqrt{s_\mathrm{NN}} = 130$ GeV) are already much more baryon–antibaryon symmetric than the SPS condition where $\sqrt{s_\mathrm{NN}} \leq 17.3$ GeV, and the highest RHIC and LHC energies will allow us to extrapolate our understanding from $\mu_\mathrm{b}/T \leq 1$ to $\mu_\mathrm{b}/T \ll 1$.

More difficult to resolve will be the differences in the physics due to the different time scales involved. The evolution of the Universe is slow on

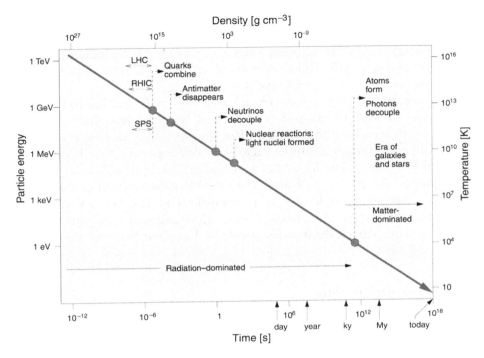

Fig. 1.2. Particle energy (temperature) as a function of time in the early Universe. Different evolutionary epochs are shown along with the accessible range of accelerator laboratory experiments.

the hadronic time scale. Given the value of τ_U, we expect that practically all unstable hadronic particles decay, all hadronic equilibria are fully attained, and there is potentially time to develop macroscopic structures in the 'mixed phase' of QGP and hadronic gas (HG), and for weak interactions to take place. All this can not occur during the life span of the dense matter created in nuclear collisions.

The temporal evolution of the Universe is depicted, in Fig. 1.2, as a function of time. Beginning with decoupling of neutrinos and nucleosynthesis at time $\mathcal{O}(1)$ s the evolution of the Universe is well understood today. In comparison, little work has gone into the detailed understanding of the earlier period when the nearly symmetric matter–antimatter hadron gas emerged from the quark–gluon phase and evolved into the baryon Universe in which we find ourselves today. This period spans the temperature interval 300 MeV $< T < 1$ MeV, separating the perturbative QGP epoch from the epoch of decoupling of neutrinos and cosmological nucleosynthesis.

We see, in Fig. 1.2, that, after about $10\,\mu$s, the deconfined phase of quarks and gluons is transformed into a hot gas of hadrons, namely mesons, baryons, and antibaryons. Just after that, the evolution of our Universe was marked by a period of baryon–antibaryon annihilation, and, possibly, separation: although we have not been able to observe antimatter in our galaxy, or in the neighborhood of our galaxy, it is far from certain that there is no antimatter in our Universe.

The laboratory study of the formation of the QGP and hadronization is expected to lead to an understanding of how the hot, baryon- and antibaryon-rich hadron gas evolved after its formation at $T \simeq 170$ MeV. Employing the statistical-physics methods developed in this book, one finds that the energy fraction of baryons and antibaryons within hadronic-gas matter is about 25% just after the QGP has hadronized in the early Universe, and nearly half of this is comprised of the heavier and un-stable strange baryons and antibaryons. It is believed that this strong antimatter component disappears from the Universe prior to the era of nucleosynthesis.

1.2 Quarks and gluons

Both quarks and gluons manifest themselves only for a short instant fol-lowing a high-energy interaction, and have never been observed as free objects at macroscopic distances from the space–time volume of the reac-tion; they are 'confined'. Gluons interact only through strong interactions and pose a great experimental challenge regarding the study of their prop-erties. The measurement of the properties of confined quarks is relatively easy, since, in addition to the strong-interaction (color) charge, they also carry the electro-weak charges.

There are six different *flavors* of quarks, see table 1.1, two practically stable flavors referred to as *up* – for the proton-like quark u, and *down* – for the neutron-like quark d. We often refer to these two light quarks by their generic name q. Light quarks q may be viewed as a single entity with two states, up or down. The semi-heavy *strange*-flavor s-quark decays due to electro-weak interaction when it is bound in hadrons, typically within 10^{-10} s, whereas the heavier *charm* c and *bottom* b flavors have approximate life spans of 3×10^{-13} s and 10^{-12} s, respectively.

These six flavors of quarks form three doublets:

$$\begin{pmatrix} u \\ d \end{pmatrix}, \quad \begin{pmatrix} c \\ s \end{pmatrix}, \quad \begin{pmatrix} t \\ b \end{pmatrix}; \quad Q_q = \begin{pmatrix} +\frac{2}{3} \\ -\frac{1}{3} \end{pmatrix}.$$

The upper component of a doublet has charge $Q_q = +\frac{2}{3}$, in units of the proton charge, whereas the lower component has one unit of charge less, as is also the case for the related lepton doublet comprising the three charged leptons (electron, muon, and tau) accompanied by their neutrinos. There is an antiquark for each quark, carrying the opposite electrical charge.

Table 1.1. Properties of quarks: flavor f symbol, flavor name, electrical charge Q_f (in units of proton charge Q_p), and mass m_f at energy scale 2 GeV; see the text for further discussion of strange-quark mass

f	Quark	$Q_f\ [Q_p]$	$m_f(2\text{ GeV})$
u	Up	$+\frac{2}{3}$	3.5 ± 2 MeV
d	Down	$-\frac{1}{3}$	6 ± 3 MeV
s	Strange	$-\frac{1}{3}$	115 ± 55 MeV
c	Charm	$+\frac{2}{3}$	1.25 ± 0.15 GeV
b	Bottom	$-\frac{1}{3}$	4.25 ± 0.15 GeV
t	Top	$+\frac{2}{3}$	174.3 ± 5.1 GeV

Quarks differ from charged leptons (electrons e, muons μ, and taus τ), and neutrinos (ν_i, i = e, μ, and τ) by a further internal quantum number they must carry, in addition to spin. The presence of this additional quantum number arises even in the simplest quark models. For example, consider hadronic particles containing three quarks of the same flavor, such as the spin-3/2 baryons:

$$\Delta^{++} = (\text{uuu}), \qquad \Delta^{-} = (\text{ddd}), \qquad \Omega^{-} = (\text{sss}).$$

The physical properties of these baryons imply that three identical quarks are present in the same S-wave with the same spin wave function. Since quarks are fermions, they are subject to the Pauli principle. Thus, there must be an additional way to distinguish the quarks, aside from spin. This additional degeneracy factor has been determined to be $g_c = 3$. It became known as the color of quarks – in analogy to the three fundamental colors: red, green, and blue.

Color is an internal quantum number, which like the electrical charge, is thought to be the source of a force [123]. It seems that there is no way to build an apparatus to distinguish the three fundamental color charges, all colors must everywhere be exactly equal physically. The theory of color forces must satisfy the principle of local nonabelian gauge invariance, e.g., invariance under arbitrary local $SU(3)$ transformations in the three-dimensional color space. In other aspects, there is considerable formal similarity with quantum electrodynamics (QED). Therefore, the theory of strong interactions based on such color forces has been called quantum chromodynamics (QCD).

The flavor structure and symmetry of quarks and leptons remains a mystery today. We also do not have a fundamental understanding of the origin of quark masses. In table 1.1 we see that quarks of various flavors differ widely in their 'current' mass m_f, that is mass which enters the elementary QCD Lagrangian $\mathcal{L}_{QCD}$. The values presented in table 1.1 are for the momentum scale 2 GeV.

Since quarks are confined inside hadrons, and the zero-point energy of confinement is much larger than the masses of light quarks, their masses could not be determined by direct measurement. However, the precise masses of light u and d quarks do not matter in the study of hadronic interactions, being generally much smaller than the pertinent energy scales,

The mass of the strange quark m_s is barely heavy enough to be determined directly in a study of hadronic structure. We adopt, in this book, the value $m_s(1\,\text{GeV}) = 200 \pm 20$ MeV [150]. In the value of m_s reference is made to the scale of energy at which the mass of the strange quark is measured: akin to the interaction strength, also the mass of quarks depends on the (energy) scale. This value of m_s corresponds to $m_s(2\,\text{GeV}) \simeq m_s(1\,\text{GeV})/1.30 = 154 \pm 15$ MeV. A somewhat smaller value $m_s(2\,\text{GeV}) = 115 \pm 55$ MeV, see table 1.1, corresponding to $m_s(1\,\text{GeV}) \simeq 150 \pm 70$ MeV, is the recommended value. The rather rapid change by 30% of the quark mass between the 1- and 2-GeV scales is well known, but often not remembered, e.g., the 'low' recommended mass of the charmed quark presented in table 1.1 in fact corresponds to $m_c(1\,\text{GeV}) = 1.6$ GeV, a rather 'high' value.

1.3 The hadronic phase transition in the early Universe

We will now show that the 'freezing' of quark–gluon 'color' deconfined degrees of freedom is the essential ingredient in determining the conditions in a transition between phases that has time to develop into equilibrium. The following discussion tacitly assumes the presence of latent heat $\mathcal{B}$ in the transition, and a discontinuity in the number of degrees of freedom, $g_2 \neq g_1$, where '1' refers to the primeval QGP phase and '2' to the final hadronic-gas state.

To find the phase-transition point, we determine the (critical) temperature at which the pressures in the two phases are equal. We allow, in a transition of first order, for a difference in energy density $\epsilon_1 \neq \epsilon_2$ associated with the appearance of latent heat $\mathcal{B}$ (the 'bag constant'), which also enters the pressure of the deconfined phase. We consider the Stefan–Boltzmann pressure of a massless photon-like gas with degeneracy g_i:

$$P_c \equiv P_1(T_c) = \frac{\pi^2}{90} g_1 T_c^4 - \mathcal{B}, \tag{1.6}$$

$$P_c \equiv P_2(T_c) = \frac{\pi^2}{90} g_2 T_c^4. \tag{1.7}$$

We obtain

$$\frac{\mathcal{B}}{T_{\mathrm{c}}^4} = \frac{\pi^2}{90} \Delta g, \qquad T_{\mathrm{c}} = \mathcal{B}^{\frac{1}{4}} \left(\frac{90}{\pi^2 \Delta g} \right)^{\frac{1}{4}}, \qquad \Delta g = g_1 - g_2. \qquad (1.8)$$

The transition temperature, in the early Universe, is slightly higher than the value seen in laboratory experiments, even though Eq. (1.8) involves only the difference in the number of degrees of freedom. For the pressure at the transition we obtain

$$P_{\mathrm{c}} = \mathcal{B} \frac{g_2}{\Delta g}. \qquad (1.9)$$

The pressure, and therefore the dynamics of the transition in the early Universe, depends on the presence of non-hadronic degrees of freedom, which are absent from laboratory experiments with heavy ions.

In summary, the phase-transition dynamics in the early Universe is determined by

(a) the effective number of confined degrees of freedom, g_2, at T_{c};
(b) the change in the number of acting degrees of freedom Δg, which occurs exclusively in the strong-interaction sector; and
(c) the vacuum pressure (latent heat) $\mathcal{B}$, a property of strong interactions.

In order to understand the early Universe, we need to measure these quantities in laboratory experiments.

Both phases involved in the hadronization transition contain effectively massless electro-weak (EW) particles. Even though the critical temperature does not depend on the background of EW particles not participating in the transition, the value of the critical pressure, Eq. (1.9), depends on this, and thus we will briefly digress to consider the active electro-weak degrees of freedom. These involve photons, γ, and all light fermions, viz., e, $\mu, \nu_{\mathrm{e}}, \nu_{\mu}$, and ν_{τ} (we exclude the heavy τ-lepton with $m_{\tau} \gg T$, and we consider the muon as being effectively a massless particle). Near to $T \simeq 200$ MeV, we obtain

$$g^{\mathrm{EW}} = g_{\gamma} + \tfrac{7}{4} g_{\mathrm{F}}^{\mathrm{EW}} = 14.25, \qquad (1.10)$$

with

$$g_{\gamma} = 2, \qquad \tfrac{7}{4} g_{\mathrm{F}}^{\mathrm{EW}} = \tfrac{7}{8} \times 2 \times (2_{\mathrm{e}} + 2_{\mu} + 3_{\nu}) = 12.25,$$

where charged, effectively massless fermions enter with spin multiplicity 2, and we have three neutrino flavors – there are only left-handed light neutrinos and right-handed antineutrinos, and thus only half as many neutrino degrees of freedom as would naively be expected.

In the deconfined QGP phase of the early Universe, we have

$$g_1 = g^{\mathrm{EW}} + g_{\mathrm{g}} + \tfrac{7}{4} g_{\mathrm{q}}. \qquad (1.11)$$

The number of effectively present strongly interacting degrees of freedom of quarks and gluons is influenced by their interactions, characterized by the strong coupling constant α_s, and this book will address this topic in depth,

$$g_g = 2_s \times 8_c \left(1 - \frac{15}{4\pi}\alpha_s\right), \quad \frac{7}{4}g_q = \frac{7}{4}2_s \times 2.5_f \times 3_c \left(1 - \frac{50}{21\pi}\alpha_s\right), \quad (1.12)$$

where the flavor degeneracy factor used is 2.5, allowing in a qualitative manner for the contribution of more massive strangeness; table 1.1. The degeneracies of quarks and gluons are indicated by the subscripts s(pin) and, c(olor), respectively. We obtain

$$g_1 = \begin{cases} 56.5, & \text{for } \alpha_s = 0, \\ \sim 37, & \text{for } \alpha_s = 0.5, \\ \sim 33, & \text{for } \alpha_s = 0.6. \end{cases} \quad (1.13)$$

For the QCD perturbative interactions with $\alpha_s = 0.5\text{--}0.6$, we see that $g_1 \simeq 35 \pm 2$.

We now consider the final HG phase of the early Universe: there is no light, strongly interacting fermion. Aside from three light bosons (pions $\pi^\pm$ and π^0), the presence of heavier hadrons contributes at $T \lesssim 170$ MeV, and one finds for the hadronic degrees of freedom $g_2^h \simeq 5$

$$g_2 \equiv g^{\text{EW}} + g_2^h \simeq 19. \quad (1.14)$$

Thus, we find from Eqs. (1.13) and (1.14),

$$g_1 - g_2 = \Delta g = \begin{cases} \sim 37, & \text{for } \alpha_s = 0, \\ \sim 18, & \text{for } \alpha_s = 0.5, \\ \sim 14, & \text{for } \alpha_s = 0.6. \end{cases} \quad (1.15)$$

For the QCD perturbative interactions with $\alpha_s = 0.5\text{--}0.6$, we see that about half of the degrees of freedom freeze across the transition in the early Universe.

For the value $\mathcal{B}^{1/4} = 190$ MeV and $\alpha_s \simeq 0.5$, we obtain from Eq. (1.8) a transition temperature $T_c \simeq 160$ MeV. At this temperature, the critical pressure Eq. (1.9) is found to be $P_c \simeq 1.4\,\mathcal{B}$, and it includes both hadronic and electro-weak partial pressure contributions. The hadronic fractional pressure present in laboratory experiments and seen in lattice simulations of gauge theories (compare with section 15.5) is $P_c^h \simeq \mathcal{B}/4$.

1.4 Entropy-conserving (isentropic) expansion

Much of the time dependence of an expanding Universe is related to the assumption of adiabatic, i.e., entropy-conserving, expansion dynamics:

$$dE + P\,dV = T\,dS = 0, \qquad dE = d(\epsilon V), \qquad \frac{dV}{V} = \frac{3\,dR}{R}. \quad (1.16)$$

Here, as usual, ϵ is the energy density in the local restframe, and the three-dimensional volume element dV scales with the third power of the distance scale R. We obtain

$$\frac{3\,dR}{R} = -\frac{d\epsilon}{\epsilon + P}. \tag{1.17}$$

We will revisit Eq. (1.17) which describes general expansion dynamics of the micro-bang.

We now relate the expansion dynamics to the velocity of sound, and use well known relations of thermodynamics which we will discuss in this book,

$$d\epsilon = \frac{d\epsilon}{dP}\frac{dP}{dT}\,dT = \frac{1}{v_{\rm s}^2}\sigma\,dT = \frac{1}{v_{\rm s}^2}\frac{\epsilon + P}{T}\,dT. \tag{1.18}$$

We will revisit the derivation Eq. (1.18) when we study the same physics occurring in the expansion of the dense-matter phase formed in heavy-ion collisions in section 6.3. Using Eq. (1.17), we obtain

$$\frac{3\,dR}{R} = -\frac{1}{v_{\rm s}^2}\frac{dT}{T}. \tag{1.19}$$

This equation allows the integral

$$RT^{1/(3v_{\rm s}^2)} = \text{constant}, \tag{1.20}$$

which describes exactly how the temperature decreases in an isentropic expansion once the equation of state $P = P(\epsilon)$, and hence the velocity of sound is known.

For a relativistic equation of state, $v_{\rm s}^2 = \frac{1}{3}$ and thus

$$R(t)T(t) = \text{constant}, \qquad V(t)T^3(t) = \text{constant}. \tag{1.21}$$

While this result applies to a three-dimensional expansion, it is easily generalized to a one-dimensional expansion, such as is expected to apply in ultra-high-energy heavy-ion collisions.

1.5 The dynamic Universe

The $(0,0)$-component of the Einstein equation,

$$\mathcal{R}_{\mu\nu} - \tfrac{1}{2}g_{\mu\nu}\mathcal{R} + \Lambda_{\rm v}g_{\mu\nu} = 8\pi GT_{\mu\nu}, \tag{1.22}$$

gives the Friedmann equation which determines the rate of expansion of the homogeneous Universe,

$$H^2(t) \equiv \left(\frac{\dot{R}}{R}\right)^2 = \frac{8\pi G}{3}\epsilon + \frac{\Lambda_{\rm v}}{3} - \frac{k}{R^2}. \tag{1.23}$$

In the last term, $k = 0, 1$, and -1 for different geometries of the Universe (flat, bubble, and hyperbolic-open); this term is negligible in our considerations. Λ_v is Einstein's cosmological term, which is playing a similar dynamic role to $\mathcal{B}$, but, in comparison, it is of irrelevant magnitude during the early time period we consider. H is the Hubble 'constant' which varies with time. Its present-day value, $H(t_0) = H_0$, is of considerable interest and is given in the range

$$H_0 = 70 \pm 15 \,\text{km s}^{-1}\,\text{Mpc}^{-1} = \frac{0.7 \pm 0.15}{10^{10}\,\text{y}} = (2.2 \pm 0.5) \times 10^{-17}\,\text{s}^{-1}.$$

Inserting Eq. (1.17), with $V \propto R^3$, into Eq. (1.23) and neglecting the last two terms in Eq. (1.23), we find for $\epsilon(t)$

$$\dot{\epsilon}^2 = 24\pi G \epsilon (\epsilon + P(\epsilon))^2. \tag{1.24}$$

We equate the particle energy density and pressure, including the vacuum term $\mathcal{B}$ in the relativistic equation of state for the particle component,

$$\epsilon - \mathcal{B} \simeq \frac{\pi^2}{30} g T^4 \simeq 3(P + \mathcal{B}), \qquad \epsilon = 3P + 4\mathcal{B}. \tag{1.25}$$

Thus,

$$\dot{\epsilon}^2 = \frac{128\pi G}{3} \epsilon (\epsilon - \mathcal{B})^2, \tag{1.26}$$

which is valid (approximately) both for QGP and for HG phases, but in the HG phase $\mathcal{B} = 0$.

Despite its highly nonlinear nature, Eq. (1.26) has an analytical solution,

$$\epsilon_1 = \mathcal{B} \coth^2(t/\tau_U), \tag{1.27}$$

where τ_U is the expansion time constant we have defined in Eq. (1.5) and the subscript '1' reminds us that Eq. (1.5) describes the evolution in the quark–gluon phase with $\mathcal{B} \neq 0$. At $t < \tau_U$, the energy density rises like $1/t^2$; for $t > \tau_U$, it would remain constant at the vacuum energy density.

Once the transition to HG phase '2' with $\mathcal{B} \to \Lambda_v \sim 0$ has occurred, the analytical form of the solution changes; we exploit the singularity $\coth x \to 1/x$ and obtain the power-law solution

$$\epsilon_2 = \mathcal{B}\left(\frac{\tau_U}{t}\right)^2 = \frac{3}{32\pi G}\frac{1}{t^2} = \frac{\pi^2}{30} g_2 T^4. \tag{1.28}$$

In the middle of Eq. (1.28), we have substituted τ_U from Eq. (1.5) to show that, in principle, $\mathcal{B}$ does not enter the solution. On the right-hand side of Eq. (1.28), we show the energy density of the (radiation-dominated; see Fig. 1.2) Universe, establishing the well-known relation

$$\boxed{\left(\frac{t_2}{t}\right)^2 = \left(\frac{T}{T_{\mathrm{c}}}\right)^4, \quad T \propto \frac{1}{\sqrt{t}}.}$$ (1.29)

Using Eq. (1.8) for T_{c}, we determine t_2, the time when the Universe entered the hadronic phase, which we use below in Eq. (1.33). Similarly we also obtain the behavior of the size of the Universe with time. By inserting Eqs. (1.27) and (1.28) into Eq. (1.23), we find how the scale R of the Universe evolves:

$$R^2 \propto \sinh\left(\frac{t}{\tau_{\mathrm{U}}}\right)\Bigg|_1 \to \left(\frac{t}{\tau_{\mathrm{U}}}\right)\Bigg|_2.$$ (1.30)

To determine when the transition between the two phases occurs, and how long it takes, we consider the pressure in both phases. In deconfined phase '1' where $\mathcal{B} \neq 0$, we have, using Eqs. (1.25) and (1.27),

$$3P_1 = \mathcal{B}\left[\coth^2(t/\tau_{\mathrm{U}}) - 4\right], \quad t \leq t_1,$$ (1.31)

while in the post-transition phase

$$3P_2 = \mathcal{B}\left(\frac{\tau_{\mathrm{U}}}{t}\right)^2, \quad t \geq t_2.$$ (1.32)

Equating these two pressures with the critical pressure, Eq. (1.9), we obtain

$$\frac{3P_{\mathrm{c}}}{\mathcal{B}} = \frac{3g_2}{\Delta g} = \coth^2\left(\frac{t_1}{\tau_{\mathrm{U}}}\right) - 4 = \left(\frac{\tau_{\mathrm{U}}}{t_2}\right)^2.$$ (1.33)

Equation (1.33) relates the time t_1, when the transition begins, to t_2, when it ends, and the fraction of degrees of freedom which are 'freezing' in the transition, in units of the τ_{U}, Eq. (1.5).

We show the pressure and temperature in the Universe near hadronization in Fig. 1.3. The solid line corresponds to $\alpha_{\mathrm{s}} = 0.6$, and $\mathcal{B}^{1/4} = 195$ MeV, for which value $\tau_{\mathrm{U}} = 36$ µs. Dotted lines are for $\mathcal{B}^{1/4} = 170$ MeV and $\mathcal{B}^{1/4} = 220$ MeV; a higher value of $\mathcal{B}^{1/4}$ leads to a shorter time scale, Eq. (1.5).

It is straightforward to obtain the values of t_1 and t_2. Using Eq. (1.14), $g_2 = 19$, and Eq. (1.15), $\Delta g = 14$, we find that the transition is complete at $t_2 = 0.5\tau_{\mathrm{U}}$. The onset of the transition is found at $t_1 = 0.37\tau_{\mathrm{U}}$, and the transition lasts $0.13\tau_{\mathrm{U}}$ in this case – the major uncertainty is related to the value of $g_1 - g_2 = \Delta g$. For the central value of $\mathcal{B}^{1/4} = 195$ MeV with $\tau_{\mathrm{U}} = 36$ µs, we find that the transition lasts $\Delta t = (t_2 - t_1) = 4.7$ µs.

The duration of the hadronization transition is comparable (35%) to the prior life span of the Universe in the deconfined phase. This time is exceedingly long compared with the time scale of hadronic interactions

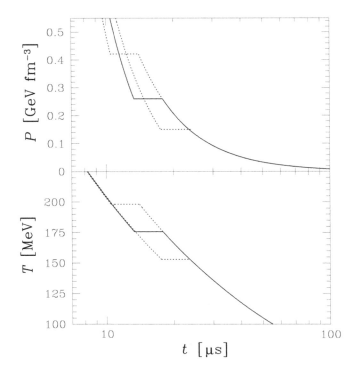

Fig. 1.3.   Pressure (upper) and temperature (lower part) in the Universe, as function of time, in the vicinity of the phase transition from the deconfined phase to the confined phase. Solid lines, $\mathcal{B}^{1/4} = 195$ MeV; dotted lines, $\mathcal{B}^{1/4} = 170$ MeV (lower part) and $\mathcal{B}^{1/4} = 220$ MeV (upper part) all for $\alpha_s = 0.6$.

$(10^{-22}\,\mathrm{s})$. It allows the decay of all unstable hadronic particles and, potentially, the development of domain structures. Moreover, the hadronization time is also three orders of magnitude longer than the characteristic time of hadronic weak decays, and is even longer than the decay time for a muon. What exactly happens to matter in this last phase transition in our Universe is not yet known. Studying dense matter in relativistic heavy-ion collisions should help us establish the physical laws governing this crucial epoch in the development of the Universe.

1.6 Looking for quark–gluon plasma: strangeness

How do we look for the phase transition of the primordial state of the Universe recreated for a short glimpse of time in the laboratory? How can we distinguish between the reactions involving confined hadronic particles only, and those in which we encounter the color-deconfined quarks and gluons?

- Is a transient new phase of matter, existing for a brief instant in time, perhaps for no more than 10^{-22} s, in principle observable? This can be possible only if time-reversibility is broken more rapidly in the collision process. This will be the tacit assumption we make. How this occurs is one of the great open issues. In some sense on the time scale of 10^{-23} s 'measurement' of the colliding system must be occurring, leading to the decoherence of the many-body quantum state.

- How can we observe a new phase of matter that exists for a short time, evolves and ultimately disintegrates (hadronizes) into usual final-state particles? At first sight, everything will always appear in the data very much akin to a reaction involving only the HG phase.

Considerable effort must be put into the understanding of the temporal and spatial evolution of the colliding system. We must identify the measurable quantities that can depend on the properties of the early and dense stage of collision, allowing us to penetrate the 'nebula' of the final hadronic state.

One observable is the quark chemical composition of the fireball of dense matter, which evolves as new quark flavors, such as strangeness, are cooked up inside the micro-bang fireball. Another observable is the entropy content: when quarks and gluons are liberated, the usually 'frozen' color bonds are broken and an entropy-rich state of matter is formed. We will address these two hadronic observables of QGP in greater detail in this book, and we offer here a first short overview of the related ideas and diagnostic methods.

The quarks 'q' (up 'u' and down 'd') from which the stable matter around us is made are easily produced as quark–antiquark pairs because they have small masses; see section 1.2. Another abundantly added quark flavor is strangeness, particularly if the deconfined QGP phase of matter is formed. Strangeness was one of the first proposed signatures of the deconfined phase [220]. The mass of strange quarks and antiquarks is of the same magnitude as the temperature T at which protons, neutrons, and other hadrons are expected to dissolve into quarks. This means that the abundance of strange quarks is sensitive to the conditions, structure, and dynamics of the deconfined-matter phase. The dominant mechanism for cooking up strangeness in quark–gluon deconfined matter was found to be the gluon-fusion reaction $gg \rightarrow s\bar{s}$ [226]. We will address this process in depth in this book. Ultimately the quarks and antiquarks produced in the fireball of dense matter find their way into a multitude of final-state particles, with different quark contents, in the process of hadronization. This situation is illustrated in Fig. 1.4.

Detection of strange particles is facilitated by the fact that the massive strange quark decays into lighter quarks. Thus, strangeness-carrying

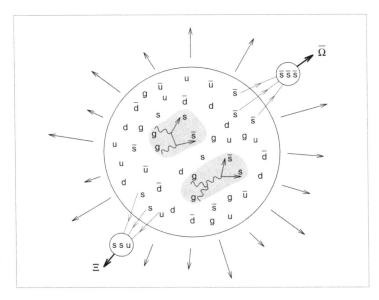

Fig. 1.4. A qualitative image of processes leading to production of (multi)-strange particles: in the QGP phase gluon collisions produce pairs of strange quarks, which are shown to assemble into otherwise rarely made multistrange baryons (for example $\Xi(ssu)$) and antibaryons (for example $\overline{\Omega}(\bar{s}\bar{s}\bar{s})$).

hadrons are naturally radioactive and decay by weak interactions that occur, in general, on a time scale that is extremely long compared with the nuclear-collision times. This makes it relatively easy to detect the strange particles through the tracks left by their decay products. It is important to remember that, unlike the light quarks, strange quarks are not brought into the reaction by the colliding nuclei. Therefore, we know for sure that any strange quarks or antiquarks observed in experiments have been made from the kinetic energy of colliding nuclei.

Should the new deconfined phase of matter be formed, we expect that final abundances of strange particles will be governed by (near) chemical equilibration of strangeness, i.e., that the yield abundance of QGP strangeness will saturate all available phase-space cells, making it into a q–$\bar{q}$–s–$\bar{s}$–g liquid. The total strangeness yielded is thus of considerable interest and is being measured as a function of the collision energy.

The excitation function of strangeness can be qualitatively studied by evaluating the ratio $K^+(\bar{s}u)/\pi^+(\bar{d}u)$ shown in Fig. 1.5. Data obtained at several experimental facilities is shown: from the KaoS experiment at the SIS/GSI; from the E917 and E866 experiments at the AGS/BNL, from NA49 and NA44 experiments at the SPS/CERN, and from the STAR experiment at the RHIC/BNL. As long as the production of strange an-

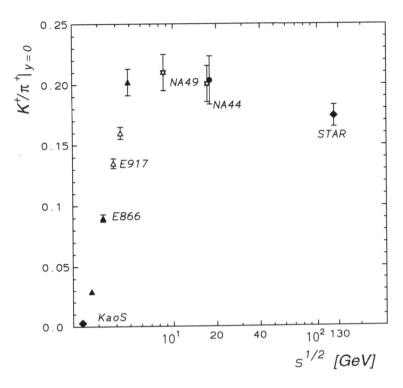

Fig. 1.5. Mid-rapidity $K^+(\bar{s}u)/\pi^+(\bar{d}u)$ in reactions of the heaviest nuclei as a function of the collision energy. See the text for details.

tibaryons is small, the $K^+(\bar{s}u)/\pi^+(\bar{d}u)$ ratio characterizes accurately the relative abundance of strangeness. This is not the case at RHIC energies, for which the strangeness content found in baryonic degrees of freedom is not negligible; see section 19.4. Using the estimate presented there, we find that

$$\frac{K^+ + \langle \bar{s} \rangle_Y}{\pi^+} \simeq 0.23,$$

where the second term $\langle \bar{s} \rangle_Y$ is the strangeness content in the baryonic degrees of freedom.

This shows that the ratio of strangeness to hadron multiplicity continues to grow as the collision energy is increased. Since this growth is here seen to occur relative to the hadron multiplicity, measured in terms of the yield of positively charged pions, this implies that the yield of strangeness increases faster than the increase in production of entropy. At low energies the increase in yield of strangeness shown in Fig. 1.5 shows the effect of the energy threshold for production of strangeness.

The specific yield of strangeness per participating baryon increases much faster. In fact the specific yield of pions increases by 50% on going from AGS to CERN energies, which implies that the yield of strangeness per baryon continues to grow rapidly in the energy range 5 GeV $< s^{1/2} < 17$ GeV.

The most interesting qualitative signature of strangeness in QGP is the yield of (multi)strange antibaryons. Given the ready supply of (strange) quarks and antiquarks, otherwise rarely produced (strange) particles will be emerging from a deconfined phase. In particular, the formation of antimatter particles comprising strangeness is of interest [215]. These particles can be more readily assembled in the high-density deconfined environment. In Fig. 1.4, we illustrate the sequence of events that leads to the formation of these particles: microscopic reactions, predominantly involving fusion of gluons, form pairs of strange quarks, of which clusters are formed and emitted.

Enhanced production of strange particles has been predicted to occur in a QGP for each strange particle species, and to increase with the strangeness content of the particle [164, 215]. Such enhancements in the number of strange particles produced per participating nucleon have now been observed in, e.g., lead–lead (Pb–Pb) collisions, compared with expectations arising from studies of proton–proton (p–p) and proton–beryllium (p–Be) collisions, as is shown in Fig. 1.6. The enhancement for a particular particle is defined as the number of that particle produced per participating nucleon in Pb–Pb collisions, divided by the number produced per participating nucleon seen in p–Be interactions [38].

In Fig. 1.6 the h$^-$ symbol denotes the yield enhancement of negatively charged hadrons, mainly negative pions. This result implies an enhancement by a factor 1.3 for all non-strange hadrons. Such an enhancement is natural if QGP is formed on account of the breaking of the color bonds, and the associated enhancement in number of accessible degrees of freedom compared with reaction scenarios involving confined hadrons. Later in this book we will discuss in depth the issues related to enhanced production of entropy in the deconfined phase.

We further see, in Fig. 1.6, that the production of particles that contain one strange quark, such as the neutral kaon K^0 and the Λ-particle, is enhanced by a factor of about three; the enhancement factor rises to about five for the doubly strange Ξ-particle (and its antiparticle, the anti-Ξ), and more than ten for the yield of $(\Omega + \overline{\Omega})$ particles, which contain three strange or antistrange quarks. The particles in the right-hand panel of Fig. 1.6 have no quarks in common with the colliding nucleons.

When one is interpreting these results as significant indicators for the formation of the deconfined state, it is important to be able to argue that both matter and antimatter particles were produced by the same

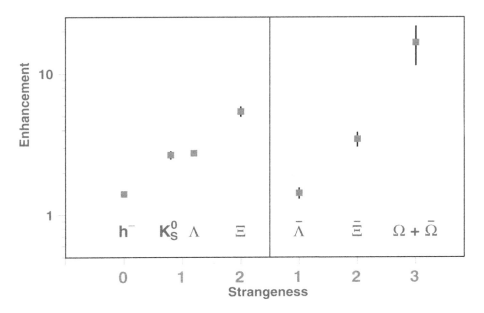

Fig. 1.6. Enhanced production of (strange) hadrons in Pb–Pb 158A-GeV heavy-ion collisions. Enhancement is defined with respect to p–Be collisions. Results were obtained by by the CERN WA97 experiment considering particles emitted by a source stationary in the CM frame of reference [38].

mechanisms, as would be the case should a deconfined soup of quarks and antiquarks break up into final-state hadrons [42]. To demonstrate this, one studies not only the abundances but also the spectra of the particles produced. In order to reduce the dependence on the flow of matter along the collision axis, which is related to the collision dynamics, it is convenient to look only at hadron spectra with momentum components transverse to the original collision axis.

In Fig. 1.7, we see the spectra of strange baryons (Λ, Ξ, and Ω) and antibaryons (anti-Λ, anti-Ξ^-, and anti-Ω^-) as functions of the transverse energy, $m_\perp = \sqrt{m^2 + p_\perp^2}$. The most significant feature of Fig. 1.7 is that the slopes of the spectra for a particle and its antiparticle are very similar. The difference between the particle and antiparticle yields is a result of the quark–antiquark yield asymmetry present. The shape identity of matter and antimatter verifies that the mechanism of production is the same, corroborating the evidence for a common, deconfined source of strange hadrons.

Much of the material of this book will be devoted to the development of ideas demonstrating that these results are a natural consequence of the formation of the deconfined QGP state. It is important to keep in mind

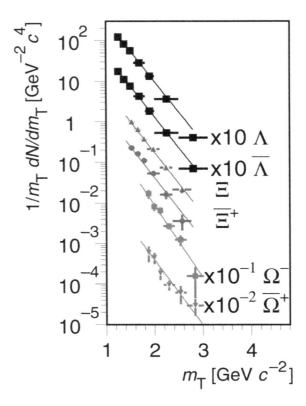

Fig. 1.7. Transverse mass spectra of strange and multistrange baryons and antibaryons. Results obtained by the CERN WA97 experiment for particles emitted at mid-rapidity [42].

that further evidence for the deconfinement of quarks in these reactions is available. The production of charmonium, i.e., particles containing a heavy charm quark and an antiquark, is another well-studied phenomenon [188].

1.7 Other probes of quark–gluon plasma

Since the charm quark is about ten times heavier than the strange quark, at SPS energies pairs of charm quarks can be formed only during the very early stages of the collision, as the nuclei begin to penetrate each other. In this early stage the colliding particles have the energy to overcome the higher energy threshold. If the QGP phase is formed, these charmed quarks have less chance of forming a charmonium state, because the gluons present within the plasma hinder their binding, or/and break the bound states. The observed strong suppression of the J/Ψ signal in 158A-GeV

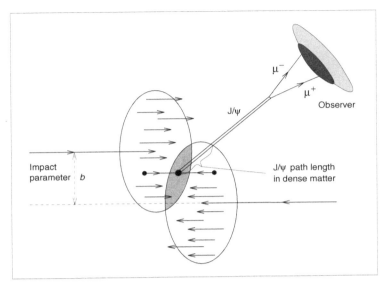

Fig. 1.8. The CM picture of initial moments of nuclear collision and an illustration of the path traveled by J/Ψ in dense matter.

Pb–Pb reactions is interpreted as being due to interactions with gluons. We will not develop this interesting topic in this book beyond a brief introduction, which follows.

Seen in the CM frame, the preformed J/Ψ must subsequently travel through a fireball of dense matter formed in the collision; see Fig. 1.8. The path in dense matter varies depending on the impact parameter, allowing one to evaluate the interaction (absorption) strength. This effect is particularly strong in most central collisions of nuclei.

The experimental J/Ψ yields, seen in Fig. 1.9, show clearly the unusual interaction strength of J/Ψ, when the path in dense matter is large as characterized by large values of the transverse energy generated by particle interactions. The J/Ψ is observed, in the NA50 experiment, by virtue of its decay into $\mu^+\mu^-$ [12]. In Fig. 1.9, the yield of J/Ψ is obtained with reference to the yield of the dimuon background, which is believed to scale just like J/Ψ with the number of participant nuclei. We see, in Fig. 1.9 at $E_\perp > 50$ GeV, a lesser increase in the suppression up to the last bins $E_\perp > 120$ GeV. It is this rapid two-fold increase in the suppression as a function of $E_\perp$ which makes an interpretation of these results, in terms of confined matter alone, difficult.

Considering quantitatively only the absorptive effect of the QGP, practically no J/Ψ should be observed for most central collisions at the higher energies available at the RHIC. However, representing the absorption of

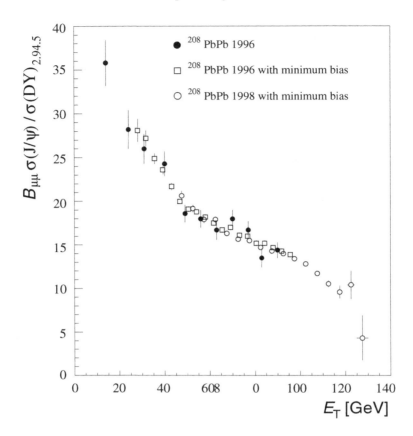

Fig. 1.9.   Suppression of production of J/Ψ as a function of AA-collision central-ity, characterized in terms of the transverse energy E_T produced in the reaction. Results of experiment NA50 [12].

the J/Ψ as a result of its interaction with the gluon content of QGP, one is led to consider the reverse reaction, and both break-up and formation become possible in the QGP medium:

$$J/\Psi + g \Longleftrightarrow c + \bar{c}.$$

Depending on the absolute abundance of open charm, there could be considerable in-plasma production of J/Ψ. A quantitative study of this effect has recently been carried out [257], resulting in the prediction of an enhancement in production of J/Ψ in central nuclear collisions in the energy range accessible at the RHIC. Observation of this new mechanism of J/Ψ formation would constitute proof of the mobility of charmed quarks in the fireball of dense matter.

If this changing pattern of J/Ψ suppression turning into enhancement is confirmed experimentally to be present relative to the directly measured yield of 'open' charm, with the expected energy dependence and centrality dependence (peripheral interactions agree with extrapolations from p-induced reactions), this will constitute convincing evidence for the formation of the deconfined form of matter.

Charmed hadrons, i.e., particles with 'open' charm, have not yet been observed directly, but considerable effort is being made today to prepare experiments at the SPS, RHIC and LHC that will have the capability to detect them. The experimental difficulty is the rapid disintegration of charmed hadrons, requiring an extremely precise tracking of particles beginning very near to the interaction vertex. The availability of direct measurements of the abundance of charm would provide a reliable baseline for understanding the phenomenon of suppression of production of J/Ψ, and perhaps at higher energies than those at the SPS, enhancement of its production.

A few further signatures of deconfinement have also been proposed. Akin to the disappearance of J/Ψ, jets of hadrons produced by hard quarks and gluon arising from initial interactions should disappear due to the enhanced strength of interaction in the 'colored' charge plasma of the deconfined medium [72]. There is very little doubt that this idea works. However, this rapid quenching of hard particles is the mechanism of thermalization of partons, aside of being a direct signature for a new phase of matter. This mechanism provides the motivation for the statistical description of the properties of dense matter. Can the thermalization of partons be seen as a signature of QGP? We will not address in this book this rapidly developing subject.

Photons γ and lepton pairs emerging from the decay of virtual photons $\gamma^* \to e^+e^-$ and $\mu^+\mu^-$ are, at first sight, the most promising probes of the dense hadronic matter [120, 154, 244]. Electromagnetic interactions are strong enough to lead to an initial detectable signal, with secondary interactions being too weak to alter substantially the shape and yield of the primary spectra. Direct photons and leptons contain information about the properties of dense matter in the initial moments of the collision. Of particular interest would be the possible exploration of the initial time period leading to the formation of thermal equilibrium.

Therefore, at first, photons and lepton pairs (often also called dileptons) originating from QGP, produced, e.g., by quark–antiquark annihilation, were considered to be the primary diagnostic tool, just as photons are a powerful diagnostic for traditional (electromagnetic) plasmas. However, thousands of particles are created in high-energy nuclear collisions. Among those the decay of neutral π^0 mesons is known to produce a strong background. It is very difficult to extract the direct QGP photon signal, which is only a small fraction of all photons produced.

Because dilepton formation, compared with formation of photons, requires one additional electromagnetic interaction, the yield of dileptons is considerably smaller, by a factor 300 or more, than the yield of direct photons. However, the dilepton background is also much reduced. While the directly produced photons are present at the level of 5% of the photon signal observed, the amount of dileptons from formation of dense matter is believed to exceed the background by a factor as large as 3–5. However, dileptons have been shown to have other origins related to properties of confined hadronic matter and the observed pattern of dilepton production, given the large systematic experimental error, is difficult to interpret [22]. We will not dwell on this complex subject in this book.

The diagnostic strength of the strangeness signature, in comparison with direct photons and dileptons, is that, aside from the overall enhancement of abundance, we also have the pattern of enhancement shown in Fig. 1.6, and the matter–antimatter symmetry seen in Fig. 1.7. The yield of strangeness is related to the initial most extreme conditions of the QGP phase, much like the photon and dilepton yields. However, strangeness has relatively little background. The source of strangeness, gluons, is by far more characteristic for the new phase of matter than is the source of photons and dileptons, which are electrical currents of quarks, present both in the confined and in the deconfined state of matter.

2 Hadrons

2.1 Baryons and mesons

As the above first and qualitative discussion of signals of QGP formation has shown, to pursue the subject we next need to understand rather well the properties of 'elementary' particles, and specifically strongly interacting (hadronic) particles. These are complex composites of the more elementary strongly interacting particles: quarks and gluons. Quark-composite hadronic particles are final products emerging from heavy-ion collisions, and play an important role in this book – there are two different types of colorless quark bound states: baryons, comprising three valence quarks, and mesons, which are quark–antiquark bound states.

Typically, each type of (composite) hadron has many resonances, i.e., states of greater mass but the same quantum number, which are generally internal structural excitations of the lowest-mass state. Normally the lowest-mass states of a given quantum number are stable (that is, if they decay, it occurs very slowly compared with the time scale of hadronic reactions), but there are notable exceptions, such as the Δ-resonances, which are subject to the rapid hadronic decay $\Delta \to N + \pi$, which occurs within a duration less than the life span of a hadronic fireball.

On the other hand, some unstable hadronic resonances can be stable on this time scale. However, by the time relatively distant experimental detectors record a particle, all possible hadronic decays and even many weak decays have occurred. To study particles produced in high-energy nuclear collisions, we need to understand their decay patterns well.

Because the two light quarks are practically indistinguishable under strong interactions, their bound states reflect a symmetry, referred to as isospin I. In addition one is normally quoting the strange-quark content (strangeness S) of a bound hadron state. Similarly, we refer to more exotic baryons by quoting the value of their charm content C or bottom content B. It is believed that the heaviest quark t can not form hadronic bound states, since its life span is too short for a (quantum) orbit to form.

We keep in mind that, for historical reasons, strange baryons, containing strange quarks s, have *negative* strangeness (hyperon number) S, and that strange mesons, with a strange quark s, are called antikaons $\overline{K}(s\bar{q})$ (we indicate in the parentheses the valence-quark content). Similar conventions were adopted for charm and bottom flavors.

- **Baryons**

In Fig. 2.1, we present baryons made of the four lightest quarks (u, d, s, and c). Were all four quark flavors degenerate in mass, we would have instead of the isospin $SU(2)$ an $SU(4)$ flavor symmetry. In Fig. 2.1, the vertical hierarchy is generated by the largest symmetry breaking in the hadron mass spectrum, introduced by the relatively large mass of the charmed quark. The two 'foundations' of the $SU(4)$ flavor-multiplet 'pyramids' are the $SU(3)$ flavor multiplets, arising when three quarks (u, d, and s) are considered: in Fig. 2.1(a) a spin-$\frac{1}{2}$ 'octet' (eight-fold multiplicity); in Fig. 2.1(b) a spin-$\frac{3}{2}$ 'decuplet' (ten-fold multiplicity). $SU(2)$ flavor (isospin) particle multiplets are located along straight horizontal lines within the $SU(3)$ flavor multiplet planes.

The requirement that quarks form antisymmetric bound states leads to classification in these two spin–flavor configurations of spin $\frac{1}{2}^{+}$ and $\frac{3}{2}^{+}$, respectively, and thus when symbols are repeated, the excited state is the higher spin state. States with positive intrinsic parity (upper index '+') are shown, since negative parity is associated with intrinsic angular-momentum excitation, or the presence of an additional particle component such as a gluon or, equivalently, a quark–antiquark pair. Therefore, generally such excited negative-parity states can decay rapidly into the more stable positive-parity states.

Moving up in the two 'pyramids', we replace at each 'floor' one of the strange quarks by a charmed quark, omitting the non-strange baryons.

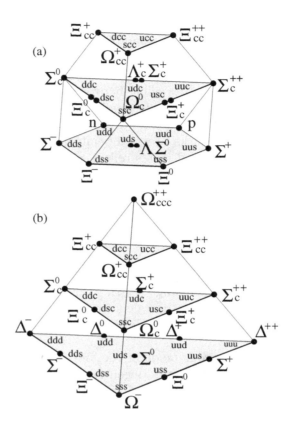

Fig. 2.1.   Baryons (positive parity, superscript +): (a) generalization of the spin-$\frac{1}{2}^{+}$ octet and (b) generalization of the spin-$\frac{3}{2}^{+}$ decuplet classified according to their valence-quark content.

In the naming convention, the name of the corresponding $SU(3)$ flavor particles is retained, with the added lower index indicating how many charmed quarks have been introduced to replace strange quarks. Similarly, when bottom-flavor-involving baryons are considered (not shown in Fig. 2.1), the lower index 'b' is used to indicate how many strange quarks have been replaced by bottom quarks.

●**Mesons**

In Fig. 2.2, we present mesons made of the four lightest quark–antiquark pairs (u, d, s, and c). Now the center-planes of the two $SU(4)$ flavor multiplets are the well-known $SU(3)$ flavor (u, d, and s) multiplets: Fig. 2.2(a), a spin-0 octet, with the ninth state $\eta_c = c\bar{c}$; Fig. 2.2(b), a spin-1 octet, which now also includes the ninth state $J/\Psi = c\bar{c}$. We recall that $\pi^0 = (u\bar{u} - d\bar{d})/\sqrt{2}$ and that the heavier η and η' share the strange-

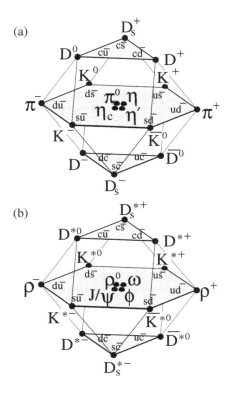

Fig. 2.2. Mesons (negative parity, superscript $-$): (a) pseudoscalar 0^- and (b) vector 1^- mesons, classified according to their valence-quark content.

quark component $s\bar{s}$; there is no pure spin-0 $s\bar{s}$ state. In the spin-1 octet, case $\phi^0 = s\bar{s}$ is a pure strange-quark pair state. Not considered in Fig. 2.2 are states containing bottom quarks:

$B = \pm 1$, bottom mesons: $B^+ = u\bar{b}$, $B^0 = d\bar{b}$, $\overline{B^0} = \bar{d}b$, and $B^- = \bar{u}b$;

$B = \pm 1$, $S = \pm 1$, bottom strange mesons: $B_s^0 = s\bar{b}$, and $\overline{B_s^0} = \bar{s}b$; and

$B = \pm 1$, $C = \pm 1$, bottom charmed mesons: $B_c^+ = \bar{b}c$ and $B_c^- = b\bar{c}$.

2.2 Strange hadrons

For the physics of interest in this book, a particularly important family of particles consists of those which contain strange quarks. Since we will address these particles by implying some of their particular properties, we will summarize in the following their key features, and we comment on the experimental methods for their detection. A similar comment applies to heavier flavors (charm and bottom), and we also present a sample of these particles below.

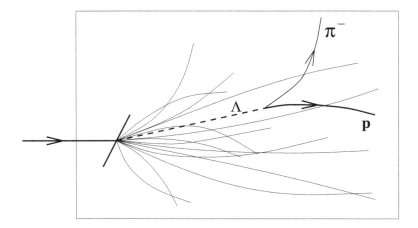

Fig. 2.3. A schematic representation of the Λ-decay topological structure showing as a dashed line the invisible Λ and the decay 'V' of the final-state charged
particles. Tracks of other directly produced charged particles propagating in a
magnetic field normal to the plane of the figure are also shown.

• **Hyperons** $Y(qqs)$ and $\overline{Y}(\bar{q}\bar{q}\bar{s})$
Sometimes all strange baryons are referred to as hyperons. We prefer to
use this term for singly strange baryons.
— The isospin singlet lambda, $\Lambda(uds)$, is a neutral particle of mass 1.116
GeV that decays weakly with proper path length $c\tau = 7.891 \pm 0.006$ cm.
The dominant and commonly observed decay is

$$\Lambda \to p + \pi^-, \quad 63.9 \pm 0.5\%.$$

The decay of a neutral particle into a pair of charged particles forms a
characteristic 'V' structure as shown in Fig. 2.3. The other important
weak decay,

$$\Lambda \to n + \pi^0, \quad 35.8 \pm 0.5\%,$$

has only (hard-to-identify) neutral particles in the final state.
In addition to the $\frac{1}{2}^+$ ground state (positive parity, spin $\frac{1}{2}$: $\frac{1}{2}^+$), we
encounter a $\frac{1}{2}^-$ resonance $\Lambda^*(1.405)$ and also a $\frac{3}{2}^-$ state $\Lambda^*(1.520)$. There
is no 'stable' positive parity $\frac{3}{2}^+$ iso-singlet Λ. $\Lambda^*(1.520)$ has a remarkably
narrow width $\Gamma = 15.6$ MeV, even though the hadronic decay into the
$N\overline{K}$ channel is open. All Λ^* excited resonance states (13 are presently
known, with mass below 2.4 GeV) decay hadronically into two principal
channels:

$$\Lambda^* \to Y + \text{meson(s)},$$
$$\Lambda^* \to N + \overline{K}.$$

Since the hadronic decays have free-space proper decay paths of 1–10 fm (widths $\Gamma = 16$–250 MeV), all these resonances can not be distinguished from the ground state of corresponding flavor content and contribute to the abundance of the observed 'stable' (on hadronic time scale) strange particles Λ and K. However, it is possible to measure their yields by testing whether the momenta of the expected decay products add to the energy–momentum product of a particle with the 'invariant' mass of the parent resonance.

The practical approach to the observation of Λ is to detect, in a tracking device, the (dominant) decay channel. The two final-state charged particles are pointing to a formation vertex remote from the collision vertex of projectile and target. The ancestor neutral particle should point to the interaction 'star' (see Fig. 2.3) and should have the correct mass and life span. This approach includes, in a certain kinematic region, the events which originate from the decay of K_s (see below). A well-established method of analysis of data allows one to distinguish the hyperon decay products from kaon decay products [211]. The procedure we describe in the following is named after Podolanski and Armenteros [208]. It exploits the difference between the phase-space distributions of the pair of particles produced.

The invisible neutral particle, reconstructed from the charged tracks forming the decay 'V', has a definitive line of flight, shown by the dashed line in Fig. 2.3. The magnitude of the transverse momentum $q_\perp$ measured with reference to this line of flight is by definition the same for both produced, oppositely charged, particles. The longitudinal momentum $p_\parallel^+ \neq p_\parallel^-$ for each particle is dependent on a random angular distribution of decay products of charges $+$ and $-$. The relative $p_\parallel$ asymmetry,

$$a = \frac{p_\parallel^+ - p_\parallel^-}{p_\parallel^+ + p_\parallel^-}, \tag{2.1}$$

can span the entire range $-1 < a < 1$ when the decay products have the same mass, as is the case for decay of kaons. When a proton is produced in Λ decay, this massive particle carries the dominant fraction of $p_\parallel$ momentum and the event appears near $a \to 1$. Similarly, the $\overline{\Lambda}$ decay populates the domain $a \to -1$. Events appear around the clearly visible kinematic $q_\perp(a)$ lines shown in Fig. 2.4. For each decay of particles K_s, Λ, and $\overline{\Lambda}$, there is a clear domain where accidental misidentification of particles is impossible. e^+–e^--pair events produce a background found near to $q_\perp = 0$. The entire $(q_\perp, a)$ plane is filled with events arising from accidental 'V's. Since the physical signal is highly concentrated along the kinematic lines seen in Fig. 2.4, the signal-to-noise ratio is very favorable. As a consequence, the measurement of production of neutral

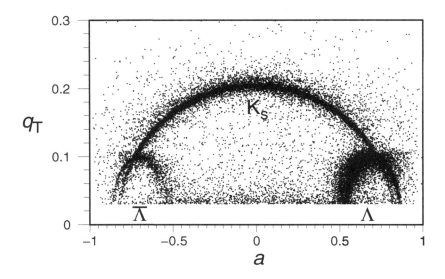

Fig. 2.4.   The Podolanski–Armenteros representation of a particle-decay event: the transverse momentum q_T in a two-particle decay, as a function of the relative longitudinal momentum asymmetry a, see Eq. (2.1).

strange hadrons using Podolanski–Armenteros analysis is highly reliable.

— The isospin triplet uds, uus, dds, i.e., Σ^0(1.193 GeV) and $\Sigma^\pm$(1.189 GeV), varies in its properties. There is only one dominant decay channel for the Σ^- decay,

$$\Sigma^- \to n + \pi^-, \qquad c\tau = 4.43 \pm 0.04 \text{ cm}.$$

Because there are two isospin-allowed decay channels of similar strength for the Σ^+,

$$\Sigma^+ \to p + \pi^0, \qquad 51.6\%,$$
$$\to n + \pi^+, \qquad 48.3\%,$$

the decay path is nearly half as long, $c\tau = 2.4$ cm. $\Sigma^\pm$ have not yet been studied in the context of QGP studies, since they are more difficult to observe than Λ – akin to the Ξ decay (see below) there is always an unobserved neutral particle in the final state of $\Sigma^\pm$ decay. Unlike Ξ decay, the kink that is generated by the conversion of one charged particle into another, accompanied by the emission of a neutral particle, is not associated with subsequent decay of the invisible neutral particle accompanied by a 'V' pair of charged particles; see Fig. 2.7 below.

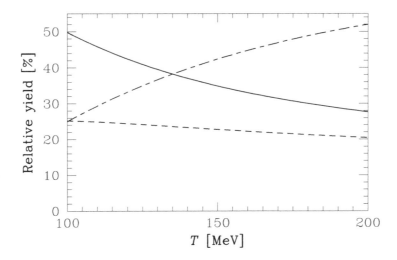

Fig. 2.5. The relative yields of the three dominant contributions to the final yield of Λ: direct production (solid line), Σ^0 contribution (dashed line), and $\Sigma^*(1385)$ contribution (chain line), as a function of T, the temperature of the hot source of these particles.

— The neutral Σ^0(uds) undergoes a rapid electromagnetic decay:

$$\Sigma^0 \to \Lambda + \gamma + 76.96 \pm 0.02 \text{ MeV}, \qquad c\tau = (2.2 \pm 0.2) \times 10^{-9} \text{ cm},$$
$$\tau = (7.4 \pm 0.7) \times 10^{-20} \text{ s}.$$

For the observer in the laboratory, this secondary Λ is practically indistinguishable from the direct production in the interaction vertex. Consequently, all measurements of Λ of interest for this book combine the abundances of Λ and Σ^0, including all the higher resonances that decay hadronically into these states, in particular $\Sigma^*(1385)$.

As with Λ, there are several (nine) heavier Σ resonances known at $m \leq 2.4$ GeV. When they are produced, all decay hadronically, producing $\overline{K}$, Λ, and Σ. Particularly important is the strong resonance $\Sigma^*(1385)$:

$$\Sigma^*(1385) \to \Lambda + \pi, \qquad 88 \pm 2\%,$$
$$\to \Sigma + \pi, \qquad 12 \pm 2\%.$$

Thus, $92 \pm 3\%$ of $\Sigma^*(1385)$ decays into Λ. Since the spin of $\Sigma^*(1385)$ is $\frac{3}{2}$, and its isospin is 1, the degeneracy of $\Sigma^*(1385)$ is six times greater than that of Λ. For this reason the final yield of Λ is in fact predominantly derived from decays of Σ^*, as is shown in Fig. 2.5, in which the chain line describes the partial contribution of Σ^* to the final yield of Λ. The solid line is the directly produced component, and the dashed line is the contribution of Σ^0.

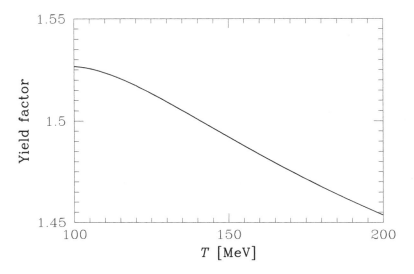

Fig. 2.6.   The multiplicative factor allowing one to infer the total yield of qqs states from the observed yield of Λ as a function of T, the temperature of the hot source of these particles.

An important practical issue is the determination of the total yield of singly strange baryons (qqs), considering that we have available the observed yield of Λ. It is generally assumed that abundances of the three isospin-1 states Σ^+, Σ^- and Σ^0 produced in relativistic nuclear collisions are equal. In the first instance, let us ignore the influence of $\Sigma^*(1385)$, which we introduce in the next paragraph. Considering the difference in mass between Λ and Σ, $\Delta m = 77$ MeV, and assuming that the relative abundance yield is appropriately described by the relation,

$$\frac{Y_{\Sigma i}}{Y_\Lambda} = \left(1 + \frac{\Delta m}{m}\right)^{3/2} e^{-\Delta m/T}, \tag{2.2}$$

derived from the relative thermal phase-space size, Eq. (10.50c), one finds for a reasonable range of values $T = 160 \pm 15$ MeV, implied by thermal hadron production models, that the total yield of hyperons (qqs) is obtained by multiplying the experimentally observed yield of Λ by a factor $F_Y = 1.8$.

This estimate has to be modified in order to allow for the important role $\Sigma^*(1385)$ plays as a contributor to the production of Λ. In Fig. 2.6, we evaluate the yield factor F_Y allowing for the contributions of Σ^0 and $\Sigma^*(1385)$ to the yield of Λ, using thermal phase space, Eq. (10.51). There is very little variation with T in a wide range, 100 MeV $< T < 200$ MeV, and we conclude that, for all practical purposes, one can use the yield factor $F_Y = 1.48 \pm 0.03$ to estimate the yield of singly strange

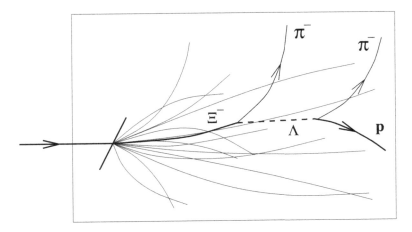

Fig. 2.7.   A schematic representation of the topological structure of decay of Ξ^- showing as a dashed line the invisible Λ emerging from the decay kink and the decay 'V' of the final-state charged particles. Tracks of other directly produced charged particles propagating in a magnetic field normal to the plane of the figure are also shown.

baryons (qqs) from the observed yield of Λ. Naturally, this result holds for antibaryons as well.   In many publications a slightly larger value, $F_Y = 1.6$, is applied.

At a smaller level the decays of multistrange Ξ and $\overline{\Xi}$ also contribute to the yield of Λ and $\overline{\Lambda}$, but this contribution is usually separated experimentally.

• **Cascades**, Ξ(qss) and $\overline{\Xi}(\overline{q}\overline{s}\overline{s})$
The doubly strange cascades, Ξ^0(uss) and Ξ^-(dss), are below the mass threshold for hadronic decays into hyperons and kaons, and also just below the weak-decay threshold for the $\pi + \Sigma$ final state. Consequently, we have one decay in each case:

$$\Xi^-(1321) \to \Lambda + \pi^-, \qquad c\tau = 4.9\,\text{cm},$$
$$\Xi^0(1315) \to \Lambda + \pi^0, \qquad c\tau = 8.7\,\text{cm}.$$

The first of these reactions can be found in charged-particle tracks since it involves conversion of the charged Ξ^- into the charged π^-, with the invisible Λ carrying the 'kink' momentum. For Ξ^- to be positively identified, it is required that the kink combines properly with an observed 'V' of two charged particles identifying a Λ decay. This decay topology is illustrated in Fig. 2.7.

There are also several Ξ^* resonances known, which normally feed down in a hadronic decay into the hyperon and kaon abundances:

$$\Xi^*(\text{qss}) \to Y(\text{qqs}) + \overline{K}(\overline{q}\text{s}).$$

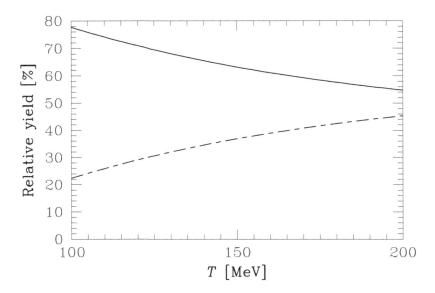

Fig. 2.8. The relative strength of the two dominant contributions to the final yield of Ξ: direct production (solid line), and $\Xi^*(1530)$ (chain line), as a function of T, the temperature of the hot source of these particles.

The main exception is the hadronic decay of the spin-$\frac{3}{2}$ recurrence of the spin-$\frac{1}{2}$ Ξ ground state:

$$\Xi^*(1530) \;\rightarrow\; \Xi + \pi, \qquad \Gamma = 9.5\,\mathrm{MeV}.$$

This relatively small width corresponds to $c\tau = 21$ fm and this decay occurs outside of the space–time region of hot matter.

Since the spin-$\frac{3}{2}$ state is populated twice as often as is the spin-$\frac{1}{2}$ ground state, its relative suppression due to the greater mass ($\Delta m \simeq 200$ MeV) is by a factor of 0.6, as is shown in Fig. 2.8. Despite this non-negligible decay contribution, the cascade spectra are, at high transverse momenta, the most representative among measured hadronic spectra of particles directly produced by the fireball of hot and dense matter.

• **Omegas,** Ω^-(sss) and $\overline{\Omega^-}(\overline{\mathrm{s}}\overline{\mathrm{s}}\overline{\mathrm{s}})$

There are several primary weak-interaction decay channels leading to the relatively short proper decay path, $c\tau = 2.46$ cm:

$$
\begin{aligned}
\Omega^-(1672) &\rightarrow \Lambda + \mathrm{K}^-, & 68\%,\\
&\rightarrow \Xi^0 + \pi^-, & 24\%,\\
&\rightarrow \Xi^- + \pi^0, & 9\%.
\end{aligned}
$$

The first of these decay channels is akin to the decay of the Ξ^-, except that the pion is replaced by a kaon in the final state. In the other two options, after cascading has finished, there is a neutral pion in the final state, which makes the detection of these channels impractical. The $\Lambda + K^-$ decay channel is not unique for the $\Omega^-(1672)$ state; there is a $\Xi^*(1690)$ that also decays in the same way. However, this is a rapid hadronic decay that occurs in the interaction vertex and can be easily separated from the weak decay of the Ω.

The highest known resonance $\Omega^*(2250)$ is rather heavy. However, present experiments do not exclude the possibility of a light resonance, with a dominant electromagnetic decay, $\Omega^*(M^* < M_\Omega + m_\pi) \to \Omega + \gamma$. The presence of this resonance would greatly affect the theoretical expectations regarding the production of Ω and $\overline{\Omega}$ in heavy-ion interactions. The statistical degeneracy due to the high spin of such resonances could easily enhance their abundance, akin to the situation with the Ξ^* state we described above.

• **Kaons**, $K(q\bar{s})$ and $\overline{K}(\bar{q}s)$

— **Neutral kaons**, K_S and K_L ($m = 497.7$ MeV)

This is not the place to describe in detail the interesting physics of the short- and long-lived neutral kaons, except to note that both are orthogonal combinations of the two electrically neutral states $(d\bar{s})$ and $(\bar{d}s)$. The short-lived combination K_S has $c\tau = 2.676$ cm and can be observed in its charged decay channel:

$$\begin{aligned}
K_S &\to \pi^+ + \pi^-, & 69\%, \\
&\to \pi^0 + \pi^0, & 31\%.
\end{aligned}$$

Care must be exercised to separate the K_S decay from Λ decay, since in both cases there are two, *a priori* unidentified, charged particles in the final state, making a 'V' originating from an invisible neutral particle; see Fig. 2.4.

The long-lived kaon K_L, with $c\tau = 1551$ cm, has not been studied in relativistic heavy-ion-collision experiments. Since both the K_L and the K_S arise from the time evolution of hadronically produced $K(d\bar{s})$ and $\overline{K}(\bar{d}s)$ their abundances are essentially equal.

— **Charged kaons**, $K^+(u\bar{s})$ and $K^-(\bar{u}s) \equiv \overline{K^+}$ ($m = 493.7$ MeV)

Charged kaons can in principle be observed directly, both in charged-particle-tracking devices and in magnetic spectrometers, since their mass differs sufficiently from those of the lighter $\pi^\pm$ and the heavier proton/antiproton. Also, $K^\pm(494)$ decays with $c\tau = 371$ cm, with three principal channels:

$$\begin{aligned}
K^+ &\to \mu^+ + \nu_\mu, & 63.5\%, \\
&\to \pi^+ + \pi^0, & 21.2\%, \\
&\to \pi^+ + \pi^+ + \pi^-, & 5.6\%.
\end{aligned}$$

The last one, with three charged hadrons in the final state, can be used to identify charged kaons within tracking devices.

Considering the quark content, one sees that the average of the production of $K^+(u\bar{s})$ and $K^-(\bar{u}s)$ satisfies $\langle K_S \rangle \simeq 0.5(\langle K^+ \rangle + \langle K^- \rangle)$.

• **The φ-meson**, $\phi(s\bar{s})$

The vector $(J = 1)$ meson φ is believed to be a 'pure' bound state of the strange-quark pair $\phi = s\bar{s}$. With mass 1019.4 MeV, it has a relatively narrow full width $\Gamma_\phi = 4.43$ MeV, since the decay into a pair of kaons is barely possible in free space. There are two open and relatively slow decay channels leading to the formation of pairs of leptons,

$$\begin{aligned} \phi \rightarrow e^+ + e^-, &\qquad 0.03\%, &\qquad \Gamma_{e^+e^-} = 1.32 \pm 0.05 \text{ keV}, \\ \rightarrow \mu^+ + \mu^-, &\qquad 0.025\%, &\qquad \Gamma_{\mu^+\mu^-} = 1.3 \ \pm 0.2 \text{ keV}, \end{aligned}$$

allowing the determination of the number of φ-mesons emerging from the hadronic-interaction region. When absolute yields of particles are difficult to determine, one can compare, using the dilepton decay channel, the yield of φ with the yield of non-strange partner mesons $\rho(770)$ $(\Gamma_{e^+e^-} = 6.77 \pm 0.32$ keV) and $\omega(782)$ $(\Gamma_{e^+e^-} = 0.60 \pm 0.02$ keV).

2.3 Charm and bottom in hadrons

• **Onium-mesons**, $c\bar{c}$, $b\bar{b}$ and $\bar{b}c$

A brief summary of the physical properties of the three main onium states follows; the first three properties, mass (note the substantial systematic and statistical uncertainty in the mass of the B_c), lifetime and its inverse the width, are determined by measurement, followed by the binding energy with regard to dissociation into mesons, and the average size determined within heavy-quark potential models.

	$J/\Psi(c\bar{c})$	$B_c(\bar{b}c)$	$\Upsilon(b\bar{b})$
Mass [MeV]	3097	$6400 \pm 390 \pm 130$	9460
τ [ps]	7.6×10^{-9}	$0.46 \pm 0.17 \pm 0.03$	12.5×10^{-9}
Γ [keV]	87 ± 5	1.4×10^{-6}	52.5 ± 1.8
Binding energy [MeV]	630	$\simeq 840$	1100
$\langle r \rangle$ [fm]	0.42	0.34	0.22

The spin-1 $J/\Psi(c\bar{c})$ has a partial width of 6% (per channel) projected into the experimentally observed lepton-pair channels $e^+ + e^-$, $\mu^+ + \mu^-$, i.e., $\Gamma_{ll}^c = 5.26$ keV; the still narrower $b\bar{b}$ has $\Gamma_{ll}^b = 0.3$ keV, which is 2.5% of the total width. We note that J/Ψ is produced very rarely in heavy-ion collisions, e.g., in Pb–Pb interactions at $158A$ GeV one J/Ψ is produced for each 3×10^6 mesons. Production of open charm has not been measured but is at the level of 0.2–0.6 c$\bar{c}$ pairs per Pb–Pb interaction at 158A GeV. Thus, only one in about 2000 c$\bar{c}$ pairs produced emerges as a bound J/Ψ(c$\bar{c}$) state. The uncertainty in this estimate is at least a factor of two and depends on the centrality of the interaction. It is hoped that further experimental information will become available soon, allowing us to understand this ratio more precisely.

The excited state Ψ' has a yield five times smaller. There has not yet been a measurement of production of the other onium states in nuclear collisions.

In the LHC energy range, one can expect that the bound state of b-quarks, the upsilonium Υ(b$\bar{b}$), will assume a similar role to that which is today being played, at SPS and RHIC energies, by J/Ψ.

The other heavy-quark bound state that is of interest is the B$_c$(b$\bar{c}$). This quarkonium state is so rarely produced that it was not discovered until very recently [9, 10]. However, it has been studied extensively theoretically, and the currently reported mass, $M = 6.4 \pm 0.39 \pm 0.13$ GeV, is in good agreement with the theoretical quark-potential model expectations. The life span, $\tau \simeq 0.5$ ps, $c\tau \simeq 150\,\mu$m, implies that the current silicon pixel detector technology allows one to distinguish the production vertex from the B$_c$(b$\bar{c}$) decay vertex.

The conventional mechanism for production of B$_c$(b$\bar{c}$) requires the formation of two pairs of heavy quarks in one elementary interaction, followed by the formation of a bound state. The probability of these three unlikely events occurring in one interaction is not large and hence neither is the relative predicted yield, (B$_c$ + B$_c^*$)/(b, $\bar{b}$) $\simeq$ (3–10) $\times$ 10^{-5} at $\sqrt{s_{\mathrm{NN}}} = 200$ GeV [169]. This small value implies that 'directly' produced B$_c$ (both in $J = 0$ and $J = 1$ channels B$_c^*$ and $\bar{\mathrm{B}}_c^*$) cannot be observed at the RHIC. On the other hand, an enhancement in production of this state is expected in the QGP-mediated recombination [239], which could lead to a measurable rate of production in nuclear interactions. Since the quark-recombination mechanism of production requires mobility of heavy color-charged quarks, observation of this new mechanism for the formation of this exotic meson would constitute another good signature of the deconfined QGP phase.

3 The vacuum as a physical medium

3.1 Confining vacuum in strong interactions

Theoretical interest in the study of relativistic heavy-ion collisions originates, in part, from the belief that we will be able to explore the vacuum structure of strong interactions and, in particular, the phenomenon of quark confinement. The picture of confinement can be summarized as follows:

1. all strongly interacting particles are made of quarks and gluons;
2. quarks q and gluons g are *color* charged [123], but all asymptotic observable physical states they can form are color neutral;
3. therefore, the *true* vacuum state $|V\rangle$ abhors color;
4. there is an excited state $|P\rangle$, referred to as *perturbative vacuum*, in which colored particles can exist as individual entities and therefore move freely;
5. $|P\rangle$ differs essentially from $|V\rangle$, the true vacuum, and in particular, it differs by a considerable amount of energy density in the regions of space–time in which the $|V\rangle$ structure is dissolved into $|P\rangle$.

In the 'true vacuum' (in which we live), color-charged quarks and gluons are 'confined'. However, under extreme conditions of density and temperature, we should reach the crossover to the color-conductive phase of the vacuum. In such a space–time domain, nearly free propagation of colored quarks and gluons is thought possible. This picture of hadronic interactions is consistent and indeed justifies the perturbative approach to quantum-chromodynamics (QCD) interactions. It is the foundation that allows us to describe hadrons as bags, i.e., confined bound states of quarks, see section 13.2. We also use these simple, but essential, features in the discussion of the physical properties of the QGP state in section 4.6. The melted color-conductive state $|P\rangle$ is a locally excited space–time domain in which quarks and gluons can move around. This state has properties that we would like intuitively to associate with a normal physical state, since it is simple, structureless. We must keep in mind that the situation is, however, inverted relative to our expectations. Since quarks and gluons are not observed individually, they cannot propagate in the true vacuum state, thus the true physical 'ground' state $|V\rangle$ must be complex and structured, and it is the excited state that is simple and structureless.

Vacuum structure keeps the colored particles bound and confined. Quark confinement has not been explained to be a direct result of quark–quark interaction, generated by the color charge and exchange of gluons. Rather, this force determines within a domain of perturbative vacuum $|P\rangle$ the structural detail: for the ground state the structure of the hadronic spectrum; at sufficiently high excitation, the properties of the color plasma of hot quarks and gluons. To be able to move color charges within a region of space, one needs to 'melt' the confining structure. For a first-order phase transition, the two phases have a difference in energy density, the latent heat $\mathcal{B}$, per unit of volume,

$$\boxed{\mathcal{B} \equiv \epsilon^{\text{QGP}}(T_{\text{cr}}, V_{\text{cr}}; b) - \epsilon^{\text{HG}}(T_{\text{cr}}, V_{\text{cr}}; b) \approx 0.5 \text{ GeV fm}^{-3}.} \qquad (3.1)$$

We would like to determine, by studying the QGP phase, the magnitude of $\mathcal{B}$. So far, only relativistic nuclear collisions can deliver (to a large region of space) the required energy and are the best and only tool we have today to study the process of melting of the QCD vacuum, see section 5.2. We will discuss the experimental methods further in chapter 5.

The vacuum properties of strong interactions can be explored only when the locally deconfined state, the QGP, is experimentally established. In our opinion, the study of the physical properties of the hadronic vacuum, in particular 'confined vacuum melting', is the fundamental challenge motivating the high-energy nuclear-collision experimental program. It is relevant to note that the key ideas and concepts underpinning the possibility of finding the vacuum 'melting' are robust against change and evolution of our knowledge: neither the questions about the existence of a true (discontinuous) phase transition between the hadronic vacuum states nor the possible quark substructures will greatly influence these considerations. All we want is to determine that the color-melted state contains particle-like quark–gluon excitations with established symmetries and interactions.

The most interesting property of the true QCD vacuum $|V\rangle$ is that it abhors the color charge of quarks and gluons. However, we are interested in determining and understanding its other physical properties. The appearance of a glue 'condensate' field, i.e., the vacuum expectation value of the 'square' of the gluon field, the so-called field-correlator in the true vacuum state [242, 243], is of particular relevance for the understanding of $|V\rangle$. With the glue fields defined as in section 13.4 we have

$$\frac{1}{2}F^2 \equiv \sum_a \frac{1}{2}F^a_{\mu\nu}F^{\mu\nu}_a = \sum_a [\vec{B}^2_a - \vec{E}^2_a], \tag{3.2}$$

where we use Einstein's summation convention for repeated Greek indices.

The value of F^2 is obtained by studying QCD sum rules [197, 198, 242, 243], and is in agreement with the results obtained numerically using lattice-gauge-theory methods [100, 101]:

$$\Delta F^2 \equiv \langle V|\frac{\alpha_\mathrm{s}}{\pi}F^2|V\rangle - \langle P|\frac{\alpha_\mathrm{s}}{\pi}F^2|P\rangle \simeq (2.3 \pm 0.3) \times 10^{-2}\,\mathrm{GeV}^4, \tag{3.3}$$

$$= [390 \pm 12\,\mathrm{MeV}]^4.$$

Here, $\alpha_\mathrm{s} = g^2_\mathrm{s}/(4\pi)$ is the coupling constant for the strong interaction. Since in empty space the vacuum state is field-free, i.e., the vacuum expectation value of the gauge field vanishes, the appearance of a non-vanishing vacuum expectation value of the square of the gauge field in Eq. (3.3) is a quantum effect without a classical analog.

3.2 Ferromagnetic vacuum

We now describe a model and discuss other properties of the vacuum state that are related to the remarkable result Eq. (3.3). Because of the non-abelian nature of color charges, the quanta that mediate the color force, gluons, can themselves interact by means of exchanging gluons. Since gluons are massless, there is no energy gap that would stabilize their number. An attractive force between them will induce a major realignment in the perturbative wave function, i.e., $|P\rangle$, of the many-body gluon system.

Upon inserting Eq. (3.2) into Eq. (3.3), we see that the color B-field (magnetic) fluctuations dominate the color E-field (electrical) fluctuations:

$$\Delta \sum_a \vec{B}_a^2 = \Delta \sum_a \vec{E}_a^2 + 2[390 \pm 12 \,\mathrm{MeV}]^4. \qquad (3.4)$$

Here, Δ is defined as on the left-hand side of Eq. (3.3). The natural interpretation of this equation is that the true vacuum structure is predominantly magnetic. Indeed, an instability of the perturbative vacuum of QCD toward the formation of a ferromagnetic structure, was discovered early on in the development of QCD [56, 187, 236]. This effect has been shown to arise due to the attractive magnetic spin–spin interaction of gluons [35, 199, 200]. This spontaneous ferromagnetic instability parallels, in many important aspects, the instability in QED vacuum in the presence of constant electro-magnetic (EM) fields.

In QED, in the presence of a constant electrical field E, there is a nonvanishing probability of spontaneous particle-pair formation, with the probability per unit time and volume given by [240]

$$w = \frac{\alpha \vec{E}^2}{\pi^2} \sum_{n=1}^{\infty} \frac{1}{n^2} \exp\left(-n\frac{\pi m^2}{|e\vec{E}|}\right). \qquad (3.5)$$

The electromagnetic fine-structure constant, $\alpha \simeq 1/137$, is relatively small, and the mass m of the lepton (electron) produced is large compared with the laboratory fields available. Thus, in fact, this process has never been observed. The physical origin of the QED vacuum instability resides in the fact that, in a constant infinitely extending field, we can always find a potential difference between two distant points that exceeds the pair mass, and thus spontaneous pair production can ensue [219]. Schwinger's rate Eq. (3.5) is arising in such a description from the process of quantum tunneling through a barrier that the potential $V = \int d\vec{x}\vec{E}$ implies, and therefore it can be adapted with ease to the study of QCD [88, 132]. This mechanism is serving as the basis for particle production within the color string models, in which breaking of the color-electrical-flux tube connecting rapidly separating quarks provides the mechanism for particle production [39].

In the case of QED, the particles produced are screening the field source, and the vacuum-state energy still has a local minimum around the perturbative vacuum-state configuration with vanishing EM fields. In this regard the situation is different in QCD, in which there is a ferromagnetic instability. To understand this QCD magnetic instability, recall that, in a constant magnetic field of magnitude B, a particle with spin projection σ and orbital momentum $l = 1, 2, 3, \ldots$, with reference to the direction of B has the Landau energy

$$E_{l\sigma}^2 = m^2 + k_\parallel^2 + 2g_s B(l + \sigma + \tfrac{1}{2}), \qquad (3.6)$$

and the effective degeneracy is

$$g_B = \frac{V^{2/3}}{2\pi} g_s B. \qquad (3.7)$$

For $\sigma = -\tfrac{1}{2}$ (leptons, quarks), the lowest energy level for $k_\parallel = l = 0$ is at $E_0^2 = m^2$, as is seen in Eq. (3.6). However, for spin-1 gluons, states with $\sigma = -1$ display an instability whereby E_0^2 becomes negative for $g_B > m^2 = g_s B_{cr}$. For gluons with $m_g = 0$, this occurs for an arbitrarily small value of B. Therefore, the spectrum of Landau states begins at a minimum momentum, $k_\parallel > \sqrt{g_s B}$ for the relevant case of gluons with $\sigma = +1$ and $l = 0$. This has a profound impact on the zero-point energy of the vacuum.

The sum over all (stable) modes of particles $(+)$ and antiparticles $(-)$ yields the vacuum energy, that is the expectation value of the Hamiltonian in the perturbative state $|B\rangle$ in the presence of the magnetic field B:

$$\langle B|\mathcal{H}|B\rangle = (-)^{2\sigma}\frac{1}{2}\left(\sum_+ E_+(B) - \sum_- E_-(B)\right) \equiv \mathcal{E}_0^\sigma(B)V. \qquad (3.8)$$

The coefficient of the zero-point energy density $\mathcal{E}_0^\sigma$ reflects the spin-statistics relation. The appearance of the lowest-angular-momentum states of the minimum allowable momentum leads for the gluon fields (after subtraction of the perturbative state and renormalization) to [35, 199, 200]

$$\mathcal{E}_0^{\mathrm{QCD}}(B) = \frac{b_0}{2}\frac{(g_s B)^2}{4\pi}\log\left(\frac{(g_s B)^2}{\Lambda^2}\right), \qquad (3.9)$$

where $b_0 = (1/2\pi)(11n_c/3 - 2n_f/3) > 0$ is as given by Eq. (14.14). Equation (3.9) proves that the vacuum state acquires an instability in the limit at $B = 0$, since the vacuum energy does not exhibit a minimum in this limit. We find a new minimum of the vacuum energy at a finite value of

$(g_s B)^2$. The scale of the 'condensation' field is determined by the renormalization scale Λ.

While these results prove the instability of the perturbative state $|P\rangle$, given the variational approach the ferromagnetic-vacuum model may be a very poor approximation to the actual vacuum structure of $|V\rangle$. Though the energy of the $|P\rangle$ vacuum is lowered, and we find a minimum at a finite value of the magnetic field B, it cannot be expected that we have, within this crude model, reached the lowest energy corresponding to the true state $|V\rangle$. Even so, Eq. (3.9) allows a first estimate of the latent energy involved in melting the (magnetic) QCD vacuum structure to be made:

$$\mathcal{B}_B \equiv -\mathcal{E}(B_{\min}) = \frac{b_0}{8\pi}(g_s B)^2_{\min} \lesssim \mathcal{B}. \tag{3.10}$$

$\mathcal{B}_B$ is seen as the variational approximation to the true value $\mathcal{B}$. The value at the minimum underestimates the true gain in energy within a more accurate vacuum structure model. To determine the scale of the magnetic field near the minimum of the energy density, we take as the average value of the square of the vacuum magnetic field the vacuum expectation value of the field operator squared, Eq. (3.3):

$$\frac{1}{2\pi^2}(g_s B)^2_{\min} \equiv \delta\frac{\alpha_s}{\pi}\langle V|F^a_{\mu\nu}F^{\mu\nu}_a|V\rangle. \tag{3.11}$$

δ is a positive number by definition. It can not be bigger than unity. The example of the quantum oscillator expression for $\langle x^2\rangle$ suggests that it is probably small relative to unity. The nonperturbative energy density of the vacuum state Eq. (3.10) is then of the magnitude

$$\mathcal{B} \gtrsim \frac{11 - \frac{2}{3}n_f}{8}\langle V|\epsilon\frac{\alpha_s}{\pi}F^2|V\rangle \simeq \delta\, 2.5 \text{ GeV fm}^{-3}. \tag{3.12}$$

We also note the Curie-point (the temperature at which the magnetic ferric structure melts) of the magnetic QCD state at temperature $T_{\mathrm{cr}} \simeq \mathcal{B}_B^{1/4}$, at which one finds a strong first-order phase transition [192].

We infer from this exploration of a magnetic-vacuum model that the perturbative QCD vacuum $|P\rangle$ is unstable for $T < T_{\mathrm{cr}}$, and that the transition to the true vacuum state involves a considerable release of latent heat. However, the quantitative results discussed here are merely providing an order-of-magnitude estimate. In fact, many other more complex semi-analytical models of the QCD vacuum structure were developed, of which the other most often addressed case is the instanton vacuum. In this approach, one draws on the (infinite) degeneracy of the unstructured state. A more thorough discussion of this model is offered in the monograph of Shuryak [245].

3.3 Chiral symmetry

The light u and d quark masses, which we have considered in table 1.1, are just slightly different when they are measured on the energy scale associated with the QCD vacuum structure, which is of the order of a few hundred MeV. This opens up an interesting interplay between the effective flavor symmetry of QCD and the vacuum properties. Recall that up and down quarks satisfy the relativistic Dirac quantum field dynamics, Eq. (13.79),

$$(i\gamma^\mu \partial_\mu - m)\Psi = 0, \qquad (3.13)$$

from which there arise two identities,

$$\partial_\mu j_+^\mu \equiv \partial_\mu (\bar{u}\gamma^\mu d) = i(m_u - m_d)\bar{u}d, \qquad (3.14)$$

$$\partial_\mu j_+^{5\mu} \equiv \partial_\mu (\bar{u}\gamma^\mu \gamma_5 d) = i(m_u + m_d)\bar{u}\gamma_5 d, \qquad (3.15)$$

where u and d are the Dirac spinor-field operators representing the two light-quark flavor fields of current-quark masses m_u and m_d, respectively. The subscript '+' reminds us that these currents 'lift' the 'down' quark to the 'up' quark; in the quantum-field-theory formulation this current is an iso-raising operator that increases the electrical charge by $+|e|$.

When the quark masses are equal, the isospin-quark current is conserved in Eq. (3.14), which implies that the Hamiltonian is symmetric under transformations that mix equal mass 'u' with 'd' quarks; this is an expression of the isospin-$SU(2)$ symmetry of strong interactions; this symmetry is broken by the electromagnetic and weak interactions, and by the difference in current-quark masses $m_u \neq m_d$, as seen in Eq. (3.14).

In case that the light quark masses were to vanish, by virtue of Eq. (3.15), the pseudo-vector isospin-quark current would also be conserved. Thus, when we are dealing with physical situations in which the current quark masses can be neglected, each isospin quark doublet operator $q \equiv (u, d)$ must be invariant under transformations that comprise two 'isospin rotations' associated with the two current-conservation laws.

When we are motivated by the physical properties of weak interactions, it is common to study the left- and right-handed quark fields

$$q_{L,R} \equiv \tfrac{1}{2}(1 \pm \gamma_5)q.$$

The reader is reminded that, for the right-handed case, the spin rotates right-handedly around the propagation axis, that is the spin and momentum vectors are pointing in the same direction; the 'helicity' is positive. It can be shown, on general grounds, that, for massless fermions, the helicity is conserved.

On forming the sum and difference of Eqs. (3.14) and (3.15), one finds that both the right- and the left-handed doublets form conserved isocurrents; thus the overall symmetry is $SU(2)_L \times SU(2)_R$. This is the so-called chiral symmetry, i.e., 'handedness' symmetry. It is important to remember that this symmetry can be exact only if the masses of u and d quarks vanish exactly, and electro-weak interactions that distinguish the light quark flavor can be neglected. Since $m_u + m_d = 5$–$15\,\text{MeV}$, we expect this nearly exact chiral symmetry to be manifesting itself strongly at the hadronic energy scale $\mathcal{O}(1)$ GeV, literally wherever we 'look'. Yet, there is no sign of the corresponding symmetry in the hadronic spectrum; there are no double doublets of hadronic parity states, e.g., we know that there is only one isospin doublet of nucleons (proton–neutron), not two: the second, chiral-symmetry-motivated opposite-parity, isospin doublet of nucleons is not observed. It would seem that chiral symmetry is badly broken by strong interactions, presumably the mass difference of quarks somehow matters.

However, the Adler–Weisberger sum rules, which relate weak and strong properties, confirm the presence of the intrinsic $SU(2) \times SU(2)$ symmetry in the elementary Hamiltonian. We refer to the recent discussion of Weinberg for a more comprehensive introduction to this rather important matter [268]. Nambu resolved this conflict between weak and strong interactions by proposing that the required symmetry-breaking mechanism is part of the *structure* of the strongly interacting vacuum state, and the physical hadron spectrum can indeed break the intrinsic (almost) chiral symmetry of the Hamiltonian [195]. Weinberg is of the opinion that the immediate acceptance of QCD as the dynamic theory of strong interactions was very much the result of a rather natural implementation in terms of practically massless 'current' u and d quarks (see Eqs. (3.14) and (3.15)) of these contradictory properties of weak and strong interactions.

The Nambu breaking of chiral symmetry in the hadronic spectrum requires that, in the limit that the quark masses vanish exactly, there would be an exactly massless Goldstone boson, a particle with quantum numbers of the broken symmetry, thus spin zero, negative parity, and isospin $I = 1$. Since the chiral symmetry of the strong-interaction Hamiltonian is not exact, the lowest-mass particle with these quantum numbers, the nearly massless pion state, expresses the properties of the massless Goldstone meson of strong interactions.

One could argue that the finite pion mass noticeable on the scale of hadronic interactions is removing from the hadronic spectrum most of the signature of chiral symmetry. The missing parity doublets of all strongly interacting particles are a 'direct product' of the Goldstone boson (pion) with all elementary hadron states. This, in turn, implies that many features of the hadronic spectrum, and possibly of the vacuum structure, should depend on the small, and seemingly irrelevant, current quark masses we see in Eqs. (3.14) and (3.15). How this could happen is not understood.

We show now that, in the limit of vanishing quark masses, we expect the pion mass also to vanish. This behavior plays an important role in the conceptual understanding of the vacuum structure of strong interactions. We consider matrix elements of the pseudo-scalar and the pseudo-vector quark currents between the vacuum state and one pion state,

$$\langle \pi^+(p)|\bar{u}(x)\gamma^\mu\gamma_5 d(x)|V\rangle \equiv -i\sqrt{2}p^\mu f_\pi e^{ip_\mu x^\mu}, \qquad (3.16)$$

$$\langle \pi^+(p)|\bar{u}(x)\gamma_5 d(x)|V\rangle \equiv i\sqrt{2}g_\pi e^{ip_\mu x^\mu}, \qquad (3.17)$$

where $p_\mu p^\mu = m_\pi^2 = (139.6\,\mathrm{MeV})^2$. The form of the right-hand sides of Eqs. (3.16) and (3.17) arises by virtue of the Lorentz symmetry properties of the (true) vacuum state $|V\rangle$ and the π^+ state $|\pi^+(p)\rangle$. We consider the divergence ∂_μ of Eq. (3.16). Using relation Eq. (3.15), the following well-known result is found:

$$\boxed{m_\pi^2 f_\pi = (m_\mathrm{u} + m_\mathrm{d})g_\pi.} \qquad (3.18)$$

The matrix element f_π is well known, since it governs the weak-inter-action decay of pions, see, e.g., Weinberg, and the value g_π is determined by sum rule methods [242, 243]:

$$f_\pi = 93.3\,\mathrm{MeV}, \qquad g_\pi \simeq (350\,\mathrm{MeV})^2. \qquad (3.19)$$

Equation (3.18) implies that

$$m_\mathrm{u} + m_\mathrm{d} = 0.1 m_\pi, \qquad (3.20)$$

a somewhat unexpected result in the present context, since the (current) quark masses are found to be much lighter even than that of the 'massless' pion.

Weinberg also presents an in-depth discussion of the exploration of the Nambu–Goldstone structure of the vacuum, in terms of symmetry relations between current-matrix elements (current algebra). In our context, the most important vacuum property involving quarks is the Gell-Mann–Oakes–Renner (GOR) relation, which, adapted to the quark language (see section 31 of [280], or [125]), implies a relation with the quark fluctuations (condensate) in the true vacuum:

$$\boxed{m_\pi^2 f_\pi^2 = 0.17 \times 10^{-3}\mathrm{GeV}^4 \simeq -\tfrac{1}{2}(m_\mathrm{u} + m_\mathrm{d})\langle \bar{u}u + \bar{d}d\rangle + \cdots.} \qquad (3.21)$$

On dividing Eq. (3.21) by Eq. (3.18), we obtain

$$-f_\pi g_\pi = \tfrac{1}{2}\langle \bar{u}u + \bar{d}d\rangle|_{1\,\mathrm{GeV}} \equiv \tfrac{1}{2}\langle \bar{q}q\rangle = -(225 \pm 9\,\mathrm{MeV})^3, \qquad (3.22)$$

where we have used the values of f_π and g_π given in Eq. (3.19). When Eq. (3.22) is combined with Eq. (3.21), and some of its generalizations, we can determine the values of current quark masses [105]. This shows how the use of matrix-element properties and sum rules allows one to establish the physical values of the light quark masses presented in table 1.1.

We have introduced, in section 3.2, the condensates of glue, and above, of quark fields as if these were two quite independent physical effects of strong interactions. There remains an important question: is there a relation between glue-condensate melting (confinement-to-deconfinement transformation of the vacuum) and quark-condensate melting (the restoration of chiral symmetry)? One could be tempted to infer that the chiral symmetry-breaking features in QCD and gluon condensation have little in common. However, studies of restoration of symmetry of the vacuum at high temperature [103] have yielded contrary evidence: the two different vacuum structures of QCD always disappear together in the numerical studies as, e.g., the temperature is varied [103]. Model calculations [106, 107, 109, 251] employing mean-field configurations of gauge fields in the QCD vacuum suggest that it is the presence of the glue-field condensate which is the driving force causing the appearance of the quark condensate.

The mechanism connecting the two structures in the QCD vacuum (glue condensate, Eq. (3.3), and chiral condensate, Eq. (3.22)) is a major unsolved theoretical problem of strong-interaction physics. We will not pursue further in this book this interesting subject, which is undergoing rapid development.

3.4 Phases of strongly interacting matter

It is expected that, in nuclear collisions at relativistic energies, we attain conditions under which the structured confining vacuum is dissolved, forming a domain of thermally equilibrated hadronic matter comprising freely movable quarks and gluons. A qualitative sketch of the phase diagram of dense hadronic matter is shown in Fig. 3.1. The different phases populate different domains of temperature T and baryon density ρ_b presented in units of the normal nuclear density in heavy nuclei, $\rho_0 \simeq \frac{1}{6}$ nucleons fm^{-3}. For high temperatures and/or high baryon density, we have the deconfined phase. If deconfinement is reached in the nuclear-collision reactions, it 'freezes' back into the state containing confined hadrons during the temporal evolution of the small 'fireball', as indicated by the arrows in Fig. 3.1. Almost the entire ρ–T region can be explored by varying the collision energy of the colliding nuclei.

The most difficult domain to reach experimentally is the one of low baryon number density, at high T, corresponding to the conditions pertaining in the early Universe. This demands extreme collision energies, which would permit the baryon number to escape from the central rapidity region, where only the collision energy is deposited; see chapter 5.

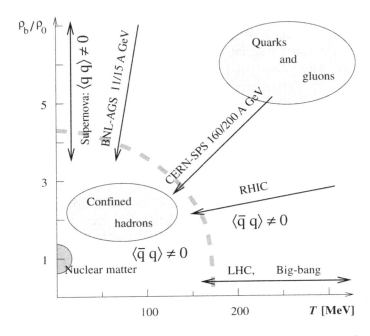

Fig. 3.1. The regions of the principal forms of hadronic matter are shown in the baryon-density–temperature plane. Their exploration with various accelerators is indicated, as are the domains relevant for cosmology and astrophysics. Also indicated is the behavior of the quark condensate.

The vertical arrow at the lowest temperatures, in Fig. 3.1, corresponds to the case of the stellar explosion of a supernova. Rather low-energy collisions at the AGS lead to such baryon-dense environments, which are more similar to nova explosions than they are to the early-Universe big-bang (horizontal arrow), which is better simulated by RHIC and future LHC experiments. In between these two extremes, we find the SPS conditions. The specific beam capabilities of the various accelerator facilities are complementary; section 5.1.

There are three regions indicated in Fig. 3.1 by the quark condensate, the expectation value of the quark fields $\langle \bar{q}q \rangle$, see Eq. (3.22), and $\langle qq \rangle$. The attractive quark–quark interaction present in some of the two-particle channels allows this di-quark color-condensate to form at low temperature and high quark density. We will not discuss the extensive work which recently addressed the properties of 'cold' quark matter, in which a 'color–flavor' locking (pairing) of quarks introduces yet other interesting structures in the deconfined state [31]. This color-superconductive phase had already been proposed early on in the study of properties of quark matter [53].

Recent studies have confirmed that the temperature at which such a color–flavor-locked phase of quark matter could exist is too low for an exploration in present-day laboratory experiments involving relativistic heavy-ion collisions [207]. At the temperature of interest in our studies, $T > 100$ MeV, the quark pairing will be largely dissolved. Work on this subject is rapidly evolving, for its current status we refer the reader to a recent review [227].

Where exactly an equilibrium transition between two phases occurs is determined from Gibbs' conditions for phase equilibrium. These establish the boundary between the physical phases considered, for bulk matter embedded in heat and particle-number 'baths'. These baths supply energy and particles to maintain given thermodynamic conditions. Although the circumstance of a nuclear-collision fireball is very different, the logic inherent in Gibbs' conditions will guide our understanding. The first condition is

$$\boxed{P_1 = P_2,} \tag{3.23}$$

which assures that there is no physical force acting on the phase boundary. We will momentarily return to discuss what happens when the phase boundary is in (relativistic) motion, see Eq. (3.28) below. The second Gibbs condition is

$$\boxed{T_1 = T_2,} \tag{3.24}$$

which assures that there is no radiative transport of energy between the phases.

How these conditions define the phase boundaries is illustrated in the P–V diagram in Fig. 3.2. The pressure in two phases (QGP and HG) is considered at fixed temperature T (and at given conserved baryon number b) as a function of volume V, at variable baryon density $\rho_b = b/V$. We distinguish three domains in Fig. 3.2:

1. the HG region for $V > V_2$ (corresponding to $\rho_b < \rho_2$), where the pressure rises modestly with the reduction of the volume;
2. the QGP region where the hadrons have disappeared at $V < V_1$ (corresponding to $\rho_b > \rho_1$); and
3. the Van der Waals regime in the intermediate region from V_1 to V_2; a way to understand this domain is, e.g., that, at V_1, the progenitors of individual hadrons begin to emerge in the QGP phase in the form of a localized cluster of quarks. Similarly, beginning at V_2 and with decreasing volume, one can, e.g., consider clustering of individual hadrons into quark-drops [220].

Because clustering of hadrons leads naturally to the formation of drops of the QGP-like phase, we refer to the coexistence region, between V_1 and

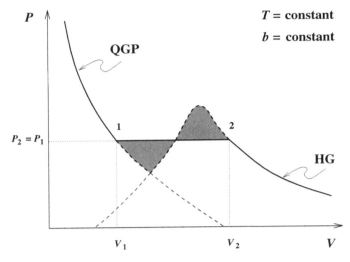

Fig. 3.2. The p–V diagram for the QGP–HG system, at fixed temperature and baryon number; dashed lines indicate unstable domains of overheated and undercooled phases.

V_2, also as the mixed phase, comprising a mixture of hadrons and drops of QGP or perhaps hadron-like clusters of quarks and free quarks. To find out at which pressure the transformation between the phases occurs, at a given temperature T (and fixed baryon number b), we find the value of the pressure, $P_1 = P_2 \equiv P_{12}$, connecting the volumes V_1 and V_2, requiring that the work done along the metastable branches vanishes:

$$\int_{V_1}^{V_2} (P - P_{12}) \, dV = 0. \tag{3.25}$$

The integrand is shown shaded in Fig. 3.2.

This Maxwell construction can be repeated for different values of b and T, and the set of resulting points 1 and 2 forms then two phase-boundary lines shown on the left-hand side in Fig. 3.3, in the $(\rho_b\text{–}T)$ plane. The Maxwell-construction line, seen in Fig. 3.2, is the vertical line connecting at fixed temperature T the two different values of baryon density found – in general, a jump in baryon density (and energy density and entropy density) is encountered if a first-order phase transition occurs. The region of high T, at fixed ρ_b, is associated with the deconfined QGP and the region of small T with HG. The shape of the phase boundary is expressing the fact that a baryonless hadronic-gas phase cannot exist at a high enough temperature, and that dense compressed cold baryon matter will transform into the deconfined quark matter phase of quark matter.

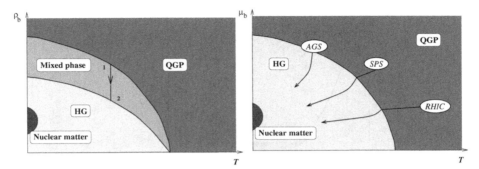

Fig. 3.3.   Left: the regions of QGP and HG in the (ρ_b-T) plane are separated by a band in which the phases coexist. The Maxwell-construction line corresponding to Fig. 3.2 is also shown, as is the path for the evolution of the Universe. Right: the same in the (μ_b-T) plane. The qualitative evolution of fireballs of dense matter created at the AGS, SPS, and RHIC is shown.

When we conserve the baryon number b on 'average' and introduce the baryochemical potential μ_b as a variable, the representation of the phase boundary changes. According to the third Gibbs condition,

$$\boxed{\mu_1 = \mu_2,} \tag{3.26}$$

the two chemical potentials must have the same value at each given T in order to assure that no transport of particles across the phase boundary occurs. Given Eq. (3.26), there is just a simple line separating any two hadronic phases in the (μ_b-T) plane, as is shown on the right-hand side in Fig. 3.3. There is a discontinuity of the energy density, baryon density, and entropy density across the phase boundary.

3.5 The expanding fireball and phase transformation

The lines shown on the right-hand side in Fig. 3.3 suggest the possible evolution of the fireball of dense matter formed in a heavy-ion collision. If a QGP fireball were indeed formed in the micro-bang, it will not expand along a fixed-temperature trajectory such as is encountered under the isothermal conditions of a heat bath. In our case, instead, entropy is the (nearly) conserved quantity for an isolated system subject to ideal flow. The evolution at constant entropy per baryon corresponds nearly to a straight line in the μ_b-T diagram in the domain of QGP (dark shaded), since the dimensionless ratio entropy per baryon is a function of other dimensionless variables – which, in the absence of significant scales other than μ_b, and T is μ_b/T. A considerable change in temperature must occur during the evolution of a fireball, as is indicated in qualitative terms by the trajectories shown in the right-hand panel of Fig. 3.3.

A phase transition that is 'strong', i.e., involves significant changes in physical properties, will be easier to find. As a strong transition, we understand a case with, e.g., a jump in magnitude of the energy, or baryon density. Should such a strong phase transition separate the two phases, the super-cooling effect of a rapidly expanding (exploding) fireball of finite size could be much more pronounced.

We recall that both the QGP and the HG phases have metastable phase branches indicated by dashed lines in Fig. 3.2. However, now we look at these at constant entropy and variable temperature. Therefore, these domains are referred to as the undercooled plasma (continuation of 1 in Fig. 3.2), or the overheated hadronic-gas states (continuation of 2 in Fig. 3.2). Thus, the pressure of the QGP phase can evolve to be well below the transition pressure.

In fact, when a fireball of dense quark–gluon matter (phase 1) rapidly explodes, driven by the high internal temperature and pressure, it is possible that it continues even beyond $P = 0$. Namely, the fluid-flow motion of quarks and gluons expands the domain of deconfinement by exercising against the vacuum component in the total pressure a force originating from the collective velocity $\vec{v}_c$.

Let P and ϵ be the pressure and energy density of the deconfined phase in the local restframe, subject to flow velocity $\vec{v}_c = (v_1, v_2, v_3)$. The pressure-tensor component in the energy–momentum tensor (compare with Eq. (6.6)) is

$$T^{ij} = P\delta_{ij} + (P + \epsilon)\frac{v_i v_j}{1 - \vec{v}^2}. \tag{3.27}$$

The rate-of-momentum-flow vector $\vec{\mathcal{P}}$ at the surface of the fireball is obtained from the energy-stress tensor T_{kl}:

$$\widehat{\mathcal{T}} \cdot \vec{n} = P\vec{n} + (P + \epsilon)\frac{\vec{v}_c(\vec{v}_c \cdot \vec{n})}{1 - \vec{v}_c^2}. \tag{3.28}$$

The pressure and energy comprise particles (subscript p) and the vacuum properties:

$$P = P_p - \mathcal{B}, \quad \epsilon = \epsilon_p + \mathcal{B}. \tag{3.29}$$

Equation (3.28) for the condition $\widehat{\mathcal{T}} \cdot \vec{n} = 0$ reads

$$\mathcal{B}\vec{n} = P_p\vec{n} + (P_p + \epsilon_p)\frac{\vec{v}_c(\vec{v}_c \cdot \vec{n})}{1 - v_c^2}, \tag{3.30}$$

and it describes the (equilibrium) condition under which the pressure of the expanding quark–gluon fluid is just balanced by the external vacuum pressure. On multiplying by $\vec{n}$, we find

$$B = P_{\mathrm{p}} + (P_{\mathrm{p}} + \epsilon_{\mathrm{p}}) \frac{\kappa v_{\mathrm{c}}^2}{1 - v_{\mathrm{c}}^2},$$

(3.31)

where we introduced the geometric factor κ:

$$\kappa = \frac{(\vec{v}_{\mathrm{c}} \cdot \vec{n})^2}{v_{\mathrm{c}}^2}.$$

(3.32)

κ characterizes the angular relation between the surface-normal vector and the direction of flow. Under the condition Eq. (3.31), the total QGP-phase pressure $P = P_{\mathrm{p}} - B$, Eq. (3.29), is negative, as we set out to show.

Expansion beyond $P \to 0$ is in general not possible. A surface region of a fireball that reaches condition Eq. (3.31) and continues to flow outward must be torn apart. This is a collective instability and the ensuing disintegration of the fireball matter should be very rapid. We find that a rapidly evolving fireball that supercools into the domain of negative pressure is in general highly unstable, and we expect that a sudden transformation (hadronization) into confined matter can ensue under such a condition. It is important to note that the situation we have described could arise only since the vacuum-pressure term is not subject to flow and always keeps the same value.

3.6 QGP and confined hadronic-gas phases

We next seek to qualitatively understand the magnitude of the temperature at which the deconfined quark–gluon phase will freeze into hadrons. The order of magnitude of this transition temperature (if a phase change occurs) or transformation temperature (if no phase transition occurs, see Fig. 3.1) is obtained by evaluating where a benchmark value for the energy density occurs:

$$\epsilon_{\mathrm{H}} \simeq 3P_{\mathrm{H}} = 1 \text{ GeV fm}^{-3}.$$

The generalized Stefan–Boltzmann law (Eqs. (1.6) and (4.66)) describes the energy density ϵ and pressure P as functions of the temperature T of a massless relativistic gas:

$$P^{\mathrm{SB}} = \frac{1}{3}\epsilon^{\mathrm{SB}} = \frac{\pi^2}{90}gT^4.$$

(3.33)

The quantity g is the number of different (relativistic) particle states available and is often called the 'number of degrees of freedom' or 'degeneracy'. In the deconfined phase,

$$g \equiv g_{\mathrm{g}} + \tfrac{7}{4}g_{\mathrm{q}},$$

(3.34)

which comprises the contribution of massless gluons (bosons) and quarks (fermions). The relative factor $2 \times \frac{7}{8} = \frac{7}{4}$ expresses the presence of particles and antiparticles (factor 2) and the smaller fermion phase space, compared with the boson case, given the exclusion principle (factor $\frac{7}{8}$, section 10.5).

We use the degrees of freedom of quarks and gluons many times in this book. Here, we ignore the role of interactions. Gluons carry color and spin, and so do quarks, which, in addition, come in two ($n_f = 2$) flavors u and d; see table 1.1. Since at high temperatures the flavor count may include the strange quark, we leave n_f as a variable. We obtain the following degeneracy in a QGP:

$$\text{gluons: } g_g = 2(\text{spin}) \times (N_c^2 - 1)(\text{color}) = 2 \times 8 = 16, \qquad (3.35a)$$

$$\text{quarks: } g_q = 2(\text{spin}) \times N_c(\text{color}) \times n_f(\text{flavor}) = 2 \times 3 \times n_f. \quad (3.35b)$$

When the semi-massive strange quarks are present, the effective number of 'light' flavors is $\simeq 2.5$. Thus, $g \simeq 40$ in Eq. (3.33), to be compared with just two directions of polarization for photons.

For a massless ideal quark–gluon gas, we find

$$T_H = 160 \text{ MeV}, \qquad \text{for} \qquad \epsilon_H = 1.1 \text{ GeV fm}^{-3}.$$

Hagedorn introduced this critical temperature in his study of the boiling point of hadronic matter [140]. Numerical simulations obtained by implementing QCD on a space–time lattice are available for zero baryon density, and these results confirm that, at approximately T_H, there is a phase transformation between confinement and deconfinement [160]. One also finds a rapid change in the behavior as the number of quarks and their masses m_s and m_q are varied.

The resulting complexity of the phase structure is shown, in Fig. 3.4, as a function of m_s and m_q. In this qualitative representation, we look at the plane spanned by the light-quark mass $m_q = m_u = m_d$ and the strange-quark mass m_s. On two boundaries of Fig. 3.4 these masses are infinite. Only near the origin is the effective number of massless flavors three, along the diagonal we have three massive flavors. Depending on the actual values of quark masses, different phase properties emerge [162].

The theoretical finding that a smooth crossover between the confined and deconfined phases is a possibility raises the question of how to understand qualitatively the gradual onset of color (quark, gluon) mobility. A gradual change implies that free quarks can coexist with confined hadrons. This then also suggests that liberation of quarks is possible since permanent confinement could be assured only at zero temperature, a mathematical limit. For any finite excitation of the system, quark mobility remains, akin to the transition of an atomic gas to an electron–ion

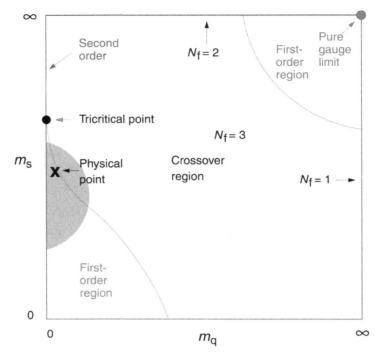

Fig. 3.4.   The nature of finite-temperature QCD phase structure as a function of quark masses m_q and m_s.

plasma. However, experimental searches for quarks have not succeeded [186]. The experimental limits which were set suggest that confinement is a fundamental physical property. This being the case, we are of the opinion that, in the physical world, the transformation from the confined to the deconfined phase is a discontinuous phase transition, most likely of first order. For this reason, we placed the physical quark-mass point within the region of first-order phase transition in Fig. 3.4. This topical area is undergoing a rapid evolution.

4 Statistical properties of hadronic matter

4.1 Equidistribution of energy

The physical tools required to describe in further detail the properties of hot hadronic matter are much like the usual ones of statistical physics, which we briefly introduce and review. A more detailed analysis will follow.

Consider a large number N of identical coupled systems, distinguishable, e.g., by their energies E_i. To simplify the matter, we assume that the energies E_i can assume only discrete values, and that there are K different 'macro' states such that $K \ll N$. Some of the energies of the macro states will be equal, i.e., most are occupied more often than once, and in general n_i times. The total energy,

$$E^{(N)} = \sum_{i=1}^{K} n_i E_i, \qquad (4.1)$$

is conserved. Further below, we will introduce also conservation of a discrete quantum number, e.g., the baryon number. We note another subsidiary condition arising from the definitions:

$$\sum_{i}^{K} n_i = N. \qquad (4.2)$$

Without an additional quantum number, systems with the same energy E_i are equivalent, i.e., in the language of quantum statistics, indistinguishable.

The distribution $\boldsymbol{n} = \{n_i\}$ having the same energy E_i can be achieved in many different ways. To find how many, consider the relation

$$K^N = (x_1 + x_2 + \cdots + \cdots x_K)^N|_{x_i=1}$$
$$= \sum_{\boldsymbol{n}} \frac{N!}{n_1! n_2! \cdots n_K!} x_1^{n_1} x_2^{n_2} \cdots x_K^{n_K}|_{x_i=1}. \qquad (4.3)$$

The normalized coefficients,

$$\boxed{W(\boldsymbol{n}) = \frac{K^{-N} N!}{\prod_{i=1}^{K} n_i!},} \qquad (4.4)$$

are the relative probabilities of realizing each state in the ensemble $\boldsymbol{n}$, with n_i equivalent elements. The practical way to find the most probable distribution $\bar{\boldsymbol{n}}$ is to seek the maximum of $\ln W$, Eq. (4.4), subject to the constraints Eqs. (4.1) and (4.2),

$$A(n_1, n_2, \ldots, n_K) = \ln W(\boldsymbol{n}) - a \sum_{i} n_i - \beta \sum_{i} n_i E_i, \qquad (4.5)$$

characterized by two Lagrange multipliers a and β. We differentiate Eq. (4.5) with respect to the n_i:

$$\frac{\partial}{\partial n_i} [-\ln(n_i!) - n_i a - \beta n_i E_i] \Big|_{\bar{\boldsymbol{n}}_m} = 0. \qquad (4.6)$$

Insofar as all $\bar{n}_i \gg 1$, we can use the relation

$$\frac{d}{dk} [\ln(k!)] \approx \frac{\ln(k!) - \ln[(k-1)!]}{(k) - (k-1)} = \ln k. \qquad (4.7)$$

One obtains for the maximum of the distribution Eq. (4.4), i.e., for the statistically most probable distribution $\bar{\boldsymbol{n}} = \{\bar{n}_i\}$, the well-known exponential

$$\bar{n}_i = \gamma e^{-\beta E_i},$$
(4.8)

where the inverse of the slope parameter β is identified below as the temperature:

$$T \equiv 1/\beta.$$
(4.9)

The supplementary condition Eq. (4.2), given the set $\bar{\boldsymbol{n}}$, now reads

$$\sum_i \bar{n}_i = \gamma \sum_{i=1}^K e^{-\beta E_i} = N.$$
(4.10)

The quantity γ,

$$\gamma \equiv e^{-a},$$
(4.11)

as seen in Eq. (4.10), controls the total number of members of the ensemble N. It is the chemical fugacity introduced in section 1.1. We will meet both statistical parameters T and γ many times again in this book.

Employing Eq. (4.8), we find for the energy $E^{(N)}$, Eq. (4.1),

$$E^{(N)} = \sum_i \bar{n}_i E_i = \gamma \sum_i E_i e^{-\beta E_i}.$$
(4.12)

On dividing $E^{(N)}$ by N, we obtain the average energy of each member of the ensemble:

$$\frac{E^{(N)}}{N} \equiv \overline{E^{(N)}} = \frac{\gamma \sum_i E_i e^{-\beta E_i}}{\gamma \sum_i e^{-\beta E_i}} \equiv -\frac{d}{d\beta} \ln Z.$$
(4.13)

We introduced here the *canonical partition function* Z:

$$Z = \sum_i \gamma e^{-\beta E_i}.$$
(4.14)

Unlike the microscopic (micro-canonical) approach in which the energy for each member of the ensemble is fixed, in the statistical 'canonical' approach, one studies the most likely distribution of energy and other physical properties among members of the ensemble. These properties are controlled solely by the statistical parameters β and γ which are the Lagrange multipliers related to the conservation of energy, and the number of members of the ensemble.

4.2 The grand-canonical ensemble

We will now relax the assumption that only energy is equipartioned as it is exchanged between macro systems. In the *grand-canonical* approach, we seek the most probable statistical distribution allowing for the flow of a discrete quantum number between the individual members of the statistical ensemble. In passing, we also show how the mathematical framework of the grand-canonical approach offers a convenient path to the evaluation of the canonical ensemble quantities, when discrete conservation laws apply.

We proceed in every detail as before, but need to characterize each macro state by an additional discrete number, and we need to introduce a further subsidiary condition to assure that this (baryon) number is conserved:

$$\sum_{i=1}^{N} n_i^b b_i = b^{(N)} \equiv N\bar{b}_i. \tag{4.15}$$

Here, $\bar{b}_i$ is the average number of baryons in each ensemble member considered. The condition Eq. (4.15) introduces a further constraining Lagrange parameter into Eq. (4.6), which we write in the form $-\ln\lambda$. In this way the generalization of Eq. (4.6) is

$$\frac{\partial}{\partial n_i^b}\left[-\ln(n_i^b!) - n_i^b a - \beta n_i^b E_i + \ln\lambda\, n_i^b b_i\right]\Bigg|_{\bar{n}_m} = 0. \tag{4.16}$$

Proceeding as in section 4.1, we obtain the most probable distribution of $\bar{n}_i$ as

$$\bar{n}_i^b = \gamma \lambda^{b_i} e^{-\beta E_i}, \tag{4.17}$$

where the number of particles is controlled by the fugacity factor λ and the factor $\gamma = e^{-a}$, see Eq. (4.11). It is common practice to introduce the chemical potential μ:

$$\boxed{\mu \equiv T\ln\lambda, \qquad \lambda = e^{\beta\mu} = e^{\mu/T}.} \tag{4.18}$$

The chemical potentials shown have physical meaning, and determine the energy required to add/remove a particle at fixed pressure, energy, and entropy. Following the method that led us to Eq. (4.13), we obtain

$$\boxed{E_{(N)} = \gamma\frac{\sum_{i;b} E_i\,\lambda^{b_i} e^{-\beta E_i}}{\gamma\sum_{i;b} \lambda^{b_i} e^{-\beta E_i}} \equiv -\frac{d}{d\beta}\ln\mathcal{Z}.} \tag{4.19}$$

We have introduced the grand-canonical partition function $\mathcal{Z}$:

$$\mathcal{Z}(V, \beta, \lambda) = \gamma \sum_{i;b} \lambda^{b_i} e^{-\beta E_i}. \tag{4.20}$$

$\mathcal{Z}$ is in fact also a 'generating' function for the canonical partition function Z_b,

$$Z_b(V, \beta) = \frac{1}{2\pi i} \oint db \frac{1}{\lambda^{b+1}} \mathcal{Z}(\beta, \lambda), \tag{4.21}$$

where $Z_b(V, \beta)$ describes a system with a fixed baryon number b. The path of integration in Eq. (4.21) leads around the singularity at $\lambda = 0$; it is often chosen to be the unit circle.

We can also evaluate the average value of b for the grand-canonical partition function:

$$\begin{aligned}
\bar{b} &= \frac{\sum_{i;b} b_i \lambda^{b_i} e^{-\beta E_i}}{\sum_{i;b} \lambda^{b_i} e^{-\beta E_i}}, \\
&= \lambda \frac{d}{d\lambda} \left(\ln \sum_{i;b} \gamma \lambda^{b_i} e^{-\beta E_i} \right) \equiv \lambda \frac{d}{d\lambda} \ln \mathcal{Z}(\beta, \lambda).
\end{aligned} \tag{4.22}$$

4.3 Independent quantum (quasi)particles

Elementary quantum physics allows a simple evaluation of the grand-canonical partition function. The discrete energies E_i of the physical systems in the statistical ensemble we introduced above are now to be understood as eigenenergies with eigenstate $|i\rangle$ of a quantum Hamiltonian $\hat{H}$:

$$\hat{H}|i\rangle = E_i|i\rangle. \tag{4.23}$$

Since the (conserved-baryon-number) operator $\hat{b}$ commutes with the Hamiltonian, $[\hat{b}, \hat{H}] = 0$, the eigenstates can furthermore be characterized by their baryon number (and strangeness, and other discrete quantum numbers that are constants of motion, but we restrict the present discussion to the baryon number only). We have

$$\hat{b}|i, b\rangle = b|i, b\rangle. \tag{4.24}$$

The grand-canonical partition function, Eq. (4.20), can be written as

$$\mathcal{Z} \equiv \sum_{i,b} \langle i, b| \gamma e^{-\beta(\hat{H} - \mu \hat{b})} |i, b\rangle = \text{Tr} \, \gamma \, e^{-\beta(\hat{H} - \mu \hat{b})}, \tag{4.25}$$

$$\equiv \sum_{n} \langle n| e^{-\beta(\hat{H} - \mu \hat{b} - \beta^{-1} \ln \gamma)} |n\rangle.$$

The great usefulness of this relation is that the trace of a quantum operator is representation-independent; that is, any complete set of microscopic basis states $|n\rangle$ may be used to find the (quantum) canonical or grand-canonical partition function. This allows us to obtain the physical properties of quantum gases in the, often useful, approximation that they consist of practically independent (quasi)particles, and, eventually, to incorporate any remaining interactions by means of a perturbative expansion.

The reference to quasi-particles is made since, e.g., in a medium, masses of particle-like objects can be different from masses of 'elementary' particles. Generally there will be collective excitation modes characterized by a mass spectrum. In this respect, dense hadronic matter behaves like any dense-matter system. As long as there is a set of well-defined excitations, it really does not matter whether we are dealing with real particles or quasi-particles, when we compute the trace of the quantum partition function Eq. (4.25). Putting it differently, even though we compute the properties of a 'free'-particle quantum gas, by choosing a suitable quasi-particle basis, we accommodate much of the effect of the strong interactions between particles.

The 'single (quasi)particle' occupation-number basis is the suitable one for the evaluation of the trace in Eq. (4.25). In this approach, each macro state $|n\rangle$ is characterized by the set of occupation numbers $n = \{n_i\}$ of the single (quasi)particle states with baryon charge b_i of energy ε_i, and the state energy is given by $E_n = \sum_i n_i \varepsilon_i$. The sum over all possible states corresponds to a sum over all allowed sets n: for fermions, each $n_i \in 0, 1$ and for bosons, $n_i \in 0, 1, 2, \ldots, \infty$:

$$\mathcal{Z} = \sum_n e^{-\sum_{i=1}^{\infty} n_i \beta(\varepsilon_i - \mu b_i - \beta^{-1} \ln \gamma)},$$

$$= \sum_n \prod_i e^{-n_i \beta(\varepsilon_i - \mu b_i - \beta^{-1} \ln \gamma)}. \tag{4.26}$$

In this case the sum and product can be interchanged:

$$\sum_n \prod_i e^{-n_i \beta(\varepsilon_i - \mu b_i - \beta^{-1} \ln \gamma)} = \prod_i \sum_{n_i = 0, 1 \ldots} e^{-n_i \beta(\varepsilon_i - \mu b_i - \beta^{-1} \ln \gamma)}. \tag{4.27}$$

To show this equality, one considers whether all the terms on the left-hand side are included on the right-hand side, where the sum is not over all the sets of occupation numbers n, but over all the allowed values of occupation numbers n_i. For fermions (F, Fermi–Dirac statistics) we can have only $n_i = 0, 1$, whereas for bosons (B, the Bose–Einstein statistics) $n_i = 0, 1, \ldots, \infty$. The resulting sums are easily carried out analytically:

$$\ln \mathcal{Z}_{\mathrm{F/B}} = \ln \prod_i \left(1 \pm \gamma e^{-\beta(\varepsilon_i - \mu b_i)}\right)^{\pm 1} = \pm \sum_i \ln(1 \pm \gamma \lambda_i^b e^{-\beta \varepsilon_i}). \tag{4.28}$$

The plus sign applies to F, and the minus sign to B; fermions have Pauli occupancy 0, γ, of each distinct single-particle state, and bosons have occupancy 0, γ^1, $2\gamma^2$, ..., ∞. The factor γ^n arises naturally since we have not tacitly set the occupancy of each single-particle level to unity as is commonly done when absolute chemical equilibrium is assumed.

For antiparticles, the eigenvalue of $\hat{b}$, in Eq. (4.28), is the negative of the particle value. Consequently, the fugacity $\lambda_{\bar{f}}$ for antiparticles $\bar{f}$ is

$$\boxed{\lambda_{\bar{f}} = \lambda_f^{-1}.}\tag{4.29}$$

It is convenient to also introduce this change in sign into the definition of the chemical potential, see Eq. (4.18), and to introduce particle and antiparticle chemical potentials such that

$$\boxed{\mu_f = -\mu_{\bar{f}}.}\tag{4.30}$$

These relations, Eqs. (4.29) and (4.30), will often be implied in what follows in this book. The microscopic (quasi)particle energy is denoted by ε in Eq. (4.28). For a homogeneous space–time, it is determined in terms of the momentum $\vec{p}$ in the usual manner:

$$\boxed{\varepsilon_i = \sqrt{m_i^2 + \vec{p}^2}.}\tag{4.31}$$

In order to make any practical evaluations of Eq. (4.28), we need to interpret the level sum $\sum_i$ with some precision. If energy is the only controlling factor then we carry out this summation in terms of the single-particle level density $\sigma_1(\varepsilon, V)$:

$$\sum_i [\ldots] = \int d\varepsilon\, \sigma_1(\varepsilon, V)[\ldots].\tag{4.32}$$

To obtain σ_1, i.e., the number of levels in a box of (infinite) volume $V = L^3$ per unit of energy ε, we note that quantum mechanics does not allow a continuous range of $\vec{p}$ in Eq. (4.31).

We consider a box L^3 with periodic boundary conditions and obtain the complete set of plane-wave states ψ having the required periodicity,

$$\psi \propto e^{i(\vec{p}_\alpha \cdot \vec{X})} = e^{i\vec{p}_\alpha(\vec{X}+\vec{n}L)},\tag{4.33}$$

where $\vec{n} = (n_1, n_2, n_3)$ with $n = 0, \pm 1, \pm 2, \ldots$. This fixes the allowed $\vec{p}_\alpha$ to

$$L\vec{p}_\alpha \cdot \vec{n} = 2\pi k, \qquad k = 0, \pm 1, \pm 2, \ldots.\tag{4.34}$$

This can be satisfied only if

$$\vec{p}_\alpha = \frac{2\pi}{L}(k_1, k_2, k_3), \qquad \text{with } k_i = 0, \pm 1, \pm 2, \ldots.\tag{4.35}$$

To sum over all single-particle states, we sum over all k_i. The number of permitted states equals the number of lattice points in the 'inverse' or 'phase-space' k-lattice. In the limit of large L,

$$[\text{number of states in } d^3k] = \left(\frac{L}{2\pi}\right)^3 d^3p = \frac{V}{(2\pi)^3}\, d^3p. \tag{4.36}$$

Thus, we obtain the single-level density, Eq. (4.32):

$$\sum_i [\ldots] = \int d\varepsilon\, \frac{V\, d^3p}{(2\pi)^3\, d\varepsilon} [\ldots]. \tag{4.37}$$

We keep in mind that, in general, the replacement of the discrete-level sum implies, in the limit of infinite volume of the system the phase-space integral

$$\boxed{\sum_i \to g \int \frac{d^3x\, d^3p}{(2\pi)^3}.} \tag{4.38}$$

Discrete quantum numbers, such as spin, isospin, and flavor, contribute an additive component of the same form in the sum over the single-particle states, which gives rise to the degeneracy coefficient g in Eq. (4.38). Aside from the volume term shown in Eq. (4.38), there is, in general, also a correction that has the form of a surface term. The magnitude and sign of the surface term depend on the physical problem considered. We will not pursue this topic further in this book; for a general discussion of this subject see, e.g., [51, 52].

4.4 The Fermi and Bose quantum gases

Allowing for the presence both of particles and of antiparticles, the quantum-statistical grand partition function Eq. (4.28) for a particle of mass m and degeneracy g can be written explicitly as

$$\boxed{\begin{aligned} \ln \mathcal{Z}_{\text{F/B}}(V, \beta, \lambda, \gamma) = \pm gV \int \frac{d^3p}{(2\pi)^3}\Big[&\ln(1 \pm \gamma\lambda e^{-\beta\sqrt{p^2+m^2}}) \\ &+ \ln(1 \pm \gamma\lambda^{-1}e^{-\beta\sqrt{p^2+m^2}})\Big]. \end{aligned}} \tag{4.39}$$

The second term in Eq. (4.39) is due to antiparticles. The well-known 'classical' Boltzmann limit arises from expansion of the logarithms, i.e., when it is possible to consider the exponential term as small relative to unity:

$$\boxed{\ln \mathcal{Z}_{\text{cl}}(V, \beta, \lambda, \gamma) = gV \int \frac{d^3p}{(2\pi)^3} \gamma(\lambda + \lambda^{-1}) e^{-\beta\sqrt{p^2+m^2}}.} \tag{4.40}$$

We will often use the (normalized) particle *spectrum*, the average relative probability of finding a particle at the energy E_i, which is the coefficient of E_i in Eq. (4.13). Using Eqs. (4.8) and (4.10) we obtain

$$\overline{w_i} \equiv \frac{\bar{n}_i}{N} = \frac{e^{-\beta E_i}}{\sum_j e^{-\beta E_j}}$$

$$= -\frac{1}{\beta} \frac{\partial}{\partial E_i} \left(\ln \sum_j \gamma e^{-\beta E_j} \right) = -\frac{1}{\beta} \frac{\partial}{\partial E_i} \ln Z. \tag{4.41}$$

The single-particle spectrum that follows from Eq. (4.28) is easily evaluated,

$$\boxed{f_{\mathrm{F/B}}(\varepsilon; \beta, \lambda, \gamma) = \frac{1}{\gamma^{-1}\lambda^{-1}e^{\beta\varepsilon} \pm 1},} \tag{4.42}$$

where the plus sign applies for fermions, and the minus sign for bosons. For antiparticles, we replace λ by λ^{-1}. The classical Boltzmann approximation arises again in the limit in which it is possible to neglect the term ± 1 in the denominator, i.e., when the phase-space abundance is small,

$$\boxed{f_{\mathrm{F/B}} \rightarrow f_{\mathrm{cl}} = \gamma\lambda e^{-\beta\varepsilon},} \tag{4.43}$$

where $\lambda \rightarrow 1/\lambda$ for antiparticles. More generally, for $\gamma\lambda e^{-\beta\varepsilon} < 1$, this Stefan–Boltzmann spectral shape can be written as an infinite series:

$$f_{\mathrm{F/B}} = \pm \sum_{n=1}^{\infty} \left(\pm\gamma\lambda e^{-\beta\varepsilon} \right)^n. \tag{4.44}$$

We consider, as an example, the spectra and yield of gluons, a special case of interest to us among bosons. Their behavior is similar to the case of photons ($g_\gamma = 2$) but gluons have an eight-fold greater color degeneracy ($g_{\mathrm{g}} = 16$). Both photons and gluons do not have an antiparticle partner, and their number is unrestricted by particle/antiparticle conservation; hence $\lambda \rightarrow 1$. We obtain

$$\ln \mathcal{Z}_{\gamma,\mathrm{g}} = -g_{\gamma,\mathrm{g}} V \int \frac{d^3p}{(2\pi)^3} \ln(1 - \gamma e^{-\beta\varepsilon}), \tag{4.45}$$

where $\varepsilon = \varepsilon(\vec{p}) = \sqrt{m^2 + \vec{p}^2} \rightarrow |\vec{p}|$, except when we consider a non-vanishing thermal gluon mass in the medium. The particle occupation probability is

$$\boxed{f_{\gamma,\mathrm{g}}(\varepsilon) = \frac{1}{\gamma^{-1}e^{\beta\varepsilon} - 1} = \sum_{n=1}^{\infty} \gamma^n e^{-n\beta\varepsilon}, \quad \gamma < e^{\beta m}.} \tag{4.46}$$

The gluon (and photon) particle densities are

$$\rho_{\gamma,g} \equiv \frac{N_{\gamma,g}}{V} = \frac{1}{V} \lim_{\lambda \to 1} \lambda \frac{d}{d\lambda} \ln Z_{\gamma,g} = g_{\gamma,g} \int \frac{d^3p}{(2\pi)^3} f_{\gamma,g}. \tag{4.47}$$

Using the series expansion from Eq. (4.46), we can explicitly evaluate this integral, substituting np/T for x term by term:

$$\rho_{\gamma,g} = \frac{g_{\gamma,g}}{2\pi^2} T^3 \sum_{n=1}^{\infty} \frac{\gamma^n}{n^3} \int_0^{\infty} dx\, x^2 e^{-\sqrt{(nm/T)^2 + x^2}}. \tag{4.48}$$

For $m \to 0$ and $\gamma \to 1$, we obtain the well-known Stefan–Boltzmann equilibrium limit:

$$\rho_{\gamma,g} = \frac{g_{\gamma,g}}{\pi^2} T^3 \zeta(3), \tag{4.49}$$

with the Riemann zeta-function $\zeta(3) \simeq 1.202$; see Eq. (10.66b). Using Eq. (10.50a), for the general case, to evaluate the integral, we obtain an infinite sum over terms containing the Bessel function K_2 (also called the McDonald function), which is discussed in section 10.4:

$$\rho_{\gamma,g} = \frac{g_{\gamma,g}}{\pi^2} T m^2 \sum_{n=1}^{\infty} \frac{\gamma^n}{n} K_2(nm/T). \tag{4.50}$$

Many other properties of the quark–gluon gas are discussed in section 10.5.

The statistical method is a powerful tool to deal with the physics we address in this book. Looking back, we recognize that we have assumed the presence of sufficiently many (weakly) interacting (quasi)particles in this discussion of basic results of statistical physics. Two important questions come to mind.

- In our context, the practical question is that of how statistical physics works when we have a few hundred (at the SPS), or a few thousand (at the RHIC) particles experiencing a limited number of collisions each. In this book, we will answer this question by consulting the experimental results, and our finding is that the thermal particle spectra describe experimental data very well.
- It seems that, perhaps, we could derive statistical-physics laws for any type of many-object system – could it be that the statistical partition function even describes the behavior of investors on Wall Street? Let us clearly identify what specific tacit physical feature makes a system of particles so much simpler to understand than a crowd of investors. An appropriate economical toy model, in our context, would consist of taking a 'conserved' number of Wall Street investors who, in view of their frequent interactions, should see their investments equipartitioned

into an exponential wealth distribution, provided that all members of the same wealth class are, basically, indistinguishable, a hypothesis many of our colleagues agree with. All the above equations apply, with E_i being now the wealth range of n_i investors. To compute anything with precision we need, however, to specify the meaning of the discrete sum, $\sum_i$; we need to know the number of 'investors' per unit of 'wealth'. In case of physical particles, this level density Eq. (4.37) is implicit in our understanding of the many particle *phase space*, which allows us to convert the symbolic expressions into quantitative equations. We are not able to generalize this naively to non-physics applications of statistical physics.

4.5 Hadron gas

Particularly important in our study is the hadronic 'gas' (HG) matter consisting of individually confined hadronic particles. Although relativistic dynamics is required, we can consider the classical (Boltzmann)-gas limit Eq. (4.40) since, in a very 'rich' multicomponent phase, each particle species has a rather low 'non-degenerate' phase-space abundance. In other words, at sufficiently high temperature, a high density of hadronic particles can arise as a consequence of many hadron species contributing, and does not in general imply a quantum degeneracy of the phase space. However, even in the HG phase, it is possible to encounter (pion) quantum degeneracy, which requires full quantum statistics, Eq. (4.39).

To see why the classical Boltzmann distribution almost always suffices in the hadronic gas phase of matter, consider the denominator of the quantum distribution, Eq. (4.42): even for the least-massive hadronic particle, the pion, the expansion of the denominator of quantum distributions makes good sense. For a range of temperatures up to $T < 150$ MeV wherein confined hadrons exist, we find $\exp(-E_\pi/T) < \exp(-m_\pi/T) < 1$. The limits of the Boltzmann approximation are tested when, e.g., the phase space is oversaturated, i.e., $\gamma_\pi > 1$, or when the baryo-chemical potential compensates for the mass term which could occur in extremely dense baryonic systems.

We present next a brief survey of the properties of a hadronic Boltzmann gas, and refer to chapter 10 for further developments. We consider a series expansion of the logarithmic function in Eq. (4.39):

$$\ln \mathcal{Z} = \sum_{n=1}^{\infty} \frac{1}{n} Z_n. \tag{4.51}$$

Each term comprises contributions from all contributing bosons B_f and fermions F_f:

$$Z_n = \sum_{B_f} g_f \gamma_f^n (\lambda_f^n + \lambda_f^{-n}) V \int \frac{d^3p}{(2\pi)^3} e^{-n\beta\varepsilon_f}$$

$$+ (-)^{n+1} \sum_{F_f} g_f \gamma_f^n (\lambda_f^n + \lambda_f^{-n}) V \int \frac{d^3p}{(2\pi)^3} e^{-n\beta\varepsilon_f}. \quad (4.52)$$

The single-particle energy ε_f entering Eq. (4.52) depends on the mass m_f of particle f, Eq. (4.31). Since the mass spectra of hadronic bosons and fermions are quite different, $n > 1$ quantum corrections do not cancel out.

In the Boltzmann limit, the first term $n = 1$ is retained in Eq. (4.51) and there is no distinction between the Bose and Fermi ideal gases in this 'classical' limit, as seen in Eq. (4.40):

$$\ln \mathcal{Z}_{cl} = \sum_f g_f \gamma (\lambda_f + \lambda_f^{-1}) V \int \frac{d^3p}{(2\pi)^3} e^{-\beta\varepsilon(\vec{p})} \equiv Z^{(1)}. \quad (4.53)$$

The last definition reminds us that the right-hand side of Eq. (4.53) is the partition function arising for a single particle enclosed in a given volume. This is not an entirely 'classical' expression. We note that

$$\mathcal{Z}_{cl} = \sum_{k=0}^{\infty} \frac{1}{k!} (Z^{(1)})^k, \quad (4.54)$$

which expresses the fact that the partition function comprises the additive contributions of terms for k microscopic particles. However, the quantum indistinguishability is retained in the factor $1/k!$ – only with this quantum factor can one obtain the correct 'classical' Maxwell distribution of atoms in a gas. This issue marked strongly the pre-quantum-mechanics development of statistical physics in the Boltzmann era since there was no easy explanation why this factor was needed.

We already know the momentum integral appearing in Eq. (4.53), from Eq. (4.48),

$$\ln \mathcal{Z}_{cl} = \frac{\beta^{-3}V}{2\pi^2} \sum_f g_f \gamma (\lambda_f + \lambda_f^{-1}) W(\beta m_f), \quad (4.55)$$

where we used the function $W(x) = x^2 K_2(x)$, shown in Fig. 10.1 on page 197.

Using Eq. (4.55), we obtain the properties of a hadronic gas in the classical (Boltzmann) limit. The 'net' (particle minus antiparticle) particle density, Eq. (4.22),

$$\rho_f = \frac{T^3}{2\pi^2} \sum_f g_f \gamma (\lambda_f - \lambda_f^{-1}) (\beta m_f)^2 K_2(\beta m_f), \quad (4.56)$$

pressure,

$$P_{\text{cl}} = \frac{T}{V} \ln \mathcal{Z}_{\text{cl}} = \frac{T^4}{2\pi^2} \sum_f g_f \gamma_f (\lambda_f + \lambda_f^{-1})(\beta m_f)^2 K_2(\beta m_f), \qquad (4.57)$$

and energy density,

$$\epsilon_{\text{cl}} = -\frac{1}{V} \frac{\partial}{\partial \beta} \ln \mathcal{Z}_{\text{cl}} = \frac{T^4}{2\pi^2} \sum_f g_f \gamma_f (\lambda_f + \lambda_f^{-1})$$

$$\times \left[3(\beta m_f)^2 K_2(\beta m_f) + (\beta m_f)^3 K_1(\beta m_f) \right]. \quad (4.58)$$

comprise the sum over all particle fractions. In Eq. (4.56) we obtained the difference between numbers of particles and antiparticles. The relation between partition function and pressure which we introduced in Eq. (4.57) is discussed in section 10.1, see Eq. (10.11). To obtain Eq. (4.58), we used $dx^2 K_2(x)/dx = -x^2 K_1(x)$.

The relativistic limits of Eqs. (4.57) and (4.58) arise in view of the properties of the Bessel function, Eqs. (10.47) and (10.50b), $K_2(x) \to 2/x^2$ and $K_1(x) \to 1/x$, and only the K_2 term contributes:

$$P_{\text{cl}} \to \frac{T^4}{\pi^2} \sum_f g_f \gamma_f (\lambda_f + \lambda_f^{-1}),$$

$$\epsilon_{\text{cl}} \to \frac{3T^4}{\pi^2} \sum_f g_f \gamma_f (\lambda_f + \lambda_f^{-1}). \qquad (4.59)$$

In the case of fermions, the Pauli exclusion principle decreases the particle degeneracy below the classical value. The energy and pressure shown in Eq. (4.59) are reduced in the relativistic limit by the Riemann η-function factor $\eta(4) = \frac{7}{8}\pi^4/90 = 0.9470$. On the other hand, since bosons are 'attracted' to each other, one finds a greater than classical degeneracy, expressed in the relativistic limit by the factor $\zeta(4) = \pi^4/90 = 1.0823$.

For a relativistic hadron gas, comprising a similar number of fermions and bosons, this quantum effect averages out. Thus, when we speak of an effective number of degrees of freedom (also effective degeneracy) in HG, we will use as a basis the classical expression Eq. (4.59):

$$g_{\text{eff}}^P \equiv \pi^2 \frac{P}{T^4}, \qquad g_{\text{eff}}^\epsilon \equiv \frac{\pi^2}{3} \frac{\epsilon}{T^4}. \qquad (4.60)$$

When $T \gg m$ for all particles, or, equivalently, when T is the only relevant energy scale, we have $g_{\text{eff}}^P \simeq g_{\text{eff}}^\epsilon$.

We consider, in Fig. 4.1, how g_{eff}^P and g_{eff}^ϵ look in a simple hadronic gas, as functions of T. Solid lines correspond to g_{eff}^ϵ, and dashed to g_{eff}^P. The thin lines are for the classical Boltzmann pion gas ($g_\pi = 3$, $m_\pi \simeq 140$

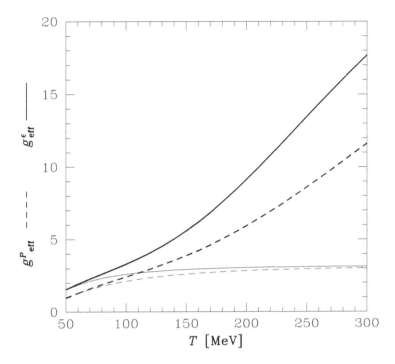

Fig. 4.1. Effective numbers of degrees of freedom from energy density (solid lines) and pressure (dashed lines), Eq. (4.60), for a Boltzmann pion gas (thin lines), and gas comprising Boltzmann pions, nucleons, kaons, and $\Delta(1232)$ for $\gamma_i = 1$ and $\lambda_i = 1$, as functions of temperature T.

MeV); thick lines also include the four kaons K, the nucleons N, and the deltas $\Delta(1232)$, and for N and Δ their antiparticles ($g_K = 4$, $m_K \simeq 495$ MeV; $g_N = 4$, $m_N \simeq 939$ MeV; and $g_\Delta = 16$, $m_\Delta \simeq 1232$ MeV) evaluated with all $\gamma = 1$, as appropriate for chemical equilibrium, and $\lambda = 1$, for a nearly baryon-free system, as appropriate for the early Universe. We see for the pion-only case (thin lines) the expected high-T limit, which is nearly reached already at $T \simeq m_\pi$. However, because of the relatively high hadron masses, the effective number of degrees of freedom keeps rising even at $T \simeq 300$ MeV toward its maximum for this example, which is near 50. We also note that the energy density approaches its relativistic limit faster than does the pressure, a point to which we shall return in Eq. (10.58).

We draw two important conclusion from results seen in Fig. 4.1.

- Since pions are several times lighter than the next heavier hadronic particle, they determine rather exactly the properties of a hadron gas at 'low' temperature below $T \simeq (m_\pi/2)$ MeV, as is seen in Fig. 4.1

(the case that the net baryon and strangeness density is zero). Even at $T \simeq m_\pi$, the pion fractional pressure is still the dominant component.
- The influence of the numerous massive hadronic particles rapidly gains in importance with rising temperature. At low temperature, the quantum corrections (not shown in Fig. 4.1) are in fact more important than the contributions of heavier particles since $2m_\pi < m_h$, h $\neq \pi$. For $2\beta m_\pi < 1$ in the HG phase, with $g_\pi = 3$, (for derivation compare Eq. (10.62)), we have

$$P_h^\pi \simeq \frac{3T^4}{2\pi^2}\left(\gamma_\pi\lambda_\pi W(\beta m_\pi) + \frac{1}{16}\gamma_\pi^2\lambda_\pi^2 W(2\beta m_\pi) + \cdots\right). \quad (4.61)$$

As the temperature increases, the small quantum correction remains a minor effect compared with a rise due to excitation of numerous heavy hadron states.

4.6 A first look at quark–gluon plasma

We consider next the properties of the QGP, modeled initially as an ideal chemically equilibrated gas of quarks and gluons, including the effect of confining vacuum structure. In the study of the quark-and-gluon gas, our task is considerably simplified by the observation that the gluons and light u and d quarks are to all intent massless particles, at least on the scale of energies available in the hot plasma, i.e., $T \approx 200$ MeV.

Since the energy density is, in general terms, given by (see Eqs. (10.7) and (10.11))

$$\epsilon = -\frac{\partial}{\partial\beta}\frac{1}{V}\ln \mathcal{Z}(\beta, \lambda), \quad (4.62)$$

in the absence of any dimensioned scales,

$$\frac{1}{V}\ln \mathcal{Z}(\beta, \lambda) = \beta^{-3}f(\lambda), \quad (4.63)$$

and we find

$$\boxed{\epsilon = 3\beta^{-4}f(\lambda) = 3\frac{T}{V}\ln \mathcal{Z}(\beta, \lambda) = 3P.} \quad (4.64)$$

The presence of masses of quarks, and in general scaled variables, breaks this perhaps most used relationship of relativistic gases. It applies to fermions, bosons, and classical gases. Equations (10.58)–(10.60) show how the presence of masses reduces the pressure below $\epsilon/3$. Put differently, massive particles are less mobile at a given temperature, and thus the pressure they can exercise is smaller than $\epsilon/3$; the energy density ϵ is 'helped' by the presence of masses, and is closer to the relativistic limit.

In the limit $\beta m = m/T \ll 1$, the phase-space integrals of ideal quantum gases are easily carried out. We can effectively neglect the particle mass m compared with the high momenta that occur. We also omit, at first, chemical potentials. We obtain for the energy density

$$\frac{E_{\rm F,B}}{V} = \frac{g}{2\pi^2} \int_0^\infty p^2\, dp\, \frac{p}{e^{\beta p} \pm 1} = \frac{g\beta^{-4}}{2\pi^2} 3! \sum_1^\infty \frac{(\pm 1)^{n-1}}{n^4}. \tag{4.65}$$

The infinite sums are the zeta and eta Riemann sums, see Eqs. (10.66a)–(10.67b), which for bosons give the well-known Stefan–Boltzmann result:

$$\boxed{P_{\rm B}|_{m=0} = \frac{T}{V}\ln \mathcal{Z}_{\rm B}|_{m=0} = \frac{g\pi^2}{90} T^4 = \frac{1}{3}\epsilon_{\rm B} \equiv \frac{E_{\rm B}}{3V}.} \tag{4.66}$$

We have made explicit the result $\epsilon = 3P$, see Eq. (4.64), which is valid when the mass of particles is small relative to their energy (massless particles or ultra-relativistic gas). For fermions, the alternating sum in Eq. (4.65) introduces a relative reduction factor, which is $\frac{7}{8}$, see Eq. (10.67b). However, allowing for the presence of antifermions, the energy density and pressure have to be multiplied by an extra factor of two, and become in fact greater by a factor $\frac{7}{4}$:

$$\epsilon_{\rm F} \equiv \frac{E_{\rm F}}{V} = \frac{g\pi^2}{30}\frac{7}{4} T^4 = 3P_{\rm F}. \tag{4.67}$$

For fermions, the inclusion of a finite chemical potential is of importance. In the limit $m \to 0$, the Fermi integrals of the relativistic quantum (degenerate) quark gas can be evaluated exactly at finite μ, see Eq. (10.73):

$$\boxed{P_{\rm F}|_{m=0} = \frac{T}{V}\ln \mathcal{Z}_{\rm F}|_{m=0} = g\frac{(\pi T)^4}{90\pi^2}\left(\frac{7}{4} + \frac{15\mu^2}{2(\pi T)^2} + \frac{15\mu^4}{4(\pi T)^4}\right).} \tag{4.68}$$

Since in the domain of freely mobile quarks and gluons the vacuum is deconfined, a finite vacuum energy density (the latent heat of the vacuum) arises within the deconfined region, as we have discussed at length in section 3.1. This also implies that there must be a (negative) associated pressure acting on the surface of this volume and attempting to reduce the size of the deconfined region. These two properties of the vacuum follow consistently from the vacuum partition function:

$$\boxed{\ln \mathcal{Z}_{\rm vac} \equiv -\mathcal{B}V\beta.} \tag{4.69}$$

On differentiating Eq. (4.69) as in Eqs. (4.57) and (4.58), we in fact find that the perturbative vacuum region is subject to the (external) pressure $-\mathcal{B}$ while the internal energy density is $+\mathcal{B}$ relative to the outside volume.

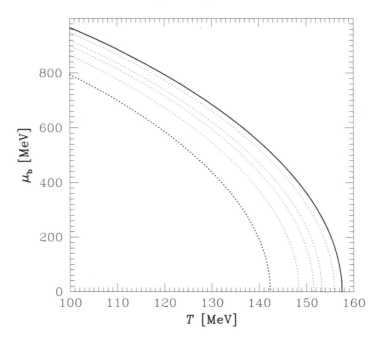

Fig. 4.2. $P = 0$ in the quark–gluon liquid in the (μ_b–T_c) plane. Dotted (from right to left): breakup conditions of the liquid for expansion velocities $v^2 = \frac{1}{10}, \frac{1}{6}, \frac{1}{5}, \frac{1}{4}$ and $\frac{1}{3}$.

The partition function (i.e., pressure) of the quark–gluon phase is obtained after we combine contributions from quarks, gluons, and vacuum:

$$\frac{T}{V} \ln \mathcal{Z}_{\text{QGP}} \equiv P_{\text{QGP}} = -\mathcal{B} + \frac{8}{45\pi^2} c_1 (\pi T)^4$$
$$+ \frac{n_f}{15\pi^2} \left[\frac{7}{4} c_2 (\pi T)^4 + \frac{15}{2} c_3 \left(\mu_q^2 (\pi T)^2 + \frac{1}{2} \mu_q^4 \right) \right]. \qquad (4.70)$$

We have inserted the quark and gluon degeneracies as shown in Eqs. (3.35a) and (3.35b). The interactions between quarks and gluons manifest their presence aside from the vacuum-structure effect, in the three coefficients $c_i \neq 1$, see Eqs. (16.1) and (16.2), [91]:

$$c_1 = 1 - \frac{15\alpha_s}{4\pi} + \cdots, \qquad (4.71a)$$

$$c_2 = 1 - \frac{50\alpha_s}{21\pi} + \cdots, \qquad (4.71b)$$

$$c_3 = 1 - \frac{2\alpha_s}{\pi} + \cdots. \qquad (4.71c)$$

One can evaluate the pressure Eq. (4.70) by choosing values for $\mathcal{B}$ and α_s. It turns out that the value of the running strong-interaction coupling constant α_s changes rather rapidly in the domain of interest to us, and hence one needs to employ a function $\alpha_s(T)$, see Fig. 14.3 on page 286. Then, also allowing for the latent heat $\mathcal{B}$, a surprisingly good agreement with lattice results in section 15.5 is found, this is shown in Fig. 16.2 on page 307. This comparison hinges strongly on an understanding of $\alpha_s(T)$, and inclusion of $\mathcal{B}$.

Drawing on these considerations, we show the QGP-phase pressure condition $P_{QGP} \to 0$ in Fig. 4.2. The solid line denotes, in the $(\mu_b$–$T_c)$ plane, where $P_{QGP} = 0$ in a stationary quark–gluon phase. The dotted lines correspond (from right to left) to the condition Eq. (3.31) for flow velocities $v^2 = \frac{1}{10}, \frac{1}{6}, \frac{1}{5}, \frac{1}{4}$ and $\frac{1}{3}$ for which an exact spherical expansion with $\kappa = 1$, see Eq. (3.32), was used. The last dotted line to the left corresponds to an expansion with the velocity of sound of relativistic (i.e., effectively massless) matter. For small baryo-chemical potentials, the equilibrium phase-transition temperature of a non-dynamically evolving system is somewhat greater than that shown here at the intercept of the solid line at $\mu_b = 0$. The actual value is $T_c \simeq 170$ MeV, as it occurs at finite pressure balanced by hadrons, compare with Fig. 3.2. Looking at the high-flow-velocity curves in Fig. 4.2, we see that an exploding QGP fireball can supercool to $T \simeq 0.9T_c$.

II
Experiments and analysis tools

5 Nuclei in collision

5.1 Heavy-ion research programs

The energy content available in the nuclear collision is the main factor in which experimental facilities differ from each other. The ultra-relativistic nuclear-collision systems we are considering are identified in table 5.1. For the maximum possible mass number up to $A_{\mathrm{max}} \simeq 200$, we show the fixed-target maximum beam energy per nucleon $E_{\mathrm{P}}^{\mathrm{max}}$ [A GeV]; for colliders, we present in this line the equivalent projectile energy. Similarly, we show the CM energy in the nucleon–nucleon system $\sqrt{s_{\mathrm{NN}}}$ [GeV], which is twice the nominal beam energy of the RHIC and LHC collider systems. We also show the total $\sqrt{s_{\mathrm{AA}}}$ [GeV] energy in the interaction region, allowing for the maximum mass number A of the beam. The final line refers to the rapidity 'gap' Δy. We will discuss these variables in the following sections.

Δy is defined as the difference between the rapidities of projectile and target. In laboratory fixed-target experiments, $y_{\mathrm{t}} = 0$, and Δy is the rapidity of the projectile y_{p}. Using the definition of rapidity Eq. (5.4), we have

$$\cosh \Delta y = E_{\mathrm{p}}/m_{\mathrm{p}}. \tag{5.1}$$

For head-on interactions occurring at rest in the laboratory, at the collider facilities, $\Delta y/2$ is the projectile (target) rapidity of each beam, which is evaluated using, e.g., Eq. (5.1) again.

A convenient way to represent the data of table 5.1 is shown in Fig. 5.1: the solid line depicts the CM energy per pair of nucleons, $\sqrt{s_{\mathrm{NN}}}$, as a function of the rapidity y. The horizontal distance between the two branches of the solid line is the projectile–target rapidity gap Δy. The shaded areas correspond to the accessible CM energies, $\sqrt{s_{\mathrm{NN}}}$, at ex-

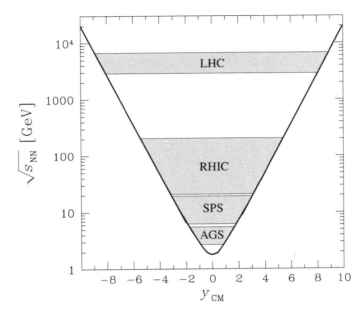

Fig. 5.1. $\sqrt{s_{NN}}$ (vertical axis) of various accelerators as a function of the projectile and target rapidities seen from the CM frame. Shaded areas: energy ranges accessible at the various accelerators.

perimental facilities that are in operation and under construction today.

As the energy increases, the rapidity gap Δy between projectile and target opens up as we see in Fig. 5.1. In the central rapidity region, we can study conditions of matter without having to account for particles spilled from the projectile and target fragments, which are known experimentally to spread over about two units of rapidity. Two extreme cases are illustrated qualitatively in Fig. 5.2, in which we sketch the distribu-

Table 5.1. Parameters of existing ultra-relativistic heavy-ion beam facilities and those under construction.

	AGS	AGS	SPS	SPS	SPS	RHIC	RHIC	LHC
Start year	1986	1992	1986	1994	1999	2000	2001	2006
A_{max}	^{28}Si	^{197}Au	^{32}S	^{208}Pb	^{208}Pb	^{197}Au	^{197}Au	^{208}Pb
$E_P^{max}[A\ \mathrm{GeV}]$	14.6	11	200	158	40	0.91×10^4	2.1×10^4	1.9×10^7
$\sqrt{s_{NN}}$ [GeV]	5.4	4.7	19.2	17.2	8.75	130	200	6000
$\sqrt{s_{AA}}$ [GeV]	151	934	614	3.6×10^3	1.8×10^3	2.6×10^4	4×10^4	1.2×10^6
$\Delta y/2$	1.72	1.58	2.96	2.91	2.22	4.94	5.37	8.77

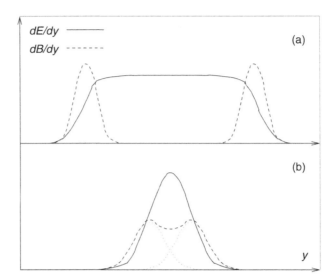

Fig. 5.2. Rapidity distributions of energy (solid lines) and baryon number (dashed lines) (in a qualitative representation): (a) for a 'Transparent' reaction mechanism; and (b) for full stopping in the collision.

tions of the energy (solid lines) and baryon number (dashed lines), as functions of the rapidity.

In the baryon-punch-through case, shown in Fig. 5.2(a), which was investigated by Bjørken [73], see section 6.4, the colliding nuclei are leaving a trail of energy between the projectile–target rapidity, but the baryon number continues to move out of the collision zone, apart from the down-shift in projectile and target rapidities necessary for conservation of energy.

The stopping limit, implicit in the work of Fermi [121] and described by Landau [173, 175], is shown in Fig. 5.2(b): both the particle multiplicity (energy) and the baryon number are centered around the central rapidity y_{CM}. The projectile and target baryons will, under the most extreme circumstance of complete stopping, lose all memory about the initial state, and in this limit there should in particular be little, if any, difference between the distributions of energy and baryon number in the longitudinal and transverse directions with respect to the collision axis.

We now survey the nuclear-collision experiments that are currently operating or under development. These include in particular the CERN–SPS heavy-ion program which continues a 15-year-long tradition in the so-called North Area (NA) in fixed-target mode with energy range up to $200A$ GeV for up to $A \simeq 100$ and dropping to $158A$ GeV for neutron-rich projectiles such as Pb. At higher energies, we have the beginning of the experimental program at the RHIC collider, and in the near future there will be one at the LHC. The RHIC allows head-on collisions of two Au ions, each carrying energy in the range $(10-100)A$ GeV. Results from the initial $65A$-GeV run will be described and a glimpse of first $100A$-GeV results is presented as this book goes to press. Compared with the SPS energies, the available CM energy per nucleon has been increased by an order of magnitude and accordingly the densities of matter reached are more extreme. Since the laboratory frame is also the CM frame in a collider experiment, the greatly increased particle density is distributed more evenly in all spatial directions.

We begin with the CERN–SPS research program*.

- The experiment NA45.2 investigates primarily the production of electron–positron pairs and of direct photons, continuing the research program of NA45 carried out with S-beams. Both experiments observe dielectron pairs and compare results with expectations based on p–p reactions. The current experimental set-up consists of a double spectrometer covering a region near mid-rapidity with full azimuthal coverage. Electrons are identified in two ring-imaging Čerenkov detectors (RICH).

- The experiment NA49, which had as its predecessor experiment NA35, uses several time-projection chambers (TPCs) for large-acceptance tracking of charged particles. Its current objective is to explore in greater detail the excitation function of strangeness near the possible threshold for the formation of the QGP phase. NA49 is at present the only experiment at the SPS capable of measuring many global observables required to characterize the nature of heavy-ion collisions as the energy is varied.

- The experiment NA57 continues the research program of experiments WA97, WA94, and WA85, all of which studied the production of (multi) strange hadrons in the central rapidity region, with particular emphasis on the production of strange antibaryons. NA57 is completely differently instrumented compared with the WA series and provides an important cross-check for all the results. It comprises silicon pixel tracking of hadrons in a magnetic field, and its results are based primarily on reconstruction of decays of strange hadrons.

- The experiment NA60 attempts detection of charmed hadrons, to complement the earlier study of suppression of production of J/Ψ by its predecessors NA50 and NA38. The NA50 muon spectrometer is complemented by a completely redesigned target area using radiation-tolerant silicon pixel detectors. The NA50 experiment studied dimuons pro-

---

* For further details consult the following CERN web pages:
  http://greybook.cern.ch/programmes/SPS.html; and see also *http://greybook.cern.ch/ programmes/EXP_NAM.html*, for all CERN experiments, including those completed.

duced in Pb–Pb and p–A collisions. The muons are measured in the former NA10 spectrometer, which is shielded from the target region by a beam stopper and absorber wall. The observed muons traverse 5 m of BeO and C.

- The completed experiment WA98 will be repeatedly mentioned in that which follows: WA98, which had as predecessors WA80 and WA87, which addressed the measurement of photons, but also measured the global production of charged hadrons. It comprised, in particular, a 10 000-module lead-glass spectrometer, which now is incorporated into the PHENIX detector (see below), yielding high-precision data on π^0 and η at mid-rapidity within a large range of transverse momenta 0.3 GeV/c > $P_\perp$ > 4.5 GeV/c for π^0. Detailed comparison of photons with the production of charged particles allowed also an evaluation of the photon enrichment potentially due to direct radiance from QGP.

We now turn to review the experimental research program at the BNL. Four experiments are at present taking data at the RHIC[†]. They are designed to allow both a survey of the reactions occurring in this hitherto unexplored condition of matter and an in-depth study of the properties of the deconfined QGP phase. We review the first results from the year-2000 run in section 9.5. The experiments currently under way are the following.

- BRAHMS (**B**road **R**ange **H**adron **M**agnetic **S**pectrometer) is designed to measure hadronic particles inclusively (that is, to measure one particle at a time irrespective of what else is happening, when the system is triggered), over a wide range of rapidity (0 < η < 4) and transverse mass (up to 30 GeV). It consists of two (forward and mid-rapidity), magnetic focusing charged-particle ($\pi^\pm$, $K^\pm$, p, $\bar{p}$) spectrometer arms, which can be set to the desired angular acceptance window.

- PHENIX (**P**ioneering **H**igh **E**nergy **N**uclear **I**nteraction **E**xperiment) is a detector optimized to observe photons and dilepton pairs (γ, $e^\pm$ and $\mu^\pm$). It comprises a central detector made of an axial field magnet and two almost identical arms placed on the left and right of the magnet, each covering a window of ±0.35 units of pseudorapidity. Each arm comprises several detector subsystems: the important goal of the central detector is observation of dielectrons at high mass resolution, allowing one to detect changes in the properties of decaying vector mesons (e.g., $J/\Psi \rightarrow e^+e^-$, $\phi \rightarrow e^+e^-$). The electro-magnetic calorimeter allows one to measure low-$p_\perp$ photons near $y = 0$. Hadron detection in the silicon vertex detector, for −2.65 < η < 2.65, will allow studies of

---

[†] For RHIC experiments, see *http://www.rhic.bnl.gov.*

the distribution of charged hadrons (without identification of particles) on an event-by-event basis. First results from this subsystem obtained in the RHIC 2000 run have recently been published [16].

- PHOBOS, a scaled down 'satellite' of MARS (**M**odular **A**rray for **R**HIC **S**pectra),

  is a very small (in comparison) arrangement of silicon-based detectors that will allow one to study low-momentum particles within the complete (pseudo)rapidity interval $-5.4 < \eta < 5.4$, aiming to explore global event structure. PHOBOS has published the first results from the RHIC 2000 run on particle multiplicity [49], as well as the RHIC 2001 run [50].

- STAR (**S**oleonoidal **T**racker **at R**HIC)

  is a (large) 4π primarily hadronic-particle detector, with a 4-m-diameter and 4-m-long solenoidal 0.5-T magnetic-field volume, comprising as the main charged-particle-tracking device a TPC with inner radius 50 cm and outer radius 200 cm, with 45 planes of tracking. This allows a pseudorapidity coverage of $-2 \leq \eta \leq +2$, and the design allows for a lower particle-momentum cutoff at 60 MeV/c. In addition, the inner silicon vertex tracker (SVT) is surrounding the interaction region between 5 and 15 cm, facilitating observation of the production of strangeness. The time-of-flight array, $\simeq 2.5$ m from the primary interaction vertex, will help identify charged particles. The outside electromagnetic calorimeter (EMC) aims to measure jets of particles, fluctuations, and high-$p_\perp$ phenomena. The high tracking resolution facilitates reconstruction of unstable hadronic resonances. First results on central production of antiprotons [19] and anisotropy of particle multiplicity (elliptical flow) [15] have been published.

Still much more extreme matter conditions will be reached when the LHC collider is completed (*http://lhc.web.cern.ch/lhc/*) and the equivalent laboratory energy of $E_{\mathrm{P}}^{\mathrm{max}} \simeq 2 \times 10^{16} A$ eV reaches into the domain of highest cosmic-particle energies, where the cosmic flux begins to decrease unusually rapidly. This 'knee' in cosmic flux as a function of the energy is below the high end of the LHC energy. At the LHC there will be initially three major detectors, ATLAS, CMS, and ALICE. ALICE is the dedicated heavy-ion experiment. CMS is intended to measure dilepton spectra under heavy-ion operation conditions. The ATLAS collaboration is exploring the potential of its detector in the heavy-ion environment.

- ALICE (**A** **L**arge **I**on **C**ollider **E**xperiment)[‡]. It comprises a TPC as a main tracking device of charged particles with an inner radius of 1 m and an outer radius of 2.5 m, and a length along the beam direction

---

[‡] See for further details the web page *http://www1.cern.ch/Alice.*

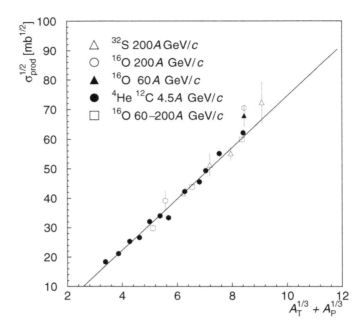

Fig. 5.3.   The square root of the inelastic reaction cross section, $\sqrt{\sigma}$, as a function of the geometric size of interacting nuclei, $A_{\mathrm{T}}^{1/3} + A_{\mathrm{P}}^{1/3}$, for various collision partners, after [37].

of 5 m, covering the pseudorapidity interval $-0.9 < \eta < 0.9$. The high-resolution inner tracking system consists of five concentric cylindrical layers with radii from 7.5 to 50 cm around the beam pipe and allows the study of decays of charmed particles. An electro-magnetic calorimeter and a dilepton arm complement this very large and universal detector.

5.2 Reaction energy and collision geometry

On intuitive grounds, we expect that, for the short-range hadronic interactions, the collision geometry determines the amount of matter participating in nuclear collisions. The collision geometry is a very important and carefully explored subject. For an in-depth discussion of the importance of collision geometry, we refer the reader to the extensive body of work for hadron–hadron and hadron–nucleus interactions [184, 190].

The earliest experimental heavy-ion results confirmed the role of this simple geometric picture of nuclear-collision reaction dynamics [201]. The reaction radius, defined as the square root of the reaction cross section, rises linearly with the geometric size of the colliding nuclei, described by the sum of their radii, which is proportional to $A^{1/3}$, as is shown in

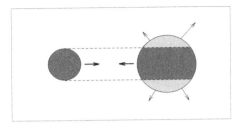
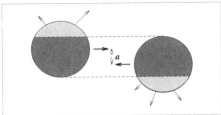

Fig. 5.4. A geometric illustration of nuclear collision. Left: small sulphur or aluminium nuclei colliding with much heavier lead or gold targets (correct relative scale). Right: symmetric collision of large nuclei at impact parameter a.

Fig. 5.3. This result confirms that the colliding nuclei need to 'touch' each other for local deposition of energy and baryon number to occur.

We show the central collision of sulphur or aluminium nuclei colliding with much heavier lead or gold targets in Fig. 5.4 (left-hand side). On the right-hand side, the symmetric slightly off-center collision with lead or gold is illustrated using the correct relative scale – we can see how important it is to assure that, in this system, the collision is geometrically as central as possible, in order to minimize the number of spectator (non-interacting, or partially interacting) nucleons. In symmetric collisions, only in a quite rare situation in which the impact parameter a is very small do we truly have the benefit of the largest possible region of interaction of the projectile and target, and do not encounter complications arising from spectator matter 'polluting' the experimental data.

A quantity of considerable importance is the energy content of the colliding system, which must be, by virtue of conservation of energy, the energy content of the final-state many-body system. The Lorentz invariant quantity we can form from the energy and momentum of the colliding projectile (p) and target (t) is

$$\sqrt{s_{\mathrm{pt}}} \equiv \sqrt{(E_{\mathrm{p}} + E_{\mathrm{t}})^2 - (\vec{p}_{\mathrm{p}} + \vec{p}_{\mathrm{t}})^2}. \tag{5.2}$$

In the CM frame where by definition $\vec{p}_{\mathrm{p}} + \vec{p}_{\mathrm{t}} = 0$, $\sqrt{s_{\mathrm{pt}}}$ is recognized as the available energy content of the projectile–target reaction, the CM energy. The quantity $\sqrt{s}$ is thus the available reaction energy. Since it is an invariant, $\sqrt{s}$ can be evaluated in any reference frame. It is natural to generalize this definition to any number of particles:

$$\sqrt{s^{(n)}} \equiv \sqrt{\left(\sum_{i=1}^{n} E_i\right)^2 - \left(\sum_{i=1}^{n} \vec{p}_i\right)^2}. \tag{5.3}$$

For $n = 1$, we see that $\sqrt{s}$ is just the mass of a particle, i.e., its energy content at rest. The conservation of energy assures that, when a particle

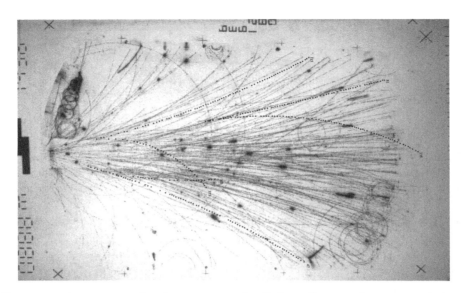

Fig. 5.5. A streamer-chamber picture of a S–Ag collision taken at 200A GeV (NA35 experiment [115]) showing the multiplicity of charged particles bent up and down in magnetic fields (with decays of neutral strange particles identified by superposed dashed lines).

decays, the final state comprising any number of particles n has the same $\sqrt{s}$. Conversely, $\sqrt{s^{(n)}}$ is also the (Lorentz invariant) mass of the ancestor system of the final-state n-body system, as determined by the momentum four vectors $p_i^\mu = (E_i, \vec{p}_i)$ of the particles produced.

This final-state energy described by Eq. (5.3) must be delivered by the colliding nuclei, see Eq. (5.2). $\sqrt{s^{(n)}}$ is also the invariant intrinsic rest energy (mass) of the fireball of dense matter, measured in terms of the participating energy and momentum of the colliding nuclei. Both these measures are jointly used in experiments to characterize a collision interaction: for example, the absence of the forward energy/momentum of the beam in the so-called zero-degree calorimeter (ZDC) can be correlated to the energy found in particles emitted in a direction transverse to the collision axis, see section 9.4, in order to define the geometric centrality of the collision. We will not follow these procedures further in this book, also since each experimental group applies a slightly different method.

In the fixed-target experiments, the longitudinal momentum is largely due to the Lorentz transformation from the CM frame to the laboratory frame. This longitudinal momentum is in general considerably greater than the transverse momentum component, and particles are focused forward along the collision axis, as seen in Fig. 5.5 [115]. In this streamer-

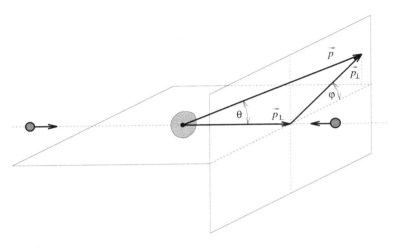

Fig. 5.6. The decomposition of particle momentum $\vec{p}$ (shown in the CM frame) into the parallel p_L and perpendicular $p_\perp$ components. Note the inclination angle θ of $\vec{p}$ and the azimuthal angle φ of $p_\perp$ (this is a qualitative presentation).

chamber picture of a S–Ag collision at $200A$-GeV of the NA35 experiment, we see all particles in a cone to the right of the interaction vertex to which the charged-particle tracks are pointing. We also see that central collisions of S–Ag nuclei at $200A$-GeV lead to the production of many secondary particles. Both positive and negative particles are bent in the applied magnetic field pointing normal to the plane of the picture. Several simultaneous photographs taken from various directions allowed precise tracking of charged particles.

Not all particle tracks go through the interaction vertex at the left-hand edge of Fig. 5.5: a few particle tracks, highlighted by dotted lines, belong to the V decays of neutral (strange) particles, see Fig. 2.3 on page 28. Low-momentum particles winding up as spirals in the high (1-T magnitude) magnetic field do not originate from the primary high-energy-vertex interactions.

5.3 Rapidity

We will now introduce the key kinematic variables that relate particle momentum to the dynamics that is occurring in the heavy-ion reaction. Each particle momentum decomposes, as shown in Fig. 5.6, into a longitudinal component (p_L) and a transverse component ($\vec{p}_\perp$) with reference to the collision axis. We note, in Fig. 5.6, the inclination angle θ of the particle against the collision axis. Also shown is the azimuthal angle φ of the two-dimensional vector $\vec{p}_\perp$.

The longitudinal momentum of a particle is an inconvenient variable, since it depends on the velocity of the CM frame with reference to the laboratory frame, as the appearance of Fig. 5.5 reminds us. For the analysis and understanding of the experimental results, it is necessary to be able to view the physical results from the CM frame, e.g., to transform the coordinate system to the CM frame of reference. The introduction of the rapidity y, replacing p_L, allows one to considerably simplify the selection and changing of the reference frame. This is due to the fact that the variable y is defined to be additive under successive Lorentz transformations along the same direction, as we shall see in Eq. (5.14): it can be understood as the 'angle' of the (hyperbolic) rotation in (3 + 1)-dimensional space. The 'angle' y is defined in terms of energy and momentum by the equations

$$\boxed{E = m_\perp \cosh y, \qquad p_L = m_\perp \sinh y,} \tag{5.4}$$

where $m_\perp$ is the transverse 'mass':

$$\boxed{m_\perp = \sqrt{m^2 + \vec{p}_\perp^{\,2}}.} \tag{5.5}$$

We note that Eqs. (5.4) and (5.5) are consistent with the relativistic dispersion relation (energy–momentum relation):

$$E = \sqrt{m_\perp^2 + p_L^2} = \sqrt{m^2 + \vec{p}_\perp^2 + p_L^2}. \tag{5.6}$$

The variable y (and $m_\perp \geq m$) replaces p_L (and $|\vec{p}_\perp|$), which are usually defining the momentum of a particle. The azimuthal angle φ of the vector $\vec{p}_\perp$, see Fig. 5.6, is the third variable required in the complete definition of $\vec{p}$.

The relation between velocity and rapidity is obtained from Eq. (5.4):

$$v_L \equiv \frac{cp_L}{E} = c \tanh y. \tag{5.7}$$

Thus, in the non-relativistic limit, $v_L \to cy$.[§] Equation (5.7) also implies that

$$\cosh y = \frac{1}{\sqrt{1 - v_L^2}} \equiv \gamma_L, \qquad \sinh y = \gamma_L v_L, \tag{5.8}$$

where γ_L is the (longitudinal) Lorentz contraction factor. Since

$$\tanh^{-1} z = \frac{1}{2} \ln\left(\frac{1 + z}{1 - z}\right), \tag{5.9}$$

[§] Even though we like to work with units that do not explicitly introduce the velocity of light c, whenever the non-relativistic limit is discussed, it is convenient to reintroduce c explicitly into the equations, as shown above.

we obtain from Eq. (5.7)

$$y = \frac{1}{2} \ln\left(\frac{1 + v_{\mathrm{L}}}{1 - v_{\mathrm{L}}}\right) = \frac{1}{2} \ln\left(\frac{E + p_{\mathrm{L}}}{E - p_{\mathrm{L}}}\right) = \ln\left(\frac{E + p_{\mathrm{L}}}{m_\perp}\right). \tag{5.10}$$

Lorentz 'boosts' are the Lorentz transformations with one of the three (x, y, z) Cartesian coordinate directions employed as the reference axis for the transformation. To verify the additivity of rapidity under a sequence of Lorentz boosts mentioned earlier, we consider the transformation of the momentum vector under a change of the reference frame along the collision axis. Under such a transformation, the transverse momentum and the transverse mass $m_\perp$ are not changed. The energy and longitudinal component of momentum transform according to

$$E' = \gamma_{\mathrm{c}}(E + v_{\mathrm{c}}\, p_{\mathrm{L}}), \qquad p'_{\mathrm{L}} = \gamma_{\mathrm{c}}(p_{\mathrm{L}} + v_{\mathrm{c}}\, E). \tag{5.11}$$

Here and below, the 'primed' quantities are seen by an observer in the laboratory system which moves with the velocity v_{c} with respect to the CM frame of reference, in which the energy E and momentum p_{L} are measured. Noting that the rapidity y_{c} of the transformation satisfies Eq. (5.8), we obtain

$$\cosh y_{\mathrm{c}} = \gamma_{\mathrm{c}}, \qquad \sinh y_{\mathrm{c}} = \gamma_{\mathrm{c}} v_{\mathrm{c}}, \tag{5.12}$$

and, upon introducing Eq. (5.4), we find for Eq. (5.11)

$$E' = m_\perp \cosh(y + y_{\mathrm{c}}), \qquad p'_{\mathrm{L}} = m_\perp \sinh(y + y_{\mathrm{c}}). \tag{5.13}$$

It is now evident that the rapidity y' seen in the laboratory system is given in terms of the CM rapidity y by

$$y' = y + y_{\mathrm{c}}. \tag{5.14}$$

It is this simple result which gives the rapidity variable its importance as a tool in the analysis of particle-production data. For example, in fixed-target experiments, we can study particle spectra using y as a variable without an explicit transformation to the CM frame of reference, and deduce from the rapidity spectra the point of symmetry corresponding to the CM rapidity. In symmetric collisions with fixed targets, the CM frame has to be in the middle between the rapidities of projectile and target; the CM rapidity is half of the rapidity of the projectile $y_{\mathrm{CM}} = y_{\mathrm{p}}/2$. In this case, the particle-rapidity spectrum must be symmetric around y_{CM}. This allows one to complement measured particle spectra: if these are available for, e.g., $y \geq y_{\mathrm{CM}}$, a reflection at the symmetry point y_{CM} gives us the part of the spectrum with $y \leq y_{\mathrm{CM}}$, or vice-versa.

Understanding the actual value of y_{CM} is of particular interest in 'asymmetric' collisions of heavy ions, i.e., those involving two different nuclei, which we continue calling 'projectile and target', though such a distinction is irrelevant in our following argument, since the result will be symmetric between the two colliding nuclei. We recollect that, considering the definition Eq. (5.4), we also have

$$E \pm p_L = m_\perp e^{\pm y}. \tag{5.15}$$

The total energy and momentum of the colliding system is obtained from the total energy and momentum of colliding nuclei:

$$E = E^p + E^t, \qquad p_L = p^p + p^t. \tag{5.16}$$

Using Eq. (5.10) for these values of E and p_L, we obtain the rapidity of the frame of reference in which the combined longitudinal momentum vanishes. For collinear collisions, the transverse momentum also vanishes, and this is the rapidity of the CM frame. Using Eq. (5.10),

$$y_{CM} = \frac{1}{2} \ln \left(\frac{E^p + E^t + p^p + p^t}{E^p + E^t - (p^p + p^t)} \right). \tag{5.17}$$

We use now Eq. (5.15) for the rapidities of projectile and target and obtain a manifestly projectile–target-symmetric expression:

$$y_{CM} = \frac{1}{2} \ln \left(\frac{m_p e^{+y_p} + m_t e^{+y_t}}{m_p e^{-y_p} + m_t e^{-y_t}} \right). \tag{5.18}$$

We now consider the asymmetric collisions both for collider and for fixed-target experiments: another way to write Eq. (5.18) offers immediate understanding of the physics involved. We take the factor e^{y_p} in the numerator and the factor e^{-y_t} in the denominator out of the logarithm and obtain

$$y_{CM} = \frac{y_p + y_t}{2} + \frac{1}{2} \ln \left(\frac{m_p + m_t e^{-(y_p - y_t)}}{m_t + m_p e^{-(y_p - y_t)}} \right). \tag{5.19}$$

In most cases of interest, we have $y_p - y_t \gg 0$ and thus

$$y_{CM} \simeq \frac{y_p + y_t}{2} - \frac{1}{2} \ln \left(\frac{m_t}{m_p} \right) + \frac{m_t^2 - m_p^2}{2 m_p m_t} e^{-(y_p - y_t)} + \cdots. \tag{5.20}$$

In general, the first two terms largely suffice. In the way we wrote Eq. (5.20), we chose the usual convention to call the more massive nucleus the 'target'. Two cases of explicit interest in Eq. (5.20) are the collider mode $y_p = -y_t$, and a stationary target $y_t = 0$ (up to Fermi motion in the stationary target nucleus).

For asymmetric collisions, the precise magnitude of m_t is determined in part by the value of the impact parameter, see Fig. 5.4. Hence the

CM rapidity, Eq. (5.20), becomes dependent on the impact parameter. The magnitude of the shift in asymmetry of rapidity arising can be easily estimated: in collisions in which the projectile with A_P emerges fully in the target A_T, all of the projectile nucleons participate: $A_p = A_P < A_T$, while the number of target participants A_t is

$$A_t \propto A_p^{2/3} A_T^{1/3}. \tag{5.21}$$

Thus,

$$y_{CM} \simeq \frac{y_p + y_t}{2} - \frac{1}{6} \ln\left(\frac{A_T}{A_P}\right). \tag{5.22}$$

For light-on-heavy-ion collisions such as of S on Pb, the expected and observed shift in mass asymmetry of rapidity (the last term in Eq. (5.22)) is noticeable (0.3 units).

5.4 Pseudorapidity and quasirapidity

In the study of production of charged hadrons, e.g., in section 9.2, we will see that observed particles are often not identified, and hence we do not know their masses, which are required in order to determine the rapidity of particles Eq. (5.10), given the momentum measured by deflection of particles within a magnetic field. On the other hand, mass can be negligible compared with the momenta carried by the particles, especially so in fixed-target experiments. Consequently, we now consider what happens with the rapidity spectra when the mass of a particle is small relative to the momentum, and the momentum alone determines the energy of a particle, e.g.,

$$E = \sqrt{p^2 + m^2} \to p. \tag{5.23}$$

In analogy to Eq. (5.4), a simpler variable, the 'pseudorapidity' η of a particle is introduced,

$$p = p_\perp \cosh \eta, \qquad p_L = p_\perp \sinh \eta, \tag{5.24}$$

which, with Eq. (5.10), leads to

$$\eta = \frac{1}{2} \ln\left(\frac{p + p_L}{p - p_L}\right) = \frac{1}{2} \ln\left(\frac{1 + \cos\theta}{1 - \cos\theta}\right) = \ln\left(\cot\frac{\theta}{2}\right). \tag{5.25}$$

Here, θ is the particle-emission angle relative to the beam axis, see Fig. 5.6.

In Fig. 5.7, we see for the range of pseudorapidity of interest to us (up to $\eta = 9$, the maximum value seen in Fig. 5.1) how the angle θ varies with the pseudorapidity. A massless particle emitted transversely at $\eta = y = 0$

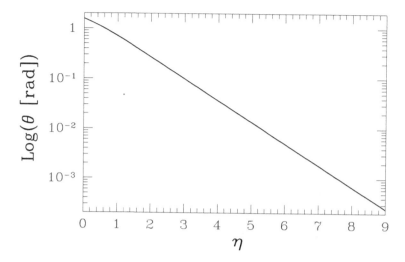

Fig. 5.7.   The emission angle θ in radians as a function of the pseudorapidity η.

has $\theta = \pi/2$; $\eta = 3$ corresponds to $\theta = 0.1$ rad, $(\equiv 0.1 \times 180°/\pi = 5.7°)$. The projectile–target fragmentation region at the LHC, where $\eta \simeq 8.5$, corresponds to $\theta = \pm 4.5 \times 10^{-4}$ rad $(\equiv \pm 0.025°)$.

From Eqs. (5.4) and (5.24), we obtain the implicit relations between pseudorapidity and rapidity:

$$\boxed{m_\perp \sinh y = p_\perp \sinh \eta, \qquad E \tanh y = p \tanh \eta.} \tag{5.26}$$

We see from these relations that the pseudorapidity is always greater than the rapidity:

$$\frac{\sinh \eta}{\sinh y} = \frac{m_\perp}{p_\perp} = \sqrt{1 + \frac{m^2}{p_\perp^2}} > 1, \quad \frac{\tanh \eta}{\tanh y} = \frac{E}{p} = \sqrt{1 + \frac{m^2}{p^2}} > 1. \tag{5.27}$$

More massive particles that have not been identified appear in a pseudorapidity particle spectrum at greater values of η than do the lighter pions.

In order to establish a precise relation between pseudorapidity and rapidity, we replace in Eq. (5.25) the (longitudinal) momentum using Eqs. (5.4) and (5.6) to obtain

$$\eta = \frac{1}{2} \ln \left(\frac{\sqrt{m_\perp^2 \cosh^2 y - m^2} + m_\perp \sinh y}{\sqrt{m_\perp^2 \cosh^2 y - m^2} - m_\perp \sinh y} \right). \tag{5.28}$$

Similarly, to determine rapidity in terms of pseudorapidity, we simply re-place the momenta in the definition of rapidity, Eq. (5.10), using definition Eq. (5.24):

$$y = \frac{1}{2} \ln \left(\frac{\sqrt{m^2 + p_\perp^2 \cosh^2 \eta} + p_\perp \sinh \eta}{\sqrt{m^2 + p_\perp^2 \cosh^2 \eta} - p_\perp \sinh \eta} \right). \tag{5.29}$$

Taking the logarithm of the first expression in Eq. (5.27), we obtain the shift in pseudorapidity relative to rapidity:

$$\delta\eta \equiv \eta - y = \frac{1}{2} \ln \left(1 + \frac{m^2}{p_\perp^2} \right) + \ln \left(\frac{1 - e^{-2y}}{1 - e^{-2\eta}} \right). \tag{5.30}$$

The leading term is the only term remaining for large η and it establishes an upper limit for the shift $\delta\eta$. The difference $\delta\eta = \eta - y$ between the pseudorapidity and the rapidity, as a function of pseudorapidity, is shown in Fig. 5.8, which was obtained by inserting Eq. (5.29) into Eq. (5.30). Thick lines are for $p_\perp = 0.3$ GeV, thin lines for $p_\perp = 0.5$ GeV; solid lines are for nucleons, chain lines for kaons, and dashed lines for pions. We see that, when $\eta \geq 3$, the first term in Eq. (5.30) in fact suffices to approximate the 'shift' in pseudorapidity which approaches a fixed maximum.

For sufficiently large $p_\perp > m$, when a particle's rest mass can be ne-glected, the shift $\delta\eta$ becomes negligible. For pions the error associated with considering the pseudorapidity instead of rapidity in hadronic re-actions can often be ignored since the mass is usually smaller than the typical momentum cut – and thus $\delta\eta < 0.1$ is seen at pseudorapidity $\eta = 3$ for $p_\perp > 0.3$ GeV. Moreover, the use of pion-quasirapidity y_π, which we discuss next, eliminates this shift completely. On the other hand, use of pseudorapidity seems not to be advisable for situations in which contribu-tions from more massive particles are of importance, unless, as Eq. (5.30) suggests, the $p_\perp$ cut is well above the mass of the particle. We see, in Fig. 5.8, that, for nucleons, taking the transverse momentum cut at 0.3 GeV, one encounters a shift of more than one rapidity unit at $\eta = 3$, the SPS value.

Since, in the upper SPS energy range (see table 5.1), the pion abun-dance dominates the hadron abundance, it has become common practice to show the distribution of hadrons as a function of pion-quasirapidity y_π, presuming that all hadrons observed are pions, as is done in Fig. 9.6 on page 166. One assumes, in lieu of the correct definition for each particle, the expression as if this particle had the mass of a pion:

$$p_{\rm L} = p_\perp \sinh \eta \rightarrow p_{\rm L} = \sqrt{p_\perp^2 + m_\pi^2} \sinh y_\pi. \tag{5.31}$$

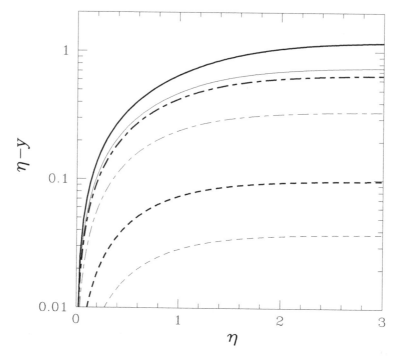

Fig. 5.8. The difference between pseudorapidity and pion-quasirapidity π (dashed lines), kaons K (chain lines), and nucleons N (solid lines) as a function of pseudorapidity η, for $p_\perp = 0.3$ GeV (thick lines) and $p_\perp = 0.5$ GeV (thin lines).

Following the derivation of Eq. (5.30), we obtain

$$\delta y \equiv y_\pi - y = \frac{1}{2}\ln\left(1 + \frac{m^2 - m_\pi^2}{p_\perp^2 + m_\pi^2}\right) + \ln\left(\frac{1 - e^{-2y}}{1 - e^{-2y_\pi}}\right). \tag{5.32}$$

In Fig. 5.9, we see the solution of the above equation as a function of the pion-quasirapidity. Lines are for $p_\perp = 0.5$ GeV(bottom line, smallest shift), $p_\perp = 0.3$ GeV (middle line), and for $p_\perp = 0.1$ GeV(top, largest shift); solid lines are for nucleons and chain lines for kaons. Again, when $\delta y \geq 3$, the first term in Eq. (5.32) nearly suffices to approximate the 'shift' in rapidity for pions as shown in Fig. 5.9 for increasing y_π, it approaches a fixed value, which for $p_\perp < m_\pi$ is significant. The quasirapidity distribution for nucleons experiences a widening by ± 1.9, and that for kaons widens by ± 1.3 units of rapidity.

As we see in Figs. 5.8 and 5.9 and Eqs. (5.30) and (5.32), the error in measurement of rapidity grows with decreasing $p_\perp$ of the particle. For kaons and nucleons, in the range of $p_\perp$ within which the pseudorapidity is

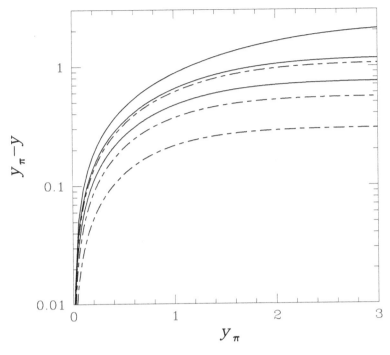

Fig. 5.9. The difference between pion-quasirapidity and rapidity as a function of quasirapidity y_π for kaons K (chain lines), and nucleons N (solid lines), from bottom to top for $p_\perp = 0.5$ GeV, $p_\perp = 0.3$ GeV, and $p_\perp = 0.1$ GeV.

failing to be a good variable, the number of particles produced increases with decreasing $p_\perp$. Thus, in fact, use of pseudorapidity or quasirapidity can be significantly misleading when one wants to understand both the spectral shape and the hadron yield.

In this context, we recall that a study of the distribution of heavy particles (nucleons and kaons) can be based on the difference between the distributions of positively and negatively charged particles, which is relatively easy to measure:

$$\frac{d(\mathrm{N}^+ - \mathrm{N}^-)}{dy_\pi} = \frac{d(\pi^+ - \pi^-)}{dy_\pi} + \frac{d(\mathrm{p} - \bar{\mathrm{p}})}{dy_\pi} + \frac{d(\mathrm{K}^+ - \mathrm{K}^-)}{dy_\pi}. \qquad (5.33)$$

The physics, in Eq. (5.33), is that the yield of pions is nearly charge symmetric (this has been observed at the SPS for $p_\perp > 0.3$ GeV [77]) and the first term cancels out. In the remainder, we have an initial measure of the quasirapidity distribution of protons and kaons. At the SPS, both protons and kaons contribute in Eq. (5.33). At the RHIC, the abundance of charged kaons is the dominating contribution, but only at the level of 1%–3% of all charged particles. As noted above, at the SPS, the canceling out of charged-pion yield is not exact at low $p_\perp < 0.25$ GeV, as a direct measurement has shown [77]: in the Pb–Pb collision system, we have 20% $(n - p)/(n + p)$ asymmetry in the number of protons and neutrons. This charge asymmetry translates into a relatively strong π^+/π^- asymmetry at small $p_\perp$, but disappears for $p_\perp > 0.2$ GeV.

5.5 Stages of evolution of dense matter

Since hadronic interactions are strong, we can hope and expect that local equilibrium conditions can be approached in experiments involving heavy ions. This is in particular the case if we characterize the essential physical properties of elementary matter in term of local, position-dependent parameters. The local average energy of each particle characterizes the local temperature T. (Local) chemical potentials μ_i need to be introduced in order to regulate the average particle and/or quark-flavor density.

These parameters express different equilibration processes in the fireball, and in general there is a considerable difference between thermal and chemical equilibrium.

- In order to establish thermal equilibrium, equipartition of energy among the different particles present has to occur in the collisional processes which lead to the statistical energy distribution. It is important to note that (local) thermal equilibrium can be achieved solely by elastic scattering. We will call the time scale on which these processes occur τ_{th}. The use of temperature T as a parameter presupposes that thermal equilibrium has (nearly) been established.

- Chemical equilibration requires reactions that change numbers of particles, and it is more difficult and thus slower to become established. There are also two quite different types of chemical equilibria.

  — *Relative* chemical equilibration, just like the commonly known case in chemistry, involves reactions that distribute a certain already existent element/property among different accessible compounds. Use of chemical potentials μ_i presupposes, in general, that the particular relative chemical equilibrium is being considered. We call the relevant time scale $\tau_{\text{chem}}^{\text{rel}}$.

  — In relativistic reactions, particles can be made as energy is converted into matter. Therefore, we can expect to approach (more slowly) the *absolute* chemical equilibrium. We call the relevant time scale $\tau_{\text{chem}}^{\text{abs}}$. We characterize the approach to absolute chemical equilibrium by a fugacity factor γ_i for particle 'i'. We often study the evolution of γ_i in the collision as a function of time, since absolute chemical equilibrium cannot generally be assumed to occur.

Some authors introduce separate chemical potentials for particles and antiparticles, $\mu_i^{\pm} = \pm\mu_i + T\ln\gamma_i$, [188, 189]. This is equivalent to our approach. However, it is still common to see in the literature that a equilibrium is assumed, $\gamma_i = 1$, with particles 'instantaneously' reaching their absolute chemical-equilibrium abundances. Such an approach cannot in general be justified. We see this on considering the relation between the relaxation times,

$$10^{-22} \text{ s} > \tau^{\text{exp}} \simeq \tau^{\text{abs}}_{\text{chem}} > \tau^{\text{rel}}_{\text{chem}} > \tau_{\text{th}}, \tag{5.34}$$

where τ^{exp} is the life span of the expanding fireball of dense matter, which is of the same magnitude as the time light needs to traverse the largest nuclei. In such a rapidly evolving system, we cannot assume that absolute chemical equilibrium, $\gamma_i = 1$, has been attained.

In order to illustrate the difference between absolute and relative chemical equilibrium better, let us consider some examples.

• We consider the baryon number, the globally conserved property of dense hadronic matter. Locally, the global conservation implies a balance of inflow against outflow, viz., there are no sources or sinks of baryon number. Generally, one always associates a conserved quantity with the presence of a chemical potential, here the chemical potential μ_b which controls the difference in number of all baryons and antibaryons. A change in the energy of the system, according to the first law of thermodynamics, is then given by

$$dE = -P\,dV + T\,dS + \mu_b\,db. \tag{5.35}$$

However, the addition of a baryon–antibaryon pair to the system will not be noted in Eq. (5.35), since the baryon number b remains unchanged!

At this point, we are not at liberty to add or remove a pair: in writing down Eq. (5.35), we *implicitly* assumed what we have above called absolute chemical equilibrium – there is a bath of baryon number in which our system is immersed, and hence a full phase-space occupancy of all available phase-space cells, and there is no place for an extra pair. By changing the chemical potential μ_b, we can regulate the difference in number of baryons and antibaryons present in the system, but densities of baryons and antibaryons move together, absolute equilibrium is assumed while relative chemical equilibrium controls the relative number of particles by virtue of the value of the chemical potential. If we change the baryon number by one at fixed volume and entropy, then according to Eq. (5.35), there is a change in energy by μ_b.

• Next, we look at the abundance of strangeness in the baryon-rich HG phase. There is no strangeness 'bath' and, initially, we have no strangeness, therefore we will be making pairs of s and $\bar{\text{s}}$ quarks; there is plenty of phase space available to fill, and we are far from absolute chemical-strangeness equilibrium. To make s$\bar{\text{s}}$ pairs in a HG, there are many possible reactions, section 18.2, classified usually as the direct- and associate-production processes. In the associate-production process, a pair of strange quarks is shared between two existent hadrons, of which one is a baryon, typically a nucleon N, which becomes a hyperon Y:

$$\pi + N \leftrightarrow K + Y.$$

In a direct-production process, a pair of strangeness-carrying particles is formed directly via annihilation of two mesons, akin to our *Gedankenexperiment* in which we are adding a pair to the system:

$$\pi + \pi \leftrightarrow K + \overline{K}.$$

Here, a pair of strange particles is made in the form of a pair of kaons, K^+K^-. With these two reaction types alone it could be that populations of strange mesons and baryons evolve differently. However, the meson carrier of the s quark, K^-, can exchange this quark rather fast, via exothermic reaction with a nucleon, forming a hyperon:

$$K^- + N \leftrightarrow \pi + Y.$$

This reaction establishes relative chemical equilibrium by being able to move the strange quark between the two different carriers, s$\bar{\text{q}}$ mesons and sqq baryons.

Reactions establishing the redistribution of existent flavor, or the abundance of some other conserved quantity, play a different role from the reactions that actually contribute to the formation of this flavor, or other quantum number, and facilitate the approach to absolute chemical equilibrium. Accordingly, the time constants for relaxation are different, since different types of reaction are involved.

Apart from the different relaxation times associated with the different types of thermal and chemical equilibria, there are different time scales associated with the different fundamental interactions involved. For example, the electro-magnetic interactions are considerably slower at reaching equilibrium than are the strong interactions governing the evolution of dense hadronic fireballs created in ultra-relativistic heavy-ion collisions. All the important time constants for relaxation in heavy-ion collisions arise from differences in mechanisms operating within the realm of strong interactions. Therefore, the separation of time scales is not as sharp as that between the different interactions, though a clear hierarchy arises, as we noted in Eq. (5.34).

Under weak interactions, there is, in comparison, an extremely slow transmutation of quark (and lepton) flavors, involving a much longer electro-weak equilibration time. Such long times are not available in the micro-bang process, in contrast to the big-bang case. We considered these electro-weak degrees of freedom in section 1.3, since the life span of the Universe at the time of the hadronic phase transition exceeds all typical relaxation times for the weak interaction.

The chemical equilibration, and hence the chemical composition of the fireball, evolve along with the temperature of the fireball. The following stages occur in heavy-ion collision.

1. The initial quantum stage.
   The formation of a thermalized state within τ_{th} is most difficult to understand, and is subject to intense current theoretical investigation. During the pre-thermal time, $0 \leq t < \tau_{th}$, the properties of the collision system require the study both of quantum transport and of decoherence phenomena, a subject reaching today far beyond the scope of this volume. We assume, in this book, that the thermal shape of a (quark, gluon) particle-momentum distribution is reached instantaneously compared with the time scales for chemical equilibration in Eq. (5.34). This allows us to sidestep questions regarding the dynamics occurring in the first moments of the heavy-ion interactions, and we explore primarily what happens after a time[¶] $\tau_0 \equiv \tau_{th} \simeq 0.25$–1 fm/$c$. The value of τ_0 decreases as the density of the pre-thermal initial state increases, e.g., as the collision energy increases. At τ_0 gluons g are, due to their greater reactivity, at or near to the local chemical equilibrium.

2. The subsequent chemical equilibration time.
   During the inter-penetration of the projectile and the target lasting no less than $\sim$1.5 fm/c, diverse particle-production reactions occur, allowing the approach to chemical equilibrium by light non-strange quarks q = u, d. As the energy is redistributed among an increasing number of accessed degrees of freedom, the temperature drops rapidly.

3. The strangeness chemical equilibration.
   A third time period, lasting up to $\simeq$5 fm/c, during which the production and chemical equilibration of strange quarks takes place. There is a reduction of temperature now mainly due to the expansion flow, though the excitation of the strange quark degree of freedom also introduces a non-negligible cooling effect.

4. The hadronization/freeze-out.
   The fireball of dense matter expands and decomposes into the final-state hadrons, possibly in an (explosive) process that does not allow re-equilibration of the final-state particles. The dynamics is strongly dependent on the size of the initial state and on the nature of the equations of state.

[¶] The time τ_{th} is often called τ_0 in the literature, and we will use this notation as well, though the subscript 'th' is more specific about the evolution step considered.

Throughout these stages, a local thermal equilibrium is rapidly established and, as noted, the local temperature evolves in time to accommodate change in the internal structure as is appropriate for an isolated physical system. We have a temperature evolution that passes through these series of stages:

T_{th} the temperature associated with the initial thermal equilibrium,
↓   *evolution dominated mainly by production of* q *and* $\bar{q}$;
T_{ch} chemical equilibrium of non-strange quarks and gluons,
↓   *evolution dominated by expansion and production of* s *and* $\bar{s}$;
T_{s} condition of chemical equilibrium of u, d and s quark flavors,
↓   *expansion, dissociation by particle radiation*;
T_{f} temperature at hadron-abundance freeze-out,
↓   *hadron rescattering, reequilibration*; and
T_{tf} temperature at thermal freeze-out, $T = T(\tau^{\text{exp}})$.

We encounter a considerable decrease in temperature. The entropy content of an evolving isolated system must increase, and this is initially related to the increase in the number of particles within the fireball and later also due to the increase in volume. However, in the later stages dominated by flow, the practical absence of viscosities in the quark–gluon fluid implies that there is little additional production of entropy. The final entropy content is close to the entropy content established in the earliest thermal stage of the collision at $t < \tau_0$, despite a drop in temperature by as much as a factor of two (under current experimental RHIC conditions) during the evolution of the fireball.

Except for the unlikely scenario of a fireball not expanding, but suddenly disintegrating, none of the temperatures discussed above corresponds to the temperature one would read off the (inverse) slopes of particle spectra. In principle, the freeze-out temperature determines the shape of emission and multiplicity of emitted particles. However, the freeze-out occurs within a local flow field of expanding matter and the thermal spectrum is to be folded with the flow which imposes a Doppler-like shift of T_{tf}: we observe a higher temperature than is actually locally present when particles decouple from flowing matter (kinetic or thermal freeze-out). The observable temperature $T_\perp$ is related to the intrinsic temperature of the source:

$$T_\perp \simeq \frac{1 + \vec{n} \cdot \vec{v}_{\text{tf}}}{\sqrt{1 - \vec{v}_{\text{f}}^2}} T_{\text{tf}} \rightarrow \sqrt{\frac{1 + v_{\text{tf}}}{1 - v_{\text{tf}}}} T_{\text{tf}}. \tag{5.36}$$

This relation must be used with caution, since it does not apply in the same fashion to all particles and has a precision rarely better than $\pm 10\%$. We study the shape of $m_\perp$-spectra in section 8.5.

5.6 Approach to local kinetic equilibrium

In the above discussion, the formation of a space–time-localized fireball of dense matter is the first key physics input. The question we wish to address now is that of how this fireball can possibly arise from a rather short sequence of individual reactions that occur when two, rather small, gas clouds of partons, clustered in nucleons, bound in the nucleus, collide. Indeed, at first sight, one would be led to believe that the small clouds comprising point-like objects would mutually disperse in the collision, and no localized, dense state of hadronic matter should be formed. At best, it was suggested in some early work, the two colliding 'eggs' should emerge from the high-energy interaction slightly 'warmed', but still largely 'unbroken'.

Two remarkable properties of hadronic interactions are responsible for just the opposite, deeply inelastic, behavior:

- the multiparticle production in hadron–hadron collisions; and
- the effective size of all hadrons expressed in term of their reaction cross sections.

What appears to be a thin system of point-like constituents is effectively already a volume-filling nucleon liquid, which will undergo, in a collision, a rapid self-multiplication with particle density rising and individual scattering times becoming progressively much shorter than the overall collision time.

Ultimately, as the energy available in collision is increased, the hadron particle/energy density will reach values at which the dissolution of the hadronic constituents into a common deconfined domain will become possible, and indeed must occur according to our knowledge about strong interactions. While we do not really know whether deconfinement of hadrons is not a general mechanism operating already at AGS energies, see table 5.1, there is today no experimental evidence that this low-energy range suffices. In contradistinction, a significant number of results obtained at the SPS energy range can be most naturally interpreted in terms of the formation of a deconfined space–time domain, section 1.6. We note that, per participant, there are as many as 7–10 further hadrons produced at SPS energies. This implies that there are thousands of quarks and gluons in the space–time domain of interest, and hence consideration of a 'local' (in space–time) equilibrium makes good sense.

There are many ways to estimate the particle number. We can use the number of final-state hadrons and evaluate the numbers of constituent quarks and antiquarks, or we can take the available energy content and divide it by the estimated energy per particle (quark, gluon). Both procedures give $\mathcal{O}(10\,000)$ particles for the case of Pb–Pb collisions at $158A$-GeV ($\sqrt{s_{\mathrm{NN}}} = 17.2$ GeV). Of these, not all particles can be causally connected, i.e., not all these particles can influence each other in classical dynamics, and local equilibrium is a feature obviously involving a causally connected region only.

A suitable measure of the causally connected size is offered by the initial decoherence time τ_0, which also determines the size of the decoherence volume, $R_0 \simeq \tau_0$. This has to be scaled up by the ensuing expansion factor, see the discussion below Eq. (6.35). For $\tau_0 = 0.5$ fm, we can expect about 5%–10% of all particles (500–1000) to be causally connected, which implies that the causal 'range' of rapidity is an interval $\Delta y \simeq 1$. Δy arises on considering the final-state rapidity distribution, see Fig. 9.6 on page 166 and Fig. 9.19 on page 184. In any case, the concept of a local equilibrium makes good sense.

When we are talking about thermal equilibria, we must first establish more precisely what these words mean. We will implicitly always refer to 'local' equilibrium. The thermalization of the momentum distributions is driven by *all* scattering processes, elastic as well as inelastic, because all of them are associated with transfer of momentum and energy between particles. The *scattering time scale*, for particles of species i, is given in terms of the collision length l by

$$\tau_{i,\text{scatt}} = \left\langle \frac{l}{v} \right\rangle_i = \frac{1}{\sum_j \langle \sigma_{ij} v_{ij} \rangle \rho_j}, \qquad (5.37)$$

where the sum in the denominator is over all particle species (with densities ρ_j) available, σ_{ij} and v_{ij} are the (energy-dependent) total cross sections and relative velocities, for a process scattering particles i and j, and the average is to be taken over the momentum distributions of the particle considered.

It is not hard to 'guestimate' the time scale governing the kinetic equilibration in the QGP. The typical particle-collision time (the inverse of the collision frequency) is obtained from Eq. (5.37) above. Given the particle densities and soft reaction cross sections, with the relative velocity of these essentially massless components being the velocity of light c, we find for the QGP scattering time,

$$\tau_i^{\text{QGP}} = 0.2\text{–}2 \, \text{fm}, \quad \text{with } \rho_i = 2\text{–}10 \, \text{fm}^{-3}, \ \sigma_i = 2\text{–}5 \, \text{mb}, \qquad (5.38)$$

as a range for different particles of type i, with the shorter time applying to the early high-density stage. This is about an order of magnitude shorter than the time scale for evolution of the fireball, which is derived from the spatial size of the colliding system: for the largest nuclei, in particular the Pb–Pb or Au–Au collisions, over a wide range of energy, we expect

$$\tau^{\text{exp}} \simeq \frac{R_A}{c} \simeq 5\text{–}8 \, \text{fm}/c. \qquad (5.39)$$

The achievement of kinetic equilibrium must be visible in the energy spectra of the particles produced, as we shall discuss below in section 8.1. This behavior, as we argue, can be understood in qualitative terms for the case of nuclear collisions. However, it remains to date a mystery why in some important aspects thermal models succeed for the case of p–p reactions. In particular, the exponential fall off of particle spectra, suggesting thermal equilibrium, has been noted with trepidation for a considerable time.

Hagedorn evaluated this behavior in the experimental data some 35 years ago [140, 145] and he developed the statistical bootstrap model (chapter 12), which assumes a statistical phase-space distribution (section 12.2). Hagedorn called it *preestablished or preformed equilibrium*: particles are produced in an elementary interaction with a probability characterized by a universal temperature. We can today only speculate about the physical mechanisms.

For example, it has been proposed that vacuum-structure fluctuations lead to color-string tension fluctuation, and thus the resulting string-breaking produces thermal hadrons [65]. Another informally discussed possibility is the presence of intrinsic chaotic dynamics capable of rapidly establishing kinetic equilibrium. We cannot pursue further in this book these ideas about the process of initial thermal equilibration.

Sometimes, the fact that we do not fully understand thermalization in the p–p case is raised as an argument against the possibility of conventional equilibration in nuclear collisions. We do not think so. In fact, if the p–p case leads to thermal hadrons, we should have a yet better thermalization in the A–A case. Thus, a microscopic model that is adopted to extrapolate from p–p to A–A collisions should respect the concept of the hadronic preestablished equilibrium, else it is not going to be fully successful, see section 6.1.

5.7 The approach to chemical equilibrium

The approach to chemical equilibrium is, in comparison with the thermal case, better understood. Firstly, we must consider which particles can be expected to have reached equilibrium and which not, and this requires a kinetic description. Though, in general, one is tempted to think of a build-up of chemical abundance of different quark flavors, the approach to absolute chemical equilibrium need not always occur from 'below', and/or the measured quark yields can be in excess of chemical equilibrium; section 19.4.

At the collision energies available at the RHIC and LHC, the more massive charm c and bottom b quarks (see table 1.1) are produced in the initial interaction, reaching and even exceeding the yield expected in absolute chemical equilibrium in thermalized deconfined matter. In QGP the chemical equilibration of these flavors occurs exceedingly slowly, and a significant excess of abundance is expected. A similar situation can arise with strangeness in presence of rapid cooling, from $T > 250$ MeV to $T \simeq 150$ MeV, which preserves the high initial thermal yield. In the early Universe, the well-known example of chemical nonequilibrium occurring, despite thermal equilibrium being established, is the freeze-out of abundances of light nuclear isotopes.

Of particular interest, in the physics of QGP, is that the saturation ('absolute' chemical equilibration) of the phase space of strange particles requires just the life span of the QGP. This is, in part, due to the relatively large threshold for the production of strange quarks and, in part, because for practical purposes most strangeness needs to be produced in thermal energy collisions – direct initial-state production of strangeness is of course quite prevalent but at the level of 10%–30% of the final-state equilibrium yield of strangeness, as long as only the normal processes of hadron collisions contribute to direct production of strangeness.

In the QGP phase, there is no need to redistribute strange quarks among different carriers and relative chemical equilibrium is automatically established. More generally, in the HG phase the relative chemical equilibrium is more easily attained than is the 'absolute' chemical equilibrium, due to the strangeness-exchange cross sections being greater than cross sections for its production.

The population master equation,

$$\frac{2\tau_{\text{chem}}^i}{\rho_i^{\text{eq}}} \frac{d\rho_i}{dt} = 1 - \left(\frac{\rho_i}{\rho_i^{\text{eq}}}\right)^2, \tag{5.40}$$

describes the population evolution of strangeness (and charm, etc.) within the scattering theory; chapter 17. τ_{chem}^i is the time constant for chemical relaxation. The quadratic term on the right-hand side, in Eq. (5.40), arises from, e.g., annihilation of strangeness, $s\bar{s} \rightarrow XX$, which rate is established by detailed balance consideration of two-body reactions. In the first instance, one has not ρ_i^2 but $\rho_i\bar{\rho}_i$, where $\bar{\rho}_i$ is the $\bar{s}$ density. However, since in heavy-ion collisions only hadronic reactions produce strangeness, we maintain the condition $\rho_i = \bar{\rho}_i$ and Eq. (5.40) follows. The solution of Eq. (5.40) approaches equilibrium exponentially for $t \rightarrow \infty$:

$$\rho_i = \rho_i^{\text{eq}} \tanh[t/(2\tau_{\text{chem}}^i)] \rightarrow (1 - e^{-t/\tau_{\text{chem}}^i})\rho_i^{\text{eq}}. \tag{5.41}$$

The chemical equilibration (relaxation) time constant τ_{chem}^i, for particle species i, is computed as an inverse of the invariant reaction rate per unit volume R_i:

$$\tau_{\text{chem}}^i = \frac{\rho_i^{\text{eq}}}{2R_i}. \tag{5.42}$$

In Eq. (5.42), ρ_i^{eq} is the chemical-equilibrium density. R_i is the rate at which the system 'chases' ρ_i^{eq}; the ratio is a characteristic time when the chase is over. The factor 2 in Eq. (5.42) is introduced to assure that the approach to equilibrium due to two-body reactions is governed by an exponential function with the time-decay parameter τ_{chem}^i, as is seen on the right-hand side of Eq. (5.41).

In terms of the reaction cross section, the invariant reaction rate per unit of time and volume is obtained from (see section 17.1)

$$R_i(x) = \sum_{a,b,X} \int_{(m_i+m_X)^2}^{\infty} 2\lambda_2(s)\, ds \int \frac{d^3k_a}{(2\pi)^3 2E_a} \int \frac{d^3k_b}{(2\pi)^3 2E_b}$$

$$\times\, f_a(k_a, x) f_b(k_b, x)\, \bar{\sigma}_{ab\to iX}(\sqrt{s})\, \delta[s - (k_a + k_b)^2]. \qquad (5.43)$$

where, see Eq. (17.10),

$$\lambda_2(s) = \big[s - (m_a + m_b)^2\big]\big[s - (m_a - m_b)^2\big].$$

In Eq. (5.43), we are neglecting Pauli or Bose quantum effects (suppression or stimulated-emission factors) in the initial and final states. Considered here, is the inelastic production process $a + b \to i + X$. $f_a(k_a, x)$ and $f_b(k_b, x)$ are the phase-space distributions of the colliding particles, and $\bar{\sigma}_{ab\to iX}(\sqrt{s})$ is the energy-dependent cross section for this inelastic channel. The 'bar' indicates that the dependence on transfer of momentum (scattering angle) is averaged over.

We will further study this integral for thermal distributions in section 17.1. However, given the importance of the final result Eq. (17.16), we record it here for the simplest case of a relativistic Boltzmann momentum distribution,

$$R_i(x) = \frac{\sum_{a,b,X} \int_{w_0}^{\infty} dw\, \lambda_2 \bar{\sigma}_{ab\to iX}(w) K_1(w/T)}{4T m_a^2 m_b^2 K_2(m_a/T) K_2(m_b/T)}, \qquad (5.44)$$

where $w = \sqrt{s}$ is the CM energy and $w_0 = m_i + m_X$ is the reaction threshold. This formula is presented in this form in [164], Eq. (5.7); it is stated there for the special case in which the reacting particles a and b are identical bosons, which, to avoid double counting of indistinguishable pairs of particles, requires an extra factor $\frac{1}{2}$, which is not included in Eq. (5.44). The interesting $m_{a,b} \to 0$ limit follows considering Fig. 10.1 and Eq. (10.47). It is implemented with a replacement of each factor $m^2 K_2(m/T)$ by $2T^2$ in Eq. (5.44), and $\lambda_2 \to s$, which reduces Eq. (5.44) to the result presented in [226], Eq. (2); [67] lacks the factor $1/T$.

We see explicitly, in Eq. (5.44), the mass threshold in the s-integration occurring for inelastic (particle-producing) rates. A high threshold combines with the exponentially small K_1 Bessel function, see Eq. (8.7), to reduce the strength of inelastic hadronic particle-production rates, which are usually much smaller than the total rates of reaction (scattering). For this reason, the time scale of chemical equilibration is, in general, considerably longer than the thermal one.

6 Understanding collision dynamics

6.1 Cascades of particles

The principal shortcomings of the near-statistical-equilibrium method, combined with ideal flow of hadronic fluid in the study of heavy-ion collisions, are the following:

- we do not have a long-lived, large region of hot hadronic matter to look at, and some features of the collision are certainly not well equilibrated;
- we need to establish the physical conditions at the initial time τ_0; and
- the system considered is subject to rapid evolution and all thermal properties are actually fields, i.e., we have a position-dependent local temperature $T(\vec{x})$, etc.

Hence, a lot of effort continues to be committed to the development of a better understanding of the initial interaction dynamics, and its subsequent description within microscopic kinetic-scattering models. The research field of the study of computer-code 'event generators' is vast and undergoing development. Consequently, in this book, we will enter into discussion of kinetic models only as matters of example and/or principle. We survey the rapidly developing field in order to offer an entry point for further study.

For a novice in this very rapidly changing panorama, the best next step is to look at the progress of the working group which has been monitoring the development of the computer codes with the objective of ensuring that reasonable quality control is attained.

OSCAR (Open Standard Codes and Routines)[||]. OSCAR begun in June 1997 to resolve the lack of common standards, documentation, version control, and accessibility in many transport codes. These transport codes for relativistic heavy-ion collisions differ from computer codes in other areas of physics, where numerical methods are only technical tools used to solve specific equations that define the physics. The source code of a nuclear-collision transport model often implements extra physical assumptions and dynamic mechanisms that go beyond the equations used to motivate its design. These algorithms often undergo evolution with time, and the very large number of phenomenological parameters also

---

[||] See: *http://www-cunuke.phys.columbia.edu/people/molnard/mirror-OSCAR/oscar.*

makes it difficult to pinpoint the relevant physical input controlling the observed computational result. Since the code itself defines the physical content of the model, it is essential to be able to closely scrutinize the actual algorithms used.

The list of codes currently either maintained or/and accessible, with the meaning of the acronyms, and principal authors is as follows**.

Correlation builders
CRAB – *Correlation After Burner*, by S. Prat.

Hydrodynamics
BJ-HYDRO – *Relativistic Hydrodynamics with Bjørken Geometry*, by A. Dumitru and D. H. Rischke.

Partonic/string transport
HIJING – *Heavy Ion Jet INteraction Generator*, by M. Gyulassy and X.-N. Wang [266].
HIJING/B-anti-B – *HIJING/Baryon Junction*, by S. Vance and M. Gyulassy.
MPC – *Molnar's Parton Cascade*, by D. Molnár.
neXus – by H.-J. Drescher and K. Werner.
PCPC – *Poincaré Covariant Parton Cascade*, by V. Boerchers, S. Gieseke, G. Martens, J. Meyer, R. Kammering and C. C. Noack.
VENUS – by K. Werner.
VNI – by K. Geiger, R. Longacre and D. Srivastava [130, 131].
VNIb – by S. A. Bass.
ZPC – *Zipping Parton Cascade*, by B. Zhang.

String/hadronic transport
ART – *Another Relativistic Transport*, by B.-A. Li and C.-M. Ko.
BEM – *Boltzmann Equation Model*, by P. Danielewicz.
BNC – *Burn and Crash*, by S. Pratt.
HSD – *Hadron String Dynamics*, by W. Cassing.
JAM –*Jet AA Microscopic Transport Model*, by Y. Nara.
JPCIAE – *Jetset Pythia CIAE (China Institute of Atomic Energy)*, by B.-H. Sa and A. Tai.
LEXUS – *Linear Extrapolation of ultra-relativistic Nucleon–Nucleon Scattering*, by S. Jeon.
LUCIAE – *Lund CIAE,*, by A. Tai and B.-H. Sa.
RQMD – *Relativistic Quantum Molecular Dynamics*, by H. Sorge [250].
UrQMD – *Ultra-relativistic Quantum Molecular Dynamics*, by S. A. Bass [274].

** For more details see *http://www-cunuke.phys.columbia.edu/people/molnard/mirror-OSCAR/oscar/models/list.html*.

Transport tools

GCP – *General Cascade Program*, by Y. Pang.

PYTHIA, JETSET, and LUND, mentioned above, are programs for the generation of high-energy-physics events, i.e., for the description of collisions at high energies between elementary particles such as e^+, e^-, p, and $\bar{p}$ in various combinations. Together, they contain theory and models for a number of aspects of physics, including hard and soft interactions, parton distributions, initial- and final-state parton showers, multiple interactions, fragmentation, and decay.

Development of JETSET, the first member of the 'Lund Monte-Carlo' family, was begun in 1978. The most extensive of these programs is PYTHIA. Over the years, these two programs have more and more come to be maintained in common. In the most recent version, they have therefore been merged into one, under the PYTHIA label. The current version is PYTHIA 6.1, by T. Sjöstrand.[††]

The common feature within transport-cascade models is that they picture a multiscattering process as a succession of binary collisions and decays, each well separated in space–time. For such an approach to have a chance of success, we must be in a physical situation dominated by well-separated collisions, the so-called collision regime. It is rather easy to see where this collision regime will occur in nuclear collisions: the de Broglie wavelength of one of the incident particles, and its (classical) mean free path in the medium, have to be compared with each other in order to identify the collision partners.

For example, at low energy, the de Broglie wavelength can be as large as the radius of the nucleus, so the dynamics will be dominated by the scattering of all nucleons, not by two-body collisions. As the energy increases, the resolving power increases and one also crosses particle-production thresholds and enters the multiple-scattering process involving elastic and inelastic nucleon–nucleon collisions, as well as collisions between the hadrons produced. We call this energy region the 'hadronic-cascade' region; as extensive studies of the data show, at AGS energies, $(10$–$15)A$ GeV, this is the dominant reaction mechanism. At higher energies, the de Broglie wavelength of the projectile becomes smaller than even a fraction of the size of a nucleon. The interaction will therefore involve the parton substructures – we reach the 'partonic-cascade' region.

In microscopic transport models describing the collision event (event generators), two primary mechanisms are used in order to describe evolution dynamics including production of particles: the nonperturbative production involving strong fields with field string-breaking, see Eq. (3.5),

---

[††] For more information, see *http://www.thep.lu.se/torbjorn/Pythia.html.*

and the production due to reactions caused by collisions of individual particles:

- Programs using primarily measured and extrapolated hadronic cross sections, e.g., RQMD [250], ARC [202, 203], QGSM [83], and UrQMD [274].
- Programs using perturbative QCD reactions, e.g., VNI [130, 131] and HIJING [266]; the main differences between these two models are the following: VNI is a Monte-Carlo implementation of a parton-cascade model (PCM) in which the time evolution of heavy-ion collisions is simulated by the parton cascading, whereas HIJING assumes the Glauber theory in the description of A–A collisions and handles the soft process on the basis of the string model.
- There are also hybrid models such as the PHC [196], a parton–hadron-cascade model, which is an extension of the hadronic-cascade model incorporating hard partonic scattering based on HIJING. However, practically all generators mentioned have, in some ways, taken the hybrid approach.

The hadronic-event generators are more suitable for lower AGS energies, and can be extended to SPS energies by introducing novel reaction mechanisms. The partonic generators are more geared to RHIC and LHC energies, but again, with some fine tuning, can be applied to the SPS energies. The SPS energy range is so difficult to cover, since p-QCD seems not to be well defined at such 'soft' energies, but the hadron cascade alone clearly cannot describe this energy range properly. The hybrid model (PHC [196]) is therefore more able to handle that energy domain.

There are major uncertainties in the hadronic-cascade models related to the impossibility of measuring reaction all relevant cross sections, section 18.2, and the necessity to introduce particle-production mechanisms well beyond the scope of the model (color ropes, for example, in RQMD). The perturbative QCD reactions in the deconfined phase are, on other hand, well determined in terms of elementary processes. The major uncertainty arises from the soft-QCD properties: for small transfers of energy, the QCD processes become very strong, and the issue of what physical mechanism is indeed responsible for the soft cutoff arises. This is reminiscent of the fact that we do not understand, in terms of QCD, the (inelastic) low-energy processes: e.g., the inelastic N–N cross section. It is for this reason that considerable attention was given to the color-string mechanism of particle production, which can be tuned to describe very well the nucleon–nucleon inelastic interactions within the LUND family adaptation (see below) to nuclear collisions, the FRITIOF model [40, 205].

Unfortunately, both the scope and the extent of this introductory book do not allow us to pursue in detail how these approaches differ. The above remarks can, however, serve as an entry point to further reading, for an up-to-date report, see [271]. The reader should be aware that a book could be written just on the subject touched on the surface in this section.

In closing this discussion, we wish to note that it is of course of interest to check how far the microscopic dynamic models are leading to near-equilibrium thermal and chemical conditions. Several studies exploring particle production and momentum distributions have revealed a very good approach to chemical and thermal equilibrium [80, 249]. This result really confirms that the large nuclear-reaction system, at the energies considered, disposes of sufficiently many degrees of freedom, and that statistical near-equilibrium methods are able to characterize the final state reached in the reaction. These results do not imply that the conditions created in the event generator are those observed experimentally.

6.2 Relativistic hydrodynamics

In the hydrodynamic description of the evolution of matter, rather than individual particles, one considers the flow of particles in a volume element. Therefore, we consider as one of the dynamic equations the conservation of (e.g., baryon) number-density current, along with the conservation of energy–momentum flow. These flows are described in terms of the local flow field $\vec{v}(\vec{x}, t)$, or equivalently in terms of the 4-velocity vector of the flow u^μ:

$$\frac{dx^\mu}{d\tau} \equiv u^\mu(x) = \gamma(1, \vec{v}), \qquad \frac{dt}{d\tau} \equiv \gamma = \frac{1}{\sqrt{1 - \vec{v}^2}}. \tag{6.1}$$

We see that, in general,

$$u_\mu u^\mu \equiv u^2 = 1, \quad u_\mu = g_{\mu\nu} u_\nu. \tag{6.2}$$

We use Einstein's summation convention for repeated Greek indices (implied summation over time '0' and space '1, 2, 3'), and work in flat space–time $g_{\mu\nu} = g^{\mu\nu}$, with the metric convention $g^{\mu\nu} \equiv \mathrm{diag}(1, -1, -1, -1)$.

There is a simple relation between the 4-divergence of the 4-velocity and the 4-divergence of the density arising from the conservation of current. We write a conserved current j_μ in terms of the local density ρ:

$$\partial_\mu(\rho u^\mu) = \rho \partial_\mu u^\mu + u^\mu \partial_\mu \rho = 0, \qquad \partial_\mu = \frac{\partial}{\partial x^\mu}. \tag{6.3}$$

The proper time τ coordinate of the local volume element and laboratory frame coordinates are related by the Euler relation:

$$\frac{d}{d\tau} = u^\mu \partial_\mu = \gamma \left(\frac{\partial}{\partial t} + \vec{v} \cdot \vec{\nabla} \right), \qquad \partial_\mu = \left\{ \frac{\partial}{\partial_t}, \vec{\nabla} \right\}. \tag{6.4}$$

We divide Eq. (6.3) by ρ and obtain, using Eq. (6.4),

$$\boxed{\partial_\mu u^\mu = -\frac{1}{\rho_i}\frac{d\rho_i}{d\tau} \equiv \frac{1}{\tau_{\exp}}.}$$

(6.5)

We suggest, at the end of condition Eq. (6.5), that it is a suitable definition of the expansion life span of the system, Eq. (5.34). In fact, as Eq. (6.32) below is showing, this is exactly true (taking freeze-out proper time) for the case of longitudinal flow in one spatial dimension.

Aside from the baryon number also, the flow of energy is considered in the hydrodynamic description of the time evolution. The hydrodynamic energy–momentum-flow equation, is

$$f^\nu \equiv \frac{\partial T^{\mu\nu}}{\partial x^\mu}, \qquad T^{\mu\nu} = (\epsilon + P)u^\mu u^\nu - g^{\mu\nu}P.$$

(6.6)

The form of $T^{\mu\nu}$ we present is suitable for adiabatic (entropy-conserving) flow of matter when the external force density vanishes, $f^\nu \to 0$, a point to which we will return momentarily.

The condition for energy–momentum conservation, Eq. (6.6), involves four equations. One of the equations can be made to look like the conservation equation Eq. (6.5): multiplication by u^ν of Eq. (6.6) yields, using Eq. (6.2),

$$u^\mu \partial_\mu \epsilon + (\epsilon + P)\partial_\mu u^\mu = 0.$$

(6.7)

If the pressure $P = 0$, this is the continuity equation Eq. (6.5) for the energy density. To make this obvious, we write

$$\boxed{\frac{\epsilon + P}{\epsilon}\partial_\mu u^\mu = -\frac{1}{\epsilon}\frac{d\epsilon}{d\tau}.}$$

(6.8)

For $P \neq 0$, the energy flow $(u^\mu \epsilon)$ is not conserved. For $P > 0$, there is a transfer of the energy content of matter to the kinetic energy of the flow of matter. The expanding matter cools. In the rare situation that $P < 0$ (see section 3.5), the transfer of energy goes from kinetic energy of flow back to the intrinsic energy density ϵ.

Equation (6.7) is equivalent to Eq. (1.17), which we recognize using Eq. (6.5),

$$\frac{1}{\epsilon + P}\frac{d\epsilon}{d\tau} = \frac{1}{\rho_i}\frac{d\rho_i}{d\tau}, \qquad \frac{d\epsilon}{\epsilon + P} = d(\ln \rho_i) = -3\frac{\dot{R}}{R},$$

(6.9)

noticing that the local density scales with $\rho \propto 1/V \propto 1/R^3$.

The other three equations which follow from Eq. (6.6) determine the velocity field $\vec{v}(\vec{x},t)$, Eq. (6.1),

$$\boxed{\frac{\partial \vec{v}}{\partial t} + (\vec{v} \cdot \vec{\nabla})\vec{v} = -\frac{1-v^2}{\epsilon+P}\left(\vec{\nabla}P + \vec{v}\frac{\partial P}{\partial t}\right),} \tag{6.10}$$

which form is obtained for the three spatial components $i = 1, 2$ and 3 in Eq. (6.6). Naturally, a solution of the hydrodynamic equations can be obtained only when the equation of state $P(\epsilon)$ is known, or equivalently, $\epsilon(T)$ and $P(T)$ are given.

As we have noted, the hydrodynamic-flow equation Eq. (6.6) conserves entropy. To show this, we consider again a contraction with u_ν of the hydrodynamic equations, but this time, we proceed in a different fashion. In the following the sum over the repeated index i is implied, which denotes more than one conserved particle number; in the simplest case, it is the baryon number,

$$\begin{aligned} u_\nu(\partial_\mu T^{\mu\nu}) &= \partial_\mu u^\mu (P+\epsilon) - u^\mu \partial_\mu P \\ &= \partial_\mu u^\mu (T\sigma + \mu_i \rho_i) - u^\mu \partial_\mu P, \end{aligned} \tag{6.11}$$

where we have used the Gibbs–Duham relation, see Eq. (10.26). After some reordering of Eq. (6.11), we obtain

$$0 = T\partial_\mu(\sigma u^\mu) + \mu_i \partial_\mu(\rho_i u^\mu) + \sigma u^\mu \partial_\mu T + \rho_i u^\mu \partial_\mu \mu_i - u^\mu \partial_\mu P. \tag{6.12}$$

The first term is the conservation of entropy flow which we are looking for,

$$\boxed{\frac{\partial \sigma^\mu}{\partial x^\mu} = 0, \quad \sigma^\mu = \sigma u^\mu,} \tag{6.13}$$

and thus other remaining terms in Eq. (6.11) should cancel out. The second term is the conservation of current flow, Eq. (6.3); it vanishes naturally.

The last three terms in Eq. (6.11) all contain the total proper time local derivative, Eq. (6.4). After multiplication with $d\tau$, we see that, for them to cancel out, we must have

$$0 = \sigma\, dT - dP + \rho_i\, d\mu_i \rightarrow 0 = S\, dT - V\, dP + b_i\, d\mu_i. \tag{6.14}$$

On multiplying by the volume V, we find the relation on the right-hand side. A more convenient way to consider Eq. (6.14) is

$$0 = d(TS - PV + b_i\mu_i) - (T\, dS - P\, dV + \mu_i\, db_i), \tag{6.15}$$

where, according to the Gibbs–Duham relation Eq. (10.26), the left parenthesis is just dE; hence, we recognize Eq. (6.15) as the first law of thermodynamics, Eq. (10.12), which proves Eq. (6.13).

For a more complete discussion of the relativistic hydrodynamics, we refer the reader to the monograph by Csernai [98], as well as introductory sections in *Gravitation and Cosmology* by Weinberg [267], who considers generalization of $T_{\mu\nu}$ with dissipative terms. A generalization of the adiabatic hydrodynamic expansion to include production of entropy has recently been proposed [114].

6.3 The evolution of matter and temperature

In our following discussion, we consider a reduction of Eq. (6.8). Introducing the velocity of sound,

$$\frac{1}{v_{\rm s}^2} \equiv \frac{d\epsilon}{dP}\bigg|_{S={\rm constant}}, \tag{6.16}$$

we obtain

$$\frac{d\epsilon}{d\tau} = \frac{d\epsilon}{dP}\frac{dP}{dT}\frac{dT}{d\tau} = \frac{1}{v_{\rm s}^2}\sigma\frac{dT}{d\tau}, \tag{6.17}$$

where we have used, in the limit of an extensive system at fixed chemical potential (here zero), for the entropy density, σ,

$$\sigma \equiv \frac{S}{V} = \frac{dP}{dT}, \tag{6.18}$$

which follows from Eq. (10.16). On substituting Eq. (6.17) into Eq. (6.8), we find

$$\partial_\mu u^\mu = -\frac{T\sigma}{\epsilon + P}\frac{1}{v_{\rm s}^2}\frac{1}{T}\frac{dT}{d\tau}. \tag{6.19}$$

The Gibbs–Duham relation, Eq. (10.26), allows us to write Eq. (6.19) in the form

$$\partial_\mu u^\mu = -\frac{1}{1 + (\mu_{\rm b}/T)b/S}\frac{1}{v_{\rm s}^2}\frac{1}{T}\frac{dT}{d\tau}. \tag{6.20}$$

We have introduced $b/S = \nu_{\rm b}/\sigma$, the inverse of the entropy per baryon, which is a constant of motion in ideal fluid dynamics: ideal flow conserves the entropy content and, of course, the baryon number is also conserved. Moreover, in an ideal gas of quarks and gluons, without a significant intrinsic dimensional scale, the ratio $\mu_{\rm b}/T$ of the two-dimensional statistical variables is also not evolving with proper time. In this case, the velocity of sound, $v_{\rm s}^2 = \frac{1}{3}$, is also exactly constant.

Using the conservation of baryon flow ($\rho_i \to \nu_{\rm b}$ in Eq. (6.5)) on the left-hand side of Eq. (6.20), we obtain:

$$\frac{1}{\nu_{\rm b}}\frac{d\nu_{\rm b}}{d\tau} = \frac{1}{1 + (\mu_{\rm b}/T)b/S}\frac{1}{v_{\rm s}^2}\frac{1}{T}\frac{dT}{d\tau}. \tag{6.21}$$

Equation (6.21) allows an exact integer for $1/v_s^2 = $ constant. Considering the relativistic quark matter $1/v_s^2 = 3$,

$$\frac{\nu_b}{\nu_b^0} = \left(\frac{T}{T_0}\right)^{3/\left(1+\frac{\mu_b}{T}\frac{b}{S}\right)}. \tag{6.22}$$

Some readers may wonder how it is possible that T, rather than μ_b, controls the evolution of the baryon density. This, of course, is just an optical illusion. Namely,

$$\frac{T}{T_0} = \frac{T}{\mu_b}\frac{\mu_b}{\mu_b^0}\frac{\mu_b^0}{T_0} = \frac{\mu_b}{\mu_b^0},$$

where the last equality arises since T/μ_b does not change during the isentropic evolution of the ideal quark–gluon gas. In this case a more palatable way of writing Eq. (6.22) is

$$\frac{\nu_b}{\nu_b^0} = \left(\frac{\mu_b}{\mu_b^0}\right)^{3/\left(1+\frac{\mu_b}{T}\frac{b}{S}\right)}. \tag{6.23}$$

It is interesting to observe that, for a baryon-dense fireball of quark matter, possibly formed in 10–40A-GeV fixed-target heavy-ion collisions, the deviations from the $\nu_b \propto \mu_b^3$ law are quite significant. However, at the SPS and RHIC, the initial conditions established assure that this relationship is valid: $S/b > 35$ (the SPS value; it is certainly larger at the RHIC) is seen to be relatively large compared with μ_b/T ($\simeq 1.4$ at the SPS and <0.1 at the RHIC).

Although we were able to extract the behavior of the baryon density from ideal-flow equations in the case of a relativistic gas of particles, the actual objective, namely the determination of the proper time variation of any of the quantities involved, has not been accomplished. We will obtain $T(\tau)$ in a very special, but interesting case, in the next section.

6.4 Longitudinal flow of matter

A special case of interest is the reaction picture invoking a rapid flow of matter along the collision axis, the so-called Bjørken scenario [73]. For this simple picture of the reaction to apply, we need to assume that

1. the colliding particles had so much energy that the flow of energy and matter after the collision remains unidirectional along the original collision axis; and

2. the transverse extent of the system is so large that the existence of the edge of matter in a direction transverse to the collision axis is of little relevance.

An interesting aspect of this 'punch'-through limit, seen in Fig. 5.2(a), is that the baryon number which is attached to the colliding valence quarks will also be leaving the interaction region, continuing to travel along the collision axis. Even though a trail of energy is deposited in the central rapidity region, the hope is that, at the highest energies, we should be able to recreate the baryon-free conditions of the early Universe.

After the time $\tau_0 = 0.25$–1 fm/c has passed since the initial contact between the Lorentz-contracted nuclear pancakes (the laboratory-frame view), the thermalized matter begins its evolution as indicated in figure Fig. 6.1. Each particle involved in the reaction has at its later freeze-out a 'proper' age τ_f since 'birth':

$$\tau_f = \int_0^f d\tau, \qquad d\tau^2 = dt^2 - d\vec{x}^{\,2}. \tag{6.24}$$

If all particles move with a constant velocity (e.g., c) along a common longitudinal direction z, and assuming that all particles have the same proper time at freeze-out, in laboratory coordinates the freeze-out time, t_f, and the freeze-out space coordinates, z_f, form a hyperbola,

$$t_f = \frac{\tau_f}{\sqrt{1 - v_f^2}}, \qquad z_f = v_f \frac{\tau_f}{\sqrt{1 - v_f^2}}, \qquad \tau_f^2 = t_f^2 - z_f^2, \tag{6.25}$$

as shown in the body of Fig. 6.1. The trajectory of each particle is a straight line $z = vt$, leading from the interaction point to the freeze-out location on the hyperbolic, $\tau_f = $ constant, surface.

The Minkowski space–time evolution of the ultra-relativistic collision is then rather simple: soon after the collision has occurred (see the CM-time snapshots to the left, beginning at the bottom in Fig. 6.1), the baryon number of the nuclei begins to separate (black lines along the light cone in Fig. 6.1), leaving in the intermediate region a trail of energy, presumed to be in the baryon-number-free QGP phase. The nuclei are trailed to right and left by the expansion of the energy they deposited at the instant of collision.

As the distance between the projectile and the target increases, the continued longitudinal expansion of every volume element reduces the local energy density/temperature until it is so low that individual had-rons can emerge (the chemical-freeze-out condition). As we shall see, the temperature will depend only on the proper time, Eq. (6.33), not on the rapidity. Therefore, a phase transition or transformation, as the case may be, and particle freeze-out occurs along given $\tau = $ constant space–time hyperbolas. In the graphic representation in Fig. 6.1, it is assumed that most of the time the QGP phase prevails, with a short period of freeze-out and hadronization, before final-state hadrons free-stream out of the interaction region.

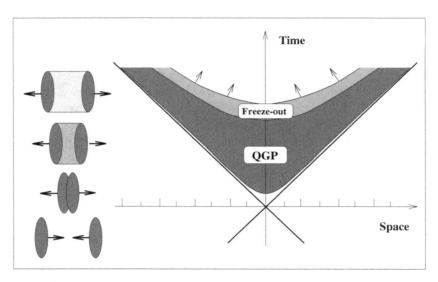

Fig. 6.1.   A space–time-image illustration of a heavy-ion collision in the ultra-relativistic (Bjørken) collision limit. Left: Lorentz-contracted nuclei collide and separate as a function of the laboratory time (vertical axis). Right: the light cone establishes the causality limit for the flow of energy, which fills the space–time domain between the separating nuclei.

The spatially central region is obviously at rest in the symmetric CM frame and will emit particles around $y_c = 0$. As we go away from the spatial center, the velocity of the local energy flow under the freeze-out condition increases, reaching the speed of the baryonic matter at the matter-trailing edge (upper right/left-hand edges of the light cone). We see that the rapidity of particles emitted and the relative position in space are correlated. Also, in laboratory time, the earliest particles to be emitted emerge at central rapidity, the latest at projectile–target rapidities, as can be seen in Fig. 6.1.

This discussion suggests that the natural variables in the study of the dynamics of longitudinally expanding matter are the proper time $\tau(t, z)$ which characterizes the parabolas in Fig. 6.1, and the space–time rapidity[‡‡] $y(t, z)$,

$$\tau = (t^2 - z^2)^{1/2}, \quad y = \frac{1}{2}\ln\left(\frac{t+z}{t-z}\right), \tag{6.26}$$

with the inverse relation

[‡‡] The reader is reminded that the variable y is not the spatial coordinate, but the space–time rapidity variable, and that only the spatial coordinate z enters into that which follows.

$$\boxed{t = \tau \cosh y, \qquad z = \tau \sinh y.}$$ (6.27)

We also record that

$$\frac{\partial y}{\partial t} = -\frac{z}{t^2 - z^2}, \quad \frac{\partial y}{\partial z} = \frac{t}{t^2 - z^2},$$

$$\frac{\partial \tau}{\partial t} = \frac{t}{(t^2 - z^2)^{1/2}}, \quad \frac{\partial \tau}{\partial z} = \frac{-z}{(t^2 - z^2)^{1/2}},$$ (6.28)

and

$$\frac{\partial t}{\partial y} = \tau \sinh y, \quad \frac{\partial t}{\partial \tau} = \cosh y,$$

$$\frac{\partial z}{\partial y} = \tau \cosh y, \quad \frac{\partial z}{\partial \tau} = \sinh y.$$ (6.29)

These transformations imply for the volume element that

$$dt\, dz = \tau\, d\tau\, dy.$$ (6.30)

The 4-velocity field of some volume element at proper time τ is

$$u^\mu \equiv \frac{dx^\mu}{d\tau} = (\cosh y, 0, 0, \sinh y), \qquad u^2 = 1.$$ (6.31)

Equation (6.31) implies that

$$\partial_\mu u^\mu = \frac{\partial u^0}{\partial t} + \frac{\partial u^3}{\partial z} = \frac{\partial y}{\partial t}\frac{\partial \cosh y}{\partial y} + \frac{\partial y}{\partial z}\frac{\partial \sinh y}{\partial y} = \frac{1}{\tau}.$$ (6.32)

All these relations become considerably more complex when one allows for flow in the transverse direction. For a velocity field including transverse cylindrical flow see Eq. (8.20).

We will now describe the Bjørken 'scaling' solution for the $(1 + 1)$-dimensional hydrodynamics [73]. The discussion above Eq. (6.26) suggests that one ought to use space–time rapidity and proper time as variables when one is considering a one-dimensional relativistic hydrodynamic model. We substitute Eq. (6.32) into Eq. (6.20) and obtain ($\mu_b \simeq 0$)

$$\boxed{\frac{v_s^2}{\tau} = -\frac{1}{T}\frac{dT}{d\tau}.}$$ (6.33)

This important result allows us to understand how fast the temperature is changing during the 'scaling' one-dimensional hydrodynamic evolution described. We encountered a related result in the study of the adiabatic (isentropic) expansion, see section 1.4.

Perhaps the most cited equation of $(1 + 1)$-dimensional hydrodynamics arises when, in Eq. (6.8), we use Eq. (6.32) (see Eq. (21) in [73]):

$$\boxed{\frac{\epsilon + P}{\tau} + \frac{d\epsilon}{d\tau} = 0.}$$ (6.34)

For $\epsilon(T)$ and $P(T)$, this implies that T is a function of τ, but not of y. This important result originates from the assumption that the proper time τ of a fluid volume element is as given in Eq. (6.26), and it is in particular independent of the transverse coordinates.

For a (nearly) relativistic gas $v_s^2 \lesssim \frac{1}{3}$, and the decrease of the temperature is slow. Explicitly, integrating Eq. (6.20), we obtain, assuming that the velocity of sound changes slowly,

$$T = T_0 \left(\frac{\tau_0}{\tau}\right)^{v_s^2},$$ (6.35)

where the initial temperature T_0 is established at an initial (proper) time τ_0, at which local thermal equilibrium has been established and the isentropic hydrodynamic expansion begins. In order to decrease the temperature by a factor two, we need the time $\tau \simeq 8\tau_0$.

In a more realistic evolution of a fireball, which allows for transverse expansion, the expansion cooling is faster [58, 163]; see section 6.2. On the other hand, one also must allow for a less than fully relativistic sound velocity. The properties of the equation of state obtained on the lattice suggest that, in the vicinity of the phase transition, i.e., for $T < 2T_c$, there are significant deviations from ideal-gas behavior. A seemingly small change in v_s matters: we note that, when $v_s \simeq 0.5$ (recall that $1/\sqrt{3} \simeq 0.58$), for the scaling solution Eq. (6.35), the time needed to decrease the temperature by a factor of two increases two-fold to $\tau \simeq 16\tau_0$.

7 Entropy and its relevance in heavy-ion collisions

7.1 Entropy and the approach to chemical equilibrium

Entropy is a quantity characterizing the arrow of time in the evolution of a physical system – in every irreversible process the entropy increases. In elementary interactions, and in particular those involving relativistic collisions of two large atomic nuclei, there is considerable production of particles and hence of entropy. A number of questions arise naturally in this context:

1. When and how is entropy produced in a quantum process, such as a nuclear collision?

2. How is production of hadronic particles related to production of entropy?

3. How does one measure the entropy produced in the reaction?

In the deconfined phase the color degree of freedom is 'melted'. Therefore, the specific entropy content per baryon (S/b), evaluated at some given (measured) values of statistical parameters, is generally greater in the deconfined state than it is in to the confined state. Entropy can only increase, and thus, once an entropy-rich state has formed, we have an opportunity, by measurement of the entropy created in the heavy-ion collision, to determine whether the color bonds of valence quarks present in the collision have been broken.

The final entropy content of the hadronic particles emerging has to exceed the initial entropy of the thermal state. In fact, quantitative studies show that very little additional entropy is produced during the entire evolution of a fireball, after the initial thermalization stage. For this reason, the final hadronic state conveys key information about the initial thermal state of dense and hot hadronic matter. For example, in the expanding quark–glue fireball the quasi-entropy-conserving evolution has been confirmed within a model study involving parton cascade [129]. The final-state entropy is largely produced in the first instant of heavy-ion collision.

The entropy can be obtained using the momentum-distribution function $f_{\mathrm{B,F}}$:

$$S_{\mathrm{B,F}} = \int d^3x \int \frac{d^3p}{(2\pi)^3} [\pm(1 \pm f_{\mathrm{B,F}}) \ln(1 \pm f_{\mathrm{B,F}}) - f_{\mathrm{B,F}} \ln f_{\mathrm{B,F}}]. \quad (7.1)$$

The upper sign $+$ is for bosons (B) and the lower sign $-$ is for fermions (F), which is somewhat counterintuitive, but in fact in agreement with Fermi and Bose statistics. We are reminded of this change by the change in the usual sequence of letters 'F, B' in the subscript.

There are two well-known ways to obtain Eq. (7.1). It follows (up to normalization) from Boltzmann's H-function in the study of momentum equilibration. It also arises naturally on rewriting Eq. (10.25) in terms of the single-particle distribution function Eq. (4.42). Since in this approach the statistical definition of entropy, which corresponds to the thermal definition, is used, the normalization is fixed by the laws of thermodynamics.

Entropies of different particles add, and the entropy of particles and antiparticles adds as well. The entropy of fermions, in Eq. (7.1), vanishes in the pure quantum-state limit for $T \to 0$, since the value of the particle-occupancy probability is either unity or zero. The 'classical' Boltzmann limit arises when $f_{\mathrm{B,F}} \ll 1$. In this case, with $f_{\mathrm{B,F}} \to f$,

$$S_{\mathrm{cl}} \equiv \int d\omega f \ln\left(\frac{e}{f}\right) = \int d\omega (f - f \ln f), \quad \int d\omega \equiv \int d^3x \int \frac{d^3p}{(2\pi)^3}. \quad (7.2)$$

Generally, expression Eq. (7.1) is presented as the generalization of the well-known classical micro-canonical definition Eq. (7.2) to quantum gases; however, this language leads to the (frequent) omission of the first term on the right-hand side, which is the number of particles present, and which comprises 25% of the total entropy of the relativistic gas.

We will study the entropy of a statistical gas in more detail in section 10.6. We note that, for massless particles (quarks, gluons), the entropy per particle following from Eq. (7.1) and Eq. (7.2) is

$$\left.\frac{S}{N}\right|^{\rm B}_{m=0} = 3.61, \quad \left.\frac{S}{N}\right|^{\rm cl}_{m=0} = 4, \quad \left.\frac{S}{N}\right|^{\rm F}_{m=0} = 4.20. \tag{7.3}$$

The effect of the finite pion mass is to increase the entropy per particle, see Eq. (10.79) and Fig. 10.4 on page 206. Each pion (a boson) emerging carries away just about 4 units of entropy from the source as long as the temperature is $T \simeq m_\pi$. These results are for a chemically equilibrated system. In general, below equilibrium at a given fixed temperature, the entropy density is lower, but the entropy per particle is higher than that in Eq. (7.3), and the opposite is true for a system above chemical equilibrium, $\gamma > 1$; section 7.5.

For an isolated system, like our hadronic fireball, a very important physical property is that relatively little entropy is generated in the approach to chemical equilibrium, both from above and from below. This happens since the change in number of particles consumes or releases thermal energy and this changes the temperature. Thus, even after chemical equilibration, the final-state entropy content is closely related to the initial entropy of the thermal state generated in the collision.

To demonstrate this, we obtain the shape of the particle-occupancy probability f of particles in the isolated fireball from the Boltzmann H-theorem result, i.e., the principle that a physical system evolves toward maximum entropy for a given energy and number of particles in the system. We seek to maximize the entropy Eq. (7.1) subject to these constraints, i.e.,

$$T(f) = \int d\omega \, [\pm(1 \pm f) \ln(1 \pm f) - f \ln f - (\alpha_f + \beta \epsilon_f) f], \tag{7.4}$$

as a functional of the distribution shape $\{f\}$. Here, $d\omega$ is the phase-space integral seen in Eqs. (7.1) and (7.2).

The results are the standard Fermi and Bose distributions Eq. (4.42), including the chemical nonequilibrium factor $\gamma = e^{-\alpha}$ which allows that the particle number is fixed independently from the temperature,

$$f_{\rm F,B} = \frac{1}{e^{\beta\epsilon + \alpha} \pm 1}. \tag{7.5}$$

This demonstrates that an isolated system in the presence of suitable internal dynamics, e.g., two-body elastic collisions, will evolve toward the kinetic-equilibrium statistical Bose/Fermi distributions, which may be in chemical nonequilibrium expressed by $\gamma \neq 1$, depending on the initial energy and number of particles. When inelastic particle-production processes are occurring, we further expect that also chemical equilibrium should be reached, but on a slower time scale; see section 5.5.

Now, we are ready to show that the entropy content of a chemically not fully equilibrated system is nearly the same as the entropy content of a system in equilibrium. For the Boltzmann approximation, we can show this analytically. The factor γ becomes a normalization factor that describes the average occupancy of the phase space relative to the equilibrium value; the additive term describes how the entropy per particle changes as the occupancy changes. We find in the Boltzmann limit

$$\mathcal{N} = \gamma \mathcal{N}|_{\text{eq}}, \tag{7.6}$$

$$\mathcal{E} = \gamma \mathcal{E}|_{\text{eq}}, \tag{7.7}$$

$$\mathcal{S} = \gamma \mathcal{S}|_{\text{eq}} + \ln\left(\gamma^{-1}\right) \gamma \mathcal{N}|_{\text{eq}}. \tag{7.8}$$

For massless particles, the phase-space integrals are easily performed and one obtains, see chapter 4,

$$\mathcal{N}^0 = aV\gamma T^3, \tag{7.9}$$

$$\mathcal{E}^0 = 3aV\gamma T^4, \tag{7.10}$$

$$\mathcal{S}^0 = 4aV\gamma T^3 + \ln\left(\gamma^{-1}\right) aV\gamma T^3, \tag{7.11}$$

where $a = g/\pi^2$ and g is the degeneracy. One easily finds how, for $\mathcal{E}^0 =$ constant, the entropy varies as a function of γ,

$$\mathcal{S} \propto \gamma^{1/4}(4 - \ln\gamma). \tag{7.12}$$

This functional has a very weak maximum at $\gamma = 1$. For example, at $\gamma = 2$ the entropy is 98.3% of the value at $\gamma = 1$.

One could imagine that an important change in number of particles is required when γ increases by say a factor of ten from 0.1 to 1. However, since the total energy and volume of the system (and hence the energy density) do not vary, we obtain a result that contradicts our intuition. Namely, at a high initial temperature the phase space is much greater and, in the Boltzmann approximation, the number of particles scales with γT^3, Eq. (7.6). Since $\mathcal{E}/V \propto \gamma T^4$, we obtain

$$\mathcal{N}|_{\mathcal{E}/V} \propto \gamma^{1/4}. \tag{7.13}$$

Thus, a ten-fold increase in γ is accompanied by a 1.8-fold increase in number of particles.

At this point, it is perhaps wise to briefly review the more familiar case of a fixed-temperature environment (a heat bath) in order to understand better the difference from the isolated-fireball system. Consider the 'black-body' radiator: a thermally insulated box with a small emission hole, for which the loss of energy due to radiation through the hole is externally compensated by keeping the temperature constant: the spectrum of the emitted radiation displays the Planck shape which minimizes the free-energy content $\mathcal{F}$ of the photon gas, at a fixed temperature of the walls – this spectrum is arising from interaction of the photons with the walls, with the spectrum and number of photons changing due to absorption and re-emission by the walls. Recalling now that $\mathcal{F} = \mathcal{E} - T\mathcal{S}$, we can combine Eqs. (7.6) and (7.8), which gives, in the Boltzmann limit,

$$\mathcal{F} = -aVT^4\gamma\left[1 + \ln\left(\gamma^{-1}\right)\right], \tag{7.14}$$

with a minimum at $\gamma = 1$. However, now a change by a factor of two in γ, at fixed T, leads to a change by 35% in the value of the free energy and an even greater change in entropy. Clearly, at fixed T, the equilibrium point $\gamma \to 1$ is much better defined than it is at fixed $\mathcal{E}$.

The lack of sensitivity of entropy to chemical equilibration for an isolated fireball assures that there is ample room to generate nonequilibrium particle yields during the dynamic evolution of the system. Given that the system we are considering is actually subject to a dynamic evolution, with expanding volume V, it is natural to expect that chemical equilibrium is an exception rather than a rule.

7.2 Entropy in a glue-ball

We are now ready to examine in detail the simplest system of dynamic interest to us. We consider an initially thermal glue–parton ball far from particle-abundance equilibrium. There are glue interactions that are producing particles, driving the system to chemical equilibrium while the temperature decreases, due to sharing of a fixed available amount of thermal energy by an ever larger number of constituents. We assume, in the example below, that, when chemical equilibrium is reached, the glue-gas state is at $T = 250$ MeV.

The intuitive expectation is that a lot of entropy is produced while this system evolves toward the particle-abundance equilibrium. However, this is not so [179]. The reason is that, as the equilibrium in particle-number abundance is approached, we must adjust the temperature of the system. There is a subtle balance between the different effects, and the result is that we find considerable constancy of the entropy of the isolated particle-producing system.

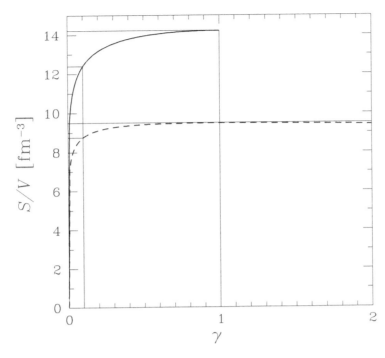

Fig. 7.1. The entropy density $\mathcal{S}/V$ (units fm^{-3}), at a fixed energy density $E/V = 2.66$ GeV fm^{-3}, for $m_g = 0$ (solid line), and at $E/V = 1.89$ GeV fm^{-3} for $m_g = 0.450$ GeV (dashed line), for a (gluon) Bose gas, as a function of the chemical occupancy γ, with $T(\gamma = 1) = 250$ MeV.

We consider both the massless-gluon case and the case of thermally massive gluons, choosing for the thermal gluon mass $m_g^{\mathrm{th}} = 0.450$ GeV, see Fig. 16.3 on page 308. A massless-gluon (Bose) gas, with $g = 16$, has an energy density $E/V = 2.66$ GeV fm^{-3}, at $T = 250$ MeV, at the chemical-equilibrium point $\gamma = 1$. The massive gas, at the chemical-equilibrium point $\gamma = 1$ at $T = 250$ MeV, has $E/V = 1.89$ GeV fm^{-3}.

In Fig. 7.1, we see the entropy density $\mathcal{S}/V$ (units fm^{-3}) as a function of γ; the solid line is for massless gluons and the dashed line is for massive gluons. The maximum in entropy at $\gamma = 1$ is shallower than would be the case for a Boltzmann gas. The curves end at the singularity of the Bose distribution function, $\gamma = 1$ for massless gluons, and $\gamma_c = e^{m_g/T} \sim 2.7$ (beyond the range shown in Fig. 7.1), which values cannot be exceeded.

The vertical line, to the left in Fig. 7.1, shows that the entropy content of the hot-glue system at $\gamma = 0.1$ is already nearly 90% of the chemical-equilibrium entropy. The 'hot' glue is at this point at $T \simeq 453$ MeV for $m_g = 0$, and at $T = 426$ MeV for $m_g = 0.450$ GeV.

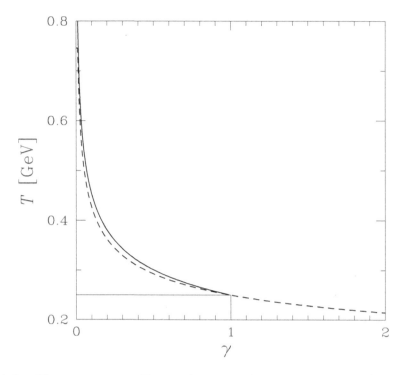

Fig. 7.2. The temperature T as a function of the chemical occupancy γ for massless gluons with $E/V = 2.66$ GeV fm^{-3} (solid line) and $m_{\mathrm{g}}^{\mathrm{th}} = 0.450$ GeV, $E/V = 1.89$ GeV fm^{-3} (dashed line). The equilibrium point $\gamma = 1$ has been chosen to occur at $T = 250$ MeV.

In order to maintain a fixed value of E/V, the temperature T and phase-space occupancy γ are not independent, and, as a function of γ, the temperature T drops rapidly, which is shown in Fig. 7.2. The dashed line, corresponding to the case of massive gluons, coincides with the solid line (massless gluons) at $\gamma = 1, T = 250$ MeV, by token of the judicious choice of E/V. As Fig. 7.2 shows, the temperature can drop rapidly in the process of chemical equilibration of the gluon gas.

It is interesting to note that, when the chemical cooling, seen for small γ in Fig. 7.2, is fastest at small γ, the collective flow of the QGP fireball should not yet be established. Therefore, it is probable that the initial-state cooling is due to chemical processes. The mechanism for a chemical equilibration of the hot initial glue phase which is faster than the volume expansion has been proposed to be inherent in the multi-glue-production reactions [247], gg $\rightarrow$ ggg, gggg, ...

The number of gluons changes relatively slowly, in particular considering massive (thermal) gluons, as can be seen in Fig. 7.3. In the case of

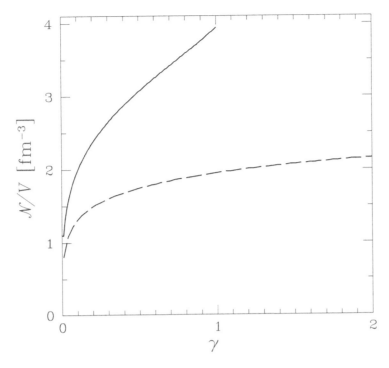

Fig. 7.3.   The particle density $\mathcal{N}/V$ (units fm^{-3}) as a function of the chemical occupancy γ. Lines are as in Fig. 7.1.

massless gluons (solid line), when γ increases by a factor of ten from 0.1 to 1, the number of gluons increases five times slower. This increase is considerably more modest for $m_{\mathrm{g}}^{\mathrm{th}} = 0.450$ GeV (dashed line in Fig. 7.3). To understand the greater change seen in Fig. 7.3 compared to Fig. 7.1 it is important to know that the entropy per particle is noticeably greater than four for a system far from chemical equilibrium, Fig. 7.8.

The process of chemical equilibration of glue involves, apart from an increase in the number of gluons, a change in the momentum distribution. In Fig. 7.4, we compare the spectra of gluons initially at $\gamma = 0.1$ with equilibrium $\gamma = 1$. At equilibrium, the temperature is $T = 250$ MeV and we see that the massless- (solid line) and massive-gluon (dashed line) spectra coincide (lines ending at $E \simeq 3.5$ GeV). The 'missing' gluons, at low energies, contribute to the difference in energy density (2.66 versus 1.89 GeV fm^{-3} for massless and massive, $m_{\mathrm{g}} = 450$ MeV, gluons, respectively). The relatively slowly falling spectra are for the early hot-glue nonequilibrium stage at $\gamma = 0.1$, at which for massless gluons $T = 453$ MeV and for massive gluons $T = 426$ MeV (values determined for fixed volume and energy of the fireball).

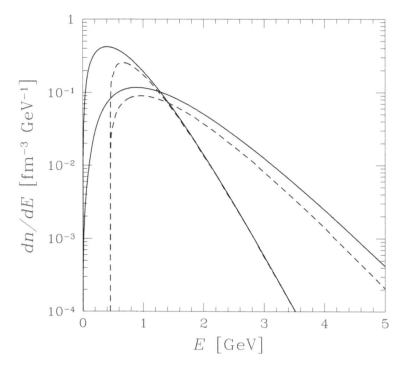

Fig. 7.4. Spectra of massless (solid lines) and $m_g = 450$ MeV (dashed lines), gluons prior to ($\gamma = 0.1$, lines reaching to right lower corner) and at ($\gamma = 1, T = 250$ MeV) chemical equilibrium.

We have shown that the dynamics of chemical equilibration is often counterintuitive. In particular, when one is considering the dynamics of an isolated fireball, we learned that the entropy varies little if chemical equilibrium is not maintained. We infer that in a rapid evolution, an isolated system can evolve away from chemical equilibrium, even if this means that the chemical entropy is not at its maximum. On the other hand, since the entropy content is not a sensitive probe of chemical-equilibrium properties we can, in the following study of the experimental entropy production, proceed as if chemical equilibrium were maintained, without loss of generality.

7.3 Measurement of entropy in heavy-ion collisions

The final-state entropy content is visible in the multiplicity of particles produced. In the HG and QGP phases of dense hadronic matter, the entropy content is in general different. The entropy content per partici-pant (specific entropy) offers a method to distinguish these two different hadronic phases. At a temperature $T > T_c \simeq 160\,\text{MeV}$, the QGP is the phase with the higher specific entropy; this difference occurs because of the liberation of the color degrees of freedom in the color-deconfined QGP phase. The question of whether it is possible to measure the entropy per baryon in the fireball arises. A measure of entropy must count the total production of particles, while the number of participants can be measured using the positive-hadron multiplicity, which comprises, in particular, protons participating in the reaction.

It has been argued that, in the SPS energy range, the ratio of net charge multiplicity to the total charged multiplicity comprises information about the specific entropy content of the matter phase in the fireball [182, 183]:

$$D_Q \equiv \frac{N^+ - N^-}{N^+ + N^-}. \tag{7.15}$$

D_Q is understood to be a function of rapidity when considering particle distribution in rapidity, rather than the total abundance number. In general, D_Q can be measured, it is an easy task if particles are not identified. The sum of positive and negative hadron multiplicity can be identified by the sign of the curvature in a magnetic field, and the emission angle; see Fig. 5.5 and Eq. (5.25).

A first estimate of this ratio is

$$D_Q^{ns} \approx \frac{A_f}{N_\pi} \frac{0.75}{1 + 0.75\,(A_f/N_\pi)}, \tag{7.16}$$

where A_f is the number of baryons in the fireball. We have considered only non-strange particles 'ns', i.e., pions and nucleons, and assumed that the system is symmetric in isospin, that is half of the participants in the fireball A_f are protons, and all yields of pions are equal: $N_{\pi^+} = N_{\pi^-} = N_{\pi^0} = N_\pi/3$.

As we see in Eq. (7.16), D_Q is indeed a measure of the baryon-to-pion ratio and thus of entropy per participant. However, this estimate Eq. (7.16) is wrong by two partially compensating factors of order 2: both higher-mass non-strange resonances and charged strange particles must also be considered. We recall the significant kaon contribution seen in Eq. (5.33).

Theoretically, such calculations require knowledge of the relative abundances of particles for higher-mass resonances as well as for strange particles. This can be determined in a statistical model of chemical freeze-out as a function of a few parameters, in particular T. Similarly, the entropy per baryon in the fireball is given as a function of the same statistical variables. Both the charge-multiplicity ratio D_Q and the specific entropy S/A_f are known functions of the statistical parameters.

As Eq. (7.17) suggests, the quantity

$$C_Q \equiv D_Q \frac{S}{A_f} \propto D_Q \frac{N_\pi}{A_f} \tag{7.17}$$

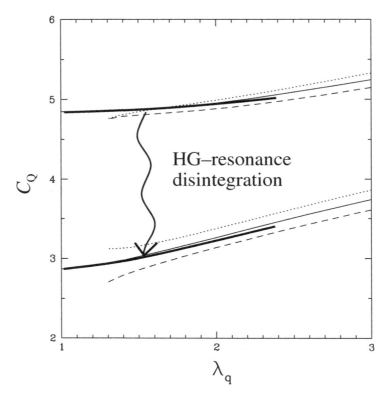

Fig. 7.5. The product $C_Q = D_Q(S/b)$ before (upper curves) and after (lower curves) resonance disintegration, as a function of λ_q, for fixed $\lambda_s = 1 \pm 0.05$ and conserved zero strangeness in an equilibrated HG.

should be a structure constant that depends somewhat on the mixture of hadronic flavors, mass spectrum and similar general hadron-spectrum properties, but should be largely independent from the statistical properties of the system. Once the value of C_Q has been established within a theoretical model, it should then apply in general – a value, $C_Q \simeq 4.5$, was found [182] for a chemically equilibrated system, see the upper line in Fig. 7.5. There is in addition the effect of hadron-resonance decays, and this increases the final-state multiplicity of charged hadrons. According to Eq. (7.15), the value of D_Q diminishes. In consequence, the observable value of $C_Q \simeq 3$ is seen to apply to the lower lines in Fig. 7.5.

7.4 The entropy content in 200A-GeV S–Pb interactions

Since D_Q is generally a small number, it has not been studied extensively. Experiment EMU05 at CERN–SPS has exposed a lead target located in

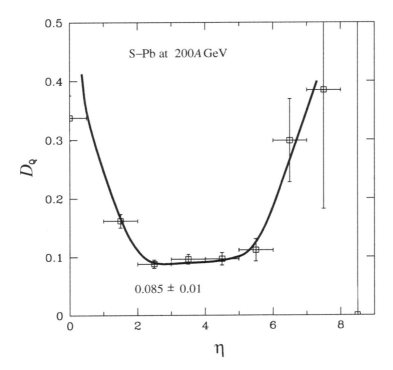

Fig. 7.6.   Emulsion data for the charged-particle-multiplicity ratio D_Q obtained in central S–Pb collisions at $200A$ GeV as a function of pseudorapidity η. The bold black line is drawn to guide the eye.

front of a photographic emulsion to a $200A$-GeV beam of sulphur atoms. Since a magnetic field was present, the charge of particles produced was determined and thus the experimental value of D_Q, as a function of pseu-dorapidity η, could be determined by evaluation of the angle of emission of charged hadrons from the interaction vertex, which fixes η, see Eq. (5.25), and the polarity of charged particles.

Selecting the most central events with charge multiplicity $N^+ + N^- >$ 300, corresponding to a total central hadron multiplicity in the range 450–1000, in the central (pseudo)rapidity region, the value

$$D_Q = 0.088 \pm 0.007, \qquad \eta \simeq 2.5 \pm 0.5,$$

is found [104, 117]. The distribution $D_Q(\eta)$ is shown in Fig. 7.6. The pseudorapidity distribution is flat in the central region, $\Delta\eta = (\eta - 2.6) \pm$ 1.5. This suggests partial transparency and the presence of longitudinal flow in the rapidity distributions of protons and K^+; see section 8.3.

Inspecting Fig. 7.5 and Fig. 7.6 we arrive at a first estimate of the entropy content of a fireball formed in these interactions:

$$\frac{S}{b} = \frac{3 \pm 0.1}{0.088 \pm 0.007} = 34 \pm 4.$$

The sources of systematic error involved in the use of D_Q to fix the entropy include the difference in the distribution between rapidity and pseudorapidity, and the uncertainty about yields of strange charged hadrons, which vary with the degree of chemical equilibration of strangeness.

This high value of specific-entropy content in highly central S–Pb 200A-GeV interactions, 40% higher than expected, suggests that an entropy-rich (deconfined) state has been created [182]. Since a high specific-entropy content can be found in the HG phase at smaller values of λ_q, i.e., smaller baryon density, it is important in comparison to the HG to have a good understanding of the baryochemical potential. The value of $\mu_b = 3\mu_q$, see Eq. (4.18), needs to be reduced by nearly a factor of two, to about $\mu_b = 100$–120 MeV, before the entropy content of HG becomes consistent with these experimental data. Such a small baryochemical potential is not in agreement with many measured yields of hadrons [176]. As this simple case shows it is the simultaneous consideration of several observables (charged-particle asymmetry combined with specific (strange) particle ratios) which allows understanding of the physics of heavy-ion collisions.

7.5 Supersaturated pion gas

The excess of hadron multiplicity (entropy) is a consistently observed phenomenon: the data obtained for Pb–Pb collisions supports this strongly. We have seen in Fig. 1.6 a significant excess of hadron multiplicity in high-energy A–A reactions compared to p–p and low-energy A–A reactions. An important question is how this excess of abundance can be theoretically described in terms of the final-state hadron phase space.

During hadronization, hadrons need to acquire the excess entropy arising from broken color bonds of QGP. We now look for the most entropy-rich hadron gas and consider the super-saturated massive (pion) Bose gas, in which the chemical potential, i.e., the abundance fugacity γ, is nearly compensating for the suppressing effect of the mass, which occurs at $\gamma e^{-m/T} \to 1$.

For pions, composed of a light-quark–antiquark pair, it is convenient to use as the abundance fugacity

$$\gamma_\pi = \gamma_q^2. \tag{7.18}$$

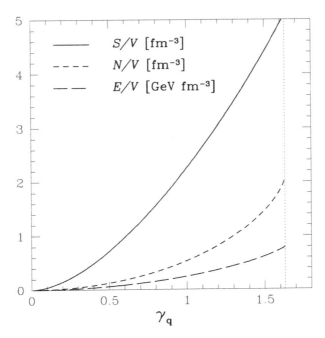

Fig. 7.7.   Pion-gas properties N/V for particles, E/V for energy, and S/V for entropy density, as functions of γ_q at $T = 142\,\mathrm{MeV}$.

This allows one to express, in terms of the abundance of pions, the relation to the abundance of quarks at hadronization, see chapter 19. We study the momentum-space distribution for pions of the form

$$f_\pi(E) = \frac{1}{\gamma_q^{-2} e^{E_\pi/T} - 1}, \qquad E_\pi = \sqrt{m_\pi^2 + p^2}. \tag{7.19}$$

The range of values for γ_q is bounded from above by the Bose singularity γ_q^c:

$$\gamma_q < \gamma_q^c = e^{m_\pi/(2T)}. \tag{7.20}$$

For $\gamma_q \to \gamma_q^c$, we encounter condensation of pions, the lowest-energy state will acquire macroscopic occupation. Formation of such a condensate 'consumes' energy without consuming entropy of the primordial high-entropy QGP phase, and thus is not likely to occur.

   In Fig. 7.7, we show the physical properties of a pion gas as functions of γ_q, for a gas temperature $T = 142\,\mathrm{MeV}$ [181]. We see (solid line) that a large range of entropy density can be accommodated by varying the parameter γ_q. A super-saturated pion gas has an entropy density rivaling that of the QGP at the point of transformation into hadrons, as we see on comparing it with Fig. 16.7 on page 315, for $T = 140\text{--}160\,\mathrm{MeV}$.

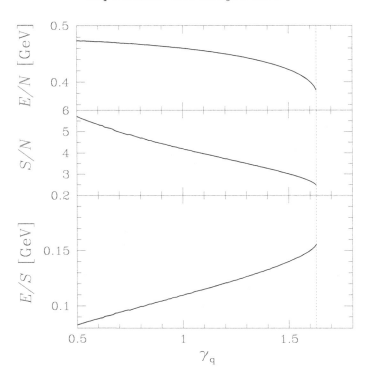

Fig. 7.8. Specific pion-gas properties E/N for energy, S/N for entropy per particle, and E/S for energy per unit of entropy, as functions of γ_{q} at $T = 142\,\mathrm{MeV}$.

The presence of chemical nonequilibrium reduces and potentially eliminates discontinuities at the phase transition, promulgating rapid evolution. This implies that, in particular, the sudden hadronization of an entropy-rich QGP should lead to the limiting value $\gamma_{\mathrm{q}} \to \gamma_{\mathrm{q}}^{\mathrm{c}}$, since other ways of increasing the entropy content involve secondary processes with relatively slow dynamics amongst hadron degrees of freedom. In fact, in an adiabatic equilibrium transformation, one allows an increase in VT^3, characterizing the entropy content, either by expanding the volume V, or invoking a rise of T (reheating), or both. Another remarkable feature of the chemical-nonequilibrium mechanism is that a first order phase transition may appear in other observables more like a phase transformation without large fluctuations.

It is important to remember that, in the hadronization of a quark–gluon phase, it is relatively easy to 'consume' excess energy density, simply by producing a few extra heavy hadrons. However, such particles, being in fact non-relativistic at the temperature considered, are not effective con-

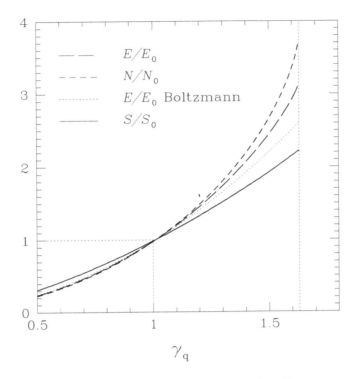

Fig. 7.9.   Pion-gas properties (N, number of particles; E, energy; and S, entropy) relative to chemical equilibrium as functions of γ_q for $T = 142 \, \mathrm{MeV}$.

tributors to pressure and entropy. As we see now, the super-saturated pion gas is just the missing element needed in order to allow rapid hadronization.

The specific properties (E/N, S/N, and E/S) of the pion gas are shown in Fig. 7.8, as functions of γ_q. We see a monotonic decrease of energy and entropy per particle while the energy per unit of entropy increases reaching the condition $E/S > T$, which plays an important role in section 19.1. We see that the entropy per pion drops as γ_q increases, and, at the condensation point $\gamma_q \to \gamma_q^c$, we can add pions without an increase in entropy. Figure 7.8 can be better understood by considering the relative (to chemical equilibrium) physical properties shown in Fig. 7.9. There is, in particular, a large growth in the number of pions, which yield is enhanced at fixed temperature by a factor 3.6 at the condensation point – it is this feature that allows one to hadronize the QGP into a super-saturated pion gas at low, supercooled temperature, accompanied by the experimentally observed excess of pions. We also see, using the energy as an example, that the Boltzmann approximation, i.e., simply a yield factor

γ_q^2 is not producing qualitatively wrong results, even though the increase in number of particles is underestimated by 50%.

7.6 Entropy in a longitudinally scaling solution

We now study the relationship between hadron multiplicity and initial conditions reached for the case of very-high-energy collisions, for which the scaling hydrodynamic solutions can be used to understand the flow of matter. The hydrodynamic expansion of an ideal fluid is entropy-conserving, Eq. (6.13). What this means, in case of the longitudinal expansion, is best seen by considering, in Eq. (6.3), the entropy current $\sigma_\mu = \sigma u_\mu$, and using Eq. (6.32) and the Euler relation Eq. (6.4):

$$0 = \frac{\partial \sigma_\mu}{\partial x_\mu} = \frac{\sigma}{\tau} + \frac{d\sigma}{d\tau} = \frac{1}{\tau} \frac{d(\tau\sigma)}{d\tau}. \tag{7.21}$$

We have

$$\tau\sigma(\tau) = \sigma(\tau_0)\tau_0 = \sigma(\tau_f)\tau_f = \text{constant}, \tag{7.22}$$

where $\tau\sigma$ is a constant of evolution independent of rapidity y.

The physical meaning of conservation of entropy, and thus conservation of $\tau\sigma(\tau)$, becomes clear on remembering the volume element Eq. (6.30). In the locally at rest frame of the fluid (the comoving frame) we have

$$\Delta S = \sigma \, dz \, dt = \sigma\tau \, dy \, d\tau, \tag{7.23}$$

whence,

$$\boxed{\frac{d}{d\tau}\left(\frac{dS}{dy}\right) = \frac{d}{d\tau}(\tau\sigma) = 0, \qquad \frac{dS}{dy} = \text{constant},} \tag{7.24}$$

where, in the last equality, we have used the result Eq. (7.22). dS/dy is independent of y and not a function of τ, i.e., it is independent of the freeze-out condition.

Since entropy is characteristic of the particle yield, Eq. (7.24) implies that the particle multiplicity is flat in rapidity, and is not evolving, i.e., it will not depend on the (uncertain) freeze-out condition. It is established during the initial period of time when the entropy density is built up as the system approaches local thermal equilibrium. Qualitatively, this result is shown in Fig. 5.2; baryons punch through and in between there is a flat distribution in y of particle abundance, since the entropy density per unit rapidity is constant.

It is important to appreciate that, as a matter of principle, the initial entropy density reached in A–A collisions remains naturally undetermined, it arises from microscopic-entropy-producing reactions occurring prior to the onset of the entropy-conserving hydrodynamic expansion. One can try (and the diverse codes we described in section 6.1 do this) to model the A–A collisions on the basis of the behavior of p–p reactions, but it is far from certain that such an approach will be successful. In other words, we cannot use scaling arguments, Eq. (7.24), to understand how big a value of particle rapidity density we can expect to find. The microscopic physics we introduce explicitly (or sometimes implicitly) into the dynamic model determines the final-state particle multiplicity.

We now relate the observed final-state particle multiplicity to the initial entropy density. Employing the volume element shown in Eq. (7.23), and using the conservation law Eq. (7.24), we obtain

$$\sigma_0 \equiv \frac{dS_0}{dV_0} = \frac{1}{F_\perp} \frac{1}{\tau_0} \frac{dS_0}{dy_0} = \frac{1}{F_\perp \tau_0} \frac{dS}{dy}\bigg|_f. \tag{7.25}$$

The transverse surface is

$$F_\perp = \pi (1.2\,\text{fm})^2 A^{2/3}, \tag{7.26}$$

as given by the geometry of the collision, at zero impact parameter.

From Eq. (7.25), we obtain, for the initial-state entropy density,

$$\boxed{\sigma_0 = \frac{1}{\pi (1.2\,\text{fm})^2 A^{2/3} \tau_0}\, 4\, \frac{3}{2} \frac{dN_{\text{ch}}}{dy},} \tag{7.27}$$

where we have assumed that a final-state particle consumes on average 4 units of entropy (see Fig. 10.5), and that the charged-particle multiplicity is two thirds of the total.

III
Particle production

8 Particle spectra

8.1 A thermal particle source: a fireball at rest

The longitudinally scaling limit in production of hadrons, section 6.4, applies at the RHIC and at higher collision energies. At the SPS and AGS energy ranges, table 5.1, it is natural to explore the other reaction picture, the full-stopping limit. In this case all *matter* and *energy* available in the collision of two nuclei is dumped into a localized fireball of hot matter. Even at the highest SPS energies many experimental results suggest that such a reaction picture is more appropriate than the $(1 + 1)$-dimensional-flow picture.

The $m_\perp$ spectra we have seen in Fig. 1.7 on page 20 provide a strong encouragement to analyze the collision region in terms of the formation of a thermalized fireball of dense hadronic matter. The high slopes seen strongly suggest that the dynamic development in the transverse direction is very important. The pattern of similarity seen for very different particles is what would be expected to occur in hadronization of a nearly static fireball, and thus this case will be the first one we explore. However, we note that this is solely an academic exercise since SPS results provide ample evidence for rather rapid $v \simeq 0.5c$ transverse expansion. One can recognize this important physical phenomenon only once the properties of the stationary fireball matter are fully understood.

We consider a space–time-localized region of thermal hadronic matter acting as a source of particles, yielding naturally a Boltzmann spectral distribution. The thermal equilibrium is strictly a local property, with different temperatures possible in different space–time domains. The necessity that there is also a local thermal pressure implies that a fireball is in general a dynamically evolving object with local flows of matter, which we shall study further below. The virtue of this model is that the spectra

130

and abundances of particles can be described in terms of a few parameters that can be measured.

The thermal analysis of the experimental results differs in many key aspects from the microscopic-transport methods introduced in section 6.1. These contain as inputs detailed reaction data and their extrapolations, including often enough hypothetical reaction cross sections and novel mechanisms without which the experimental results cannot be described completely. The attainment of thermal equilibrium is, in these calculations, a result of many complex reactions. For the N–N collisions the appearance of the thermal particle distributions in the final state is still inexplicable in terms of such dynamic microscopic models. For this reason alone, a microscopic dynamic approach cannot lead to an understanding of the thermalization of fireball matter. Moreover, since the physical thermalization processes are faster than those operating in present-day numerical transport codes, microscopic transport theory delays the thermalization of collision energy available in heavy ion reactions and thus will in general fail at predicting observables of interest which depend on (early and rapid) thermalization. As long as these issues are being studied, an empirical thermal model allowing for flow of matter and non-equilibrium abundances of particles offers considerable advantages for the understanding of experimental data.

We first aim to relate the experimental rapidity and transverse-mass spectra to the particle distribution of the fireball. We consider the differential particle-momentum distribution, e.g., near the surface of the fireball,

$$E \frac{d^3 N}{d^3 p} \equiv f(E, p_{\mathrm{L}}), \tag{8.1}$$

where the presence of p_{L} in the argument reminds us that an emitted particle could remember the collision axis; the distribution need not be intrinsically spherically symmetric as is implied when only the energy of the particle is considered. The coefficient E is introduced in Eq. (8.1) for convenience, it assures that the quantity f is invariant under Lorentz transformations. This is understood on reexpressing the left-hand side of Eq. (8.1) in terms of the invariant variables $m_\perp$ and y. Given Eq. (5.4), at constant $p_\perp$, we find

$$dy = \frac{dp_{\mathrm{L}}}{E}. \tag{8.2}$$

Since $p_\perp \, dp_\perp = m_\perp \, dm_\perp$, considering Eq. (5.5) the Lorentz-invariant momentum-space volume element is

$$\frac{d^3 p}{E} = dy \, m_\perp \, dm_\perp \, d\varphi. \tag{8.3}$$

It is important to remember, looking at the spectra, that, while $p_\perp > 0$, we have $m_\perp > m$.

We consider at first as the intrinsic distribution the simplest exponential Boltzmann-type thermal shape:

$$f(E, p_L) \rightarrow C E e^{-\beta E} = C m_\perp \cosh y\, e^{-\beta m_\perp \cosh y}, \tag{8.4}$$

with $C = gV/(2\pi^3)$ and Eq. (5.4) is used on the right-hand side to replace the particle energy by transverse mass and rapidity. Since we have a Lorentz-invariant distribution, a change of the frame of reference along the p_L axis, e.g., from the laboratory frame to the CM frame, is amounting to a shift along the rapidity y axis of the particle spectrum considered to be centered around the CM rapidity $y_{CM} = 0$, see section 5.3. The differential particle spectrum which we obtain is

$$\frac{d^2 N(y, m_\perp)}{m_\perp^2\, dm_\perp\, dy} = C \int d\varphi \cosh y\, e^{-\beta m_\perp \cosh y}. \tag{8.5}$$

To obtain the transverse-mass spectra, we need to integrate Eq. (8.5) over the applicable rapidity acceptance (often referred to as 'rapidity window'):

$$\frac{1}{m_\perp} \frac{dN(y, m_\perp)}{dm_\perp} = C \int d\varphi \int_{y^-}^{y^+} dy\, m_\perp \cosh y\, e^{-\beta m_\perp \cosh y}. \tag{8.6}$$

For a wide (see below) rapidity window, we can extend the limit of the integration to infinity, since the argument is a rapidly decreasing exponential function. We use Eqs. (10.44) and (10.45) and obtain

$$K_1(z) = \int_0^\infty dt\, e^{-z \cosh t} \cosh t,$$

$$\rightarrow \left(\frac{\pi}{2z}\right)^{1/2} e^{-z} \left(1 + \frac{3}{8z} - \frac{15}{128z^2} \cdots\right). \tag{8.7}$$

We obtain for the full rapidity coverage

$$\frac{1}{m_\perp^{3/2}} \frac{dN(y, m_\perp)}{dm_\perp} \propto e^{-\beta m_\perp} \left(1 + \frac{3}{8\beta m_\perp} \cdots\right). \tag{8.8}$$

For a narrow rapidity window, $\delta y = y^+ - y^-$, surrounding y_{CM}, one simply substitutes, in the integral Eq. (8.6), $\cosh(\delta y/2)$ by 1 and the result is

$$\frac{1}{m_\perp} \frac{dN(y, m_\perp)}{dm_\perp} \propto e^{-\beta m_\perp}. \tag{8.9}$$

In both cases, Eqs. (8.8) and (8.9), we have a (nearly) exponential transverse-mass spectrum, provided that the preexponential factors in the spectra

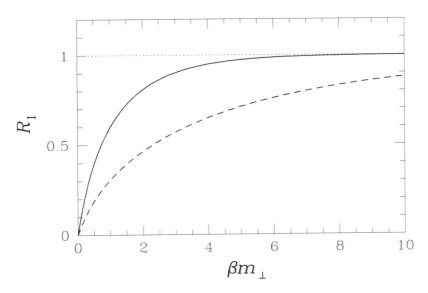

Fig. 8.1. The saturation of particle yield for a fireball at rest within a rapidity window: dashed line, $y^{\pm} = y_{\rm CM} \pm 0.5$ and solid line $y^{\pm} = y_{\rm CM} \pm 1$, as a function of $\beta m_{\perp}$; see the text for details.

are properly chosen. The result is not at all what one is naively tempted to use when one is fitting invariant cross sections, i.e., to simply take an exponential fit of the cross-section data: the choice is either to include the factor $1/m_{\perp}^{1/2}$ (compare with Fig. 8.9 and also Fig. 8.8), or to multiply by $1/m_{\perp}$ for a truly narrow rapidity window, as we see in Fig. 1.7 on page 20.

The question thus is that of how narrow the 'narrow' rapidity window must be for the factor to be as given in Eq. (8.9). We note that, in addition to the width of the typical experimental acceptance of 0.5–1 rapidity unit, one has to keep in mind that there is, in principle, a superposition of contributions to the spectra occurring due to longitudinal flow in the source, which effectively widens the rapidity acceptance domain. We show, in Fig. 8.1, the ratio

$$R_{\rm I} \equiv \frac{\int_{y_-}^{y_+} dy \, \cosh(y - y_{\rm CM}) \, e^{-\beta m_{\perp} \cosh(y - y_{\rm CM})}}{\int_{-\infty}^{+\infty} dy \, \cosh(y - y_{\rm CM}) \, e^{-\beta m_{\perp} \cosh(y - y_{\rm CM})}}, \tag{8.10}$$

of the rapidity integral Eq. (8.6) with the full rapidity coverage, as a function of $\beta m_{\perp}$. Results shown are for a rapidity window of one unit (dashed line, $y^{\pm} = y_{\rm CM} \pm 0.5$) and two units (solid line, $y^{\pm} = y_{\rm CM} \pm 1$) of rapidity, centered around $y_{\rm CM}$.

We see that, for an experimental rapidity window of one unit (e.g., $-0.5 < y - y_{\rm CM} < 0.5$) (dashed line) and for a typical 'high' transverse mass $m_\perp \simeq 1.5$ GeV at $T \to T_\perp = 1/\beta = 230\text{–}300$ MeV, we would have reached nearly 80%–90% of the full rapidity integral, justifying use of a result with a full rapidity window coverage – note that adding in smearing of flow (the solid line) means that more than 99% of the spectral strength will be effectively included. Consequently, we find that the test for applicability of Eq. (8.8) is $\beta m_\perp \cosh(\delta y/2) > 8$, with $\delta y/2$ comprising an estimate of the flow.

We next consider the thermal rapidity spectra. We now integrate Eq. (8.4) over the full range of transverse mass,

$$\frac{dN(y, m_\perp)}{dy} = C \int d\varphi \int_m^\infty dm_\perp\, m_\perp^2 \cosh y\; e^{-\beta m_\perp \cosh y}, \tag{8.11}$$

to obtain the rapidity distribution shown in the top portion of Fig. 8.2, for the case $m_\pi \lesssim \beta^{-1} < m_K$ (here $\beta^{-1} \equiv T = 160$ MeV).

The thin lines in Fig. 8.2 apply to spectra of massless particles, dashed lines are for pions ($m = 138$ MeV), chain lines are for kaons ($m = 497$ MeV), and the thick solid line depicts data for nucleons ($m = 938$ MeV). Since the experimental acceptance in $p_\perp$ cannot, for practical reasons, begin with $p_\perp = 0$, we have also shown in Fig. 8.2 what happens to these spectra when only particles with $p_\perp > p_\perp^{\min}$ are included, with minimum momentum cutoffs shown at $p_\perp^{\min} = 0.3, 0.5$, and 1 GeV. We notice that, when $p_\perp^{\min} < 0.5$ GeV, the maximum peak for massless particles is nearly half as high as that for nucleons, and, correspondingly, the widths of the distributions vary considerably with particle mass.

This change in relative abundance of the different particles increases with $p_\perp^{\min}$ (we changed the scale of the drawing by a factor of three to make the small remaining particle abundance more visible for $p_\perp^{\min} = 1$ GeV). The lighter particles disappear more rapidly and the relative abundance of the heavier ones is increased in the sample. Moreover, all shapes become increasingly more similar, resembling more and more the nucleon spectrum. We note that for $p_\perp^{\min} = 0.3$, the half-width parameter, for most particles, is within the range $0.6 < \sigma < 0.7$.

We see that, in the ideal situation of a thermal Boltzmann-like emitter, the rapidity spectra of identified particles are very narrowly distributed around the 'central' rapidity, with the distribution of the massive particles being narrower than that of lighter particles when all $p_\perp$ are included, which difference disappears when the minimum $m_\perp$ for the different particles are (nearly) the same. Since the width of the rapidity spectra is just half as large as was seen for the width for negative hadrons, h^-, see Fig. 9.6, there must be some other contribution to the width, that is in general believed to be the flow: the small source is not stationary, and its size and all other properties evolve rapidly in time, an effect we address in section 8.4.

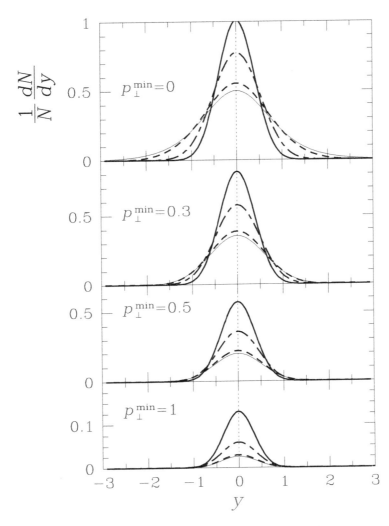

Fig. 8.2. Normalized thermal rapidity particle spectra (quantitative) for a Boltzmann (exponential) energy distribution with $\beta^{-1} = 160$ MeV and setting $y_{\mathrm{CM}} = 2.92$ as appropriate for the highest SPS energy: massless particles (thin lines), pions (dashed lines), kaons (chain lines), and nucleons (solid lines). The effect of the minimum transverse-momentum cutoff on particle yield and shape of distribution is illustrated: we show $p_{\perp}^{\mathrm{min}} > 0$, > 0.3, > 0.5, and > 1 GeV. Note the change of scale for the last (bottom) case.

Even though the above example of a thermal source is no more than a case study, we have learned much about the possible shape of the rapidity spectra of a well-defined, localized thermal source. This leads to the practical question of what spectral shape arises when identification of particles is not possible. In such a case, one generally studies pseudorapidity distributions. We keep in mind that the relatively easily measurable pseudorapidity, Eqs. (5.24) and (5.25), arises from the rapidity, Eqs. (5.4) and (5.10), in the limit $m \to 0$, and thus, in cases when an appreciable yield of nucleons and even kaons is present, significant distortions in the spectra occur.

We now discuss this quantitatively and evaluate the shape of the thermal pseudorapidity distribution. Since η is not a good Lorentz variable, we study the specific example of the spectra in the laboratory frame in which the target is at rest and the projectile had $158A$ GeV. We take as the input spectrum the rapidity shape of the thermal source defined above in Eq. (8.11). To proceed with the change of variables, we need to express the CM energy and momentum of the distributions in terms of the laboratory pseudorapidity. Using Eq. (5.4),

$$E = m_\perp \cosh(y' - y_{CM}) = E' \cosh y_{CM} - p'_L \sinh y_{CM}, \qquad (8.12a)$$

$$p_L = m_\perp \sinh(y' - y_{CM}) = p'_L \cosh y_{CM} - E' \sinh y_{CM}. \qquad (8.12b)$$

With the help of Eq. (5.24), we eliminate E' and p'_L, using the pseudorapidity η' with reference to the laboratory frame, and $p_\perp$,

$$E = \sqrt{m^2 + p_\perp^2 \cosh^2 \eta'} \cosh y_{CM} - p_\perp \sinh \eta' \sinh y_{CM}, \qquad (8.13a)$$

$$p_L = p_\perp \sinh \eta' \cosh y_{CM} - \sqrt{m^2 + p_\perp^2 \cosh^2 \eta'} \sinh y_{CM}. \qquad (8.13b)$$

We also obtain

$$\frac{1}{p_\perp} \frac{dp_L}{d\eta'} = \cosh \eta' \cosh y_{CM} - \frac{p_\perp \sinh \eta' \sinh y_{CM} \cosh \eta'}{\sqrt{m^2 + p_\perp^2 \cosh^2 \eta'}}, \qquad (8.14)$$

for the (frame-of-reference-dependent) integration Jacobian relating the CM longitudinal momentum and laboratory pseudorapidity (see Eq. (8.2) for comparison).

We now are ready to evaluate the pseudorapidity particle distribution. Proceeding in the same way as when we obtained Eq. (8.11), i.e., integrating over the azimuthal angle and the transverse momentum, and effecting the change of the integration variable from longitudinal momentum to pseudorapidity, we obtain, in the laboratory frame, the pseudorapidity distribution

$$\frac{dN}{d\eta'} \equiv 2\pi \int dp_\perp \, p_\perp^2 \left(\frac{dp'_L}{p_\perp \, d\eta'} \right) \frac{f(E, p_L)}{E}. \qquad (8.15)$$

The arguments of the distribution are as given in Eqs. (8.13a) and (8.13b) and the volume element is given by Eq. (8.14).

Equipped with this result, we can explore quantitatively the case of the exponential, thermal-like distribution, Eq. (8.4). The explicit form of the laboratory pseudorapidity distribution, including a necessary Lorentz contraction factor $(\cosh y_{\rm CM})^{-1}$ arising from the Lorentz transformation of the volume V of the source, takes the form

$$\frac{dN}{d\eta'} = 2\pi C \int_{p_\perp^{\rm min}}^{\infty} dp_\perp \; p_\perp^2 e^{-\beta\left[\cosh y_{\rm CM}\sqrt{m^2+p_\perp^2\cosh^2\eta'}-p_\perp\sinh y_{\rm CM}\sinh\eta'\right]}$$

$$\times \cosh y_{\rm CM}\left(\cosh\eta' - \tanh y_{\rm CM}\sinh\eta'\;\frac{p_\perp\cosh\eta'}{\sqrt{m^2+p_\perp^2\cosh^2\eta'}}\right). \quad (8.16)$$

A simple test of this not-so-simple expression is its normalization, which is easily (numerically) verified by integrating over η' at given m and β, for various values of $y_{\rm CM}$.

This distribution is shown in Fig. 8.3, which parallels Fig. 8.2 with the same conventions and parameters. On comparing Figs. 8.2 and 8.3, we see that the rapidity and pseudorapidity spectra agree exactly for massless particles, since the pseudorapidity is the rapidity, in this case. The nearly massless pions are visibly little changed in spectral shape. With progressively increasing mass, the pseudorapidity spectra differ more from the rapidity spectra and, in particular, their center shifts to higher pseudorapidity.

There is a notable deformation of the symmetric shape accompanied by considerable widening – the peak is only 60% of the height of the original rapidity spectrum for $p_\perp^{\rm min} = 0$. In practical situations, the small-$p_\perp$ particles are eliminated, which we allow for by means of a cutoff in $p_\perp^{\rm min}$. We see that now the spectral shapes appear progressively less shifted from their rapidity form; the pseudorapidity and rapidity shapes become more similar, although a residual shift remains for the heaviest particles (nucleons). Thus the pseudorapidity–rapidity difference is primarily a low-momentum phenomenon. As the $p_\perp$ cutoff increases, the relative strength of the particle spectra changes and, in particular, there is considerable relative enrichment of the contributions of the heaviest particles compared with those of light particles.

8.2 A dynamic fireball

Naturally, a fireball at rest is not what we are likely to encounter in the highly dynamic situation of colliding nuclei. We now look at the modifications introduced by the presence of a local flow of matter. As before a volume element in a fireball is the particle source, but now this volume is in motion, typically due to a local flow originating from a (hydrodynamic)

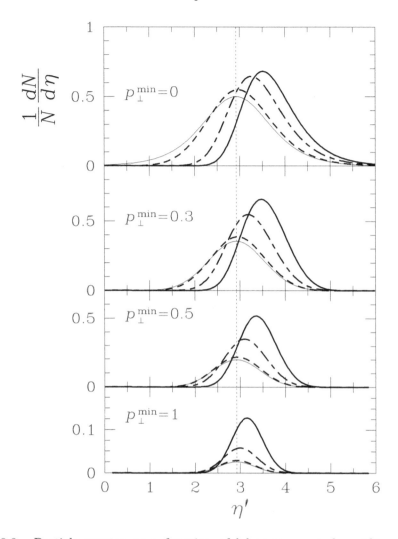

Fig. 8.3. Particle spectra as a function of laboratory pseudorapidity η', for $\beta^{-1} = 160$ MeV, $y_{\mathrm{CM}} = 2.92$; massless particles (thin lines), π (dashed lines), K (chain lines), and N (solid lines). Results for various minimum transverse-momentum cutoffs are shown: $p_\perp^{\mathrm{min}} > 0$, > 0.3, > 0.5, and > 1 GeV. Note the change of scale for the last (bottom) case.

expansion. We will refer to this collective flow velocity below simply as $\vec{v}$. We would like to know how the statistical distribution appears to a laboratory observer. However, when we refer to a statistical phase-space distribution, we always imply an observer at rest in the local 'intrinsic' frame of reference. The 'intrinsic i' particle energy E^i and momentum

$\vec{p}^{\,i}$ are measured in the local non-flowing frame of reference of a moving volume element of the fireball.

We determine now how the intrinsic thermal spectrum appears to an arbitrary Lorentz observer, such as a laboratory observer is. The physical idea is to express the intrinsic thermal phase space in terms of Lorentz co-variant quantities, and than to use variables associated with any observer, e.g., a laboratory-frame observer. There are several approaches possible, and we proceed in the first instance to consider as in section 12.3 the Touschek invariant phase-space measure [143, 261]:

$$\frac{V_0\, d^3 p^i}{(2\pi)^3} e^{-E^i/T} \longrightarrow \frac{V_\mu p^\mu}{(2\pi)^3} d^4 p\, 2\delta_0(p^2 - m^2) e^{-p_\mu u^\mu/T}. \tag{8.17}$$

The flowing volume element $V_\mu = V_0 u_\mu$ is as observed in the laboratory frame. V_0 is the comoving volume element in the local rest frame. δ_0 is the Dirac delta function for the positive (energy) roots only, . It is an invariant function for all proper Lorentz transformations. The left-hand side of Eq. (8.17), is written in terms of the intrinsic variables, but the right-hand side is not frame-dependent and we can read it in the frame of reference of the laboratory observer.

We see the invariant measure introduced in Eq. (8.3):

$$2\delta_0(p^2 - m^2)\, d^4 p = \frac{d^3 p}{E} = m_\perp\, dm_\perp\, dy\, d\phi_p = p_\perp\, dp_\perp\, dy\, d\phi_p. \tag{8.18}$$

The particle momentum defined with reference to the collision axis has the explicit form

$$p^\mu = (m_\perp \cosh y,\ p_\perp \cos \phi^p,\ p_\perp \sin \phi^p,\ m_\perp \sinh y), \tag{8.19}$$

(in cylindrical coordinates) where we omit superscript p for the variables y and $m_\perp = \sqrt{m^2 + p_\perp^2}$, which, as usual, are understood to refer to the observed particle. We recall the usual relations,

$$\frac{m_\perp}{m} = \gamma_\perp = \cosh y_\perp, \qquad \frac{p_\perp}{m} = v_\perp \gamma_\perp = \sinh y_\perp,$$

which allow us to write Eq. (8.19) in the form

$$\frac{p^\mu}{m} = (\cosh y_\perp \cosh y,\ \sinh y_\perp \cos \phi^p,\ \sinh y_\perp \sin \phi^p,\ \cosh y_\perp \sinh y).$$

This suggests that we introduce such a cylindrical representation of the velocity field as well,

$$u^\mu = (\cosh y_\perp^v \cosh y_\parallel^v,\ \sinh y_\perp^v \cos \phi^v,\ \sinh y_\perp^v \sin \phi^v,\ \cosh y_\perp^v \sinh y_\parallel^v),$$

$$u^2 = 1. \tag{8.20}$$

It is straightforward, in these coordinates, to obtain $u_\mu p^\mu$ required to construct the spectra in Eq. (8.17),

$$u_\mu p^\mu = \gamma_\perp^v \left[m_\perp \cosh(y - y_\parallel^v) - p_\perp v_\perp \cos\phi \right],$$
(8.21)

where $\gamma_\perp^v = \cosh y_\perp^v = 1/\sqrt{1 - v_\perp^2}$, $\phi = \phi^p - \phi^v$, and the variables y, $m_\perp$, and $p_\perp$ refer to the rapidity, the transverse momentum, and the transverse mass of the observed particle.

The explicit form of the invariant spectrum which generalizes Eq. (8.5), is

$$\frac{d^2 N}{m_\perp^2 \, dm_\perp \, dy} = \int \frac{d\phi \, \gamma_\perp^v}{(2\pi)^3} \left(\cosh(y - y_\parallel) - \frac{p_\perp}{m_\perp} v_\perp \cos\phi \right)$$
$$\times \exp\left\{ -\gamma_\perp^v \left[m_\perp \cosh(y - y_\parallel) - p_\perp v_\perp \cos\phi \right]/T \right\}.$$
(8.22)

For a suitable choice of the coordinate system in which the x axis is pointing in the direction of the transverse flow vector, $\phi^v = 0$, the particle-emission angle is the azimuthal angle of integration $\phi = \phi^p$. We use the range $0 < \phi \le \pi$, which has to be counted twice to include the part $\pi < \phi \le 2\pi$. The ϕ integrals we encounter are analytical:

$$\frac{1}{\pi} \int_0^\pi e^{\pm a \cos\phi} \, d\phi = I_0(a), \quad \frac{1}{\pi} \int_0^\pi e^{\pm a \cos\phi} \cos\phi \, d\phi = \pm I_1(a). \quad (8.23)$$

It is helpful to remember that I_0 'looks like' a cosh function, and I_1 like a sinh function, and the analogy goes further with

$$I_1(a) = \frac{\partial I_0(a)}{\partial a}.$$
(8.24)

However,

$$\cosh a = I_0(a) + 2I_2(a) + 2I_4(a) + 2I_6(a) + \cdots, \quad (8.25)$$
$$\sinh a = 2I_1(a) + 2I_3(a) + 2I_5(a) + \cdots, \quad (8.26)$$

where

$$I_n(a) = \frac{1}{\pi} \int_0^\pi e^{a \cos\phi} \cos n\phi \, d\phi = \sum_{k=0}^\infty \frac{(a/2)^{2k+n}}{k!(n+k)!}.$$
(8.27)

Using Eq. (8.23) in Eq. (8.22), we obtain

$$\frac{d^2N}{m_\perp^2\,dm_\perp\,dy} = \frac{\gamma_\perp^v}{(2\pi)^2}\left(\cosh(y-y_\parallel)\,I_0(p_\perp v_\perp/T) - \frac{p_\perp}{m_\perp}v_\perp I_1(p_\perp v_\perp/T)\right)$$
$$\times \exp\left[-\gamma_\perp^v m_\perp \cosh(y-y_\parallel)/T\right]. \tag{8.28}$$

This is the statistical particle spectrum seen in the laboratory frame and originating in a volume element of a fireball having two velocity components $y_\parallel$ and $v_\perp$, and emitting particles at the local temperature T. If the laboratory frame is not the CM frame, we need to shift the rapidity $y \to y - y_{\rm CM}$. This is the particle spectrum of final-state hadrons arising if the matter in the entire volume of the fireball froze out suddenly.

This volume-style statistical phase-space hadronization based on work carried out by Touschek [261], differs from the approach of Cooper and Frye [96], which allows for the dynamics of the particle-emitting surface. One imagines an opaque fireball, and each surface element is the particle source. The physical idea is thus to couple the intrinsic (statistical) particle spectrum to the Lorentz-covariant surface dynamics. The developments till 1993 are well documented in [238]. It was subsequently discovered that the radiation formula was non-positive definite and a generalization was proposed [119].

The particle phase space is written using a covariant surface in $4-1=3$ space–time dimensions. Consequently, apart from the flow, there is yet another velocity that describes how the hadronization surface moves, e.g., the surface may be flowing outward, but a rapid 'peeling' of matter may move the boundary of the particle-producing volume inward. Moreover, over the history of the particle production (freeze-out from the surface), the surface may have both positive and negative velocities, and thus it can be difficult to make sure that a particle is actually emitted rather than absorbed in the fireball.

We now illustrate the difficulty inherent in dealing with the problem of emission of particles, which continues to be actively studied. First, we recall how, in the Touschek approach, the particle density in phase space has been written in a Lorentz-invariant way as

$$\frac{d^6N}{d^3x\,d^3p} = \frac{g}{(2\pi)^3}e^{-u_\mu p_i^\mu/T}, \tag{8.29}$$

where u^μ is the 4-flow velocity. In the volume-hadronization approach, we have made the following qualitative steps:

$$E\frac{d^3N}{d^3p} = \frac{dN}{d\phi\,dy\,m_\perp\,dm_\perp}$$
$$= \frac{g}{(2\pi)^3}e^{-u_\mu p_i^\mu/T}\left[E_i\,d^3x \to p_i^\mu V_\mu \propto p_i^\mu u_\mu\right]. \tag{8.30}$$

When surface emission dominates the particle spectra, the other way to proceed is

$$E \frac{d^3N}{d^3p} = \frac{g}{(2\pi)^3} e^{-u_\mu p_i^\mu/T} \left[E_i \, d^3x \rightarrow \right.$$

$$\left. \int d\tau \, d^3\Sigma_f^\mu p_{\mu i} \, \delta\left(d\tau - \sqrt{dt_f^2 - d\vec{x}_f^2}\right) \right], \quad (8.31)$$

where $x_f^\mu = (t_f, \vec{x}_f)$ are the freeze-out surface coordinates, and Σ_f^μ is the three-dimensional hypersurface of the Minkowski volume element, characterized by a unit 4-vector normal to the surface u_f^μ, i.e., the 4-velocity of the freeze-out surface:

$$n_s^\mu = \frac{dx_f^\mu}{d\tau} = \frac{dt_f}{d\tau}\left(1, \frac{d\vec{x}_f}{dt_f}\right). \quad (8.32)$$

Equation (8.31) arises since we wish to sum the emission spectrum over the contributions made by each surface element $d^3\Sigma$ over its (proper time τ) history. For a fireball at rest, we have $n_f^\mu = (1,0,0,0)$, $d^3\Sigma_f^\mu = dt_f \, d^2x_f$, and $\vec{v}_f = d\vec{x}_f/dt_f = 0$. Noting that the δ-function simply sets the proper time to the freeze-out time, we obtain

$$E \frac{d^3N}{d^3p} = \frac{g}{(2\pi)^3} e^{-E_i/T} E_i S_f \, \Delta t_f, \quad (8.33)$$

where Δt_f is the length of (proper) time during which the emission of particles occurs, and S_f is the size of the surface.

For simple geometries, we can use

$$d\tau \, \delta\left(d\tau - dt_f \sqrt{1 - d\vec{x}_f^2/dt_f^2}\right) = \delta\left(1 - \frac{dt_f}{d\tau} \sqrt{1 - d\vec{x}_f^2/dt_f^2}\right) \rightarrow 1,$$

and we find the conventional Cooper–Frye formula:

$$E \frac{d^3N}{d^3p} = \frac{dN}{d\phi \, dy \, m_\perp \, dm_\perp} = \frac{g}{(2\pi)^3} \int_{\Sigma_f} e^{-E^i/T} p_\mu^i \, d^3\Sigma^\mu. \quad (8.34)$$

$d^3\Sigma^\mu$ is the normal surface vector for the four-dimensional space–time volume boundary, from which the emission of particles occurs,

$$d^3\Sigma^\mu \equiv \epsilon_{\mu\nu\lambda\rho} \frac{\partial\Sigma^\nu}{\partial u} \frac{\partial\Sigma^\lambda}{\partial v} \frac{\partial\Sigma^\rho}{\partial w} \, du \, dv \, dw, \quad (8.35)$$

where u, v, w is a suitable set of three locally orthogonal coordinates. In cylindrical coordinates $(u, v, w) = (r_f, \phi_f, z_f)$ and

$$\Sigma_f^\mu = (t_f, r_f \cos\phi_f, r_f \sin\phi_f, z_f). \quad (8.36)$$

The freeze-out time t_f is independent of the angle ϕ_f due to the assumption of cylindrical symmetry and thus we have $t_f(r_f, z_f)$. The covariant volume element is

$$d^3\Sigma_\mu^{cyl} = \left(1, -\frac{\partial t_f}{\partial r_f}\cos\phi_f, -\frac{\partial t_f}{\partial r_f}\sin\phi_f, -\frac{\partial t_f}{\partial z_f}\right) r_f \, dr_f \, d\phi_f \, dz_f. \quad (8.37)$$

We use the momentum vector of a particle p_i^μ in cylindrical coordinates, Eq. (8.19), and obtain

$$\frac{d\tau}{dt_f} p^\mu \, d^3\Sigma_\mu^{cyl} = r_f \, dr_f \, d\phi_f \, dz_f \left[m_\perp \left(\frac{1}{u_f^0}\cosh y - \frac{1}{u_f^\parallel}\sinh y \right) \right.$$

$$\left. -\frac{1}{u_f^\perp} p_\perp \cos\phi \right], \quad (8.38)$$

where as before $\phi = \phi^f - \phi^p$.

The 4-velocity of the freeze-out surface is

$$dx^\mu/d\tau = u_f^\mu = (u_f^0, u_f^\perp \cos\phi_f, u_f^\perp \sin\phi_f, u_f^\parallel).$$

This additional surface dynamics influences the observed spectra, even though we are dealing with a preexponential factor only. The transverse-mass spectra contain an additional factor, compared to Eq. (8.22),

$$\frac{d^2N}{m_\perp \, dm_\perp \, dy} \rightarrow \int \frac{d\phi\, \gamma_\perp^v}{(2\pi)^3} \left(1 - \frac{\vec{v}_{fr}^{-1} \cdot \vec{p}}{E} \right)$$

$$\times \left(\cosh(y - y_\parallel) - \frac{p_\perp}{m_\perp} v_\perp \cos\phi \right) \quad (8.39)$$

$$\times \exp\{ -\gamma_\perp^v \left[m_\perp \cosh(y - y_\parallel) - p_\perp v_\perp \cos\phi \right]/T \},$$

where

$$\vec{v}_{fr}^{-1} \cdot \vec{p} \equiv \frac{\partial t_f}{\partial r_f} p_\perp \cos\phi + \frac{\partial t_f}{\partial z_f} p_z. \quad (8.40)$$

Particles are emitted from the surface of fireball volume, and thus the phase space is $(2 + 2)$-dimensional when the number of particles is counted. For this dimensional reason there is one power of $m_\perp$ less in Eq. (8.39), than there is in Eq. (8.22). When $v_{fr} \rightarrow c$ the prefactor in Eq. (8.39) is able to compensate for this effect and both methods can describe the experimental hadron spectra with similar precision.

We proceed to show how the longitudinal flow, and, in section 8.4, the transverse flow, influence particle spectra. Numerical study shows that the two flows are practically independent from each other, and it has

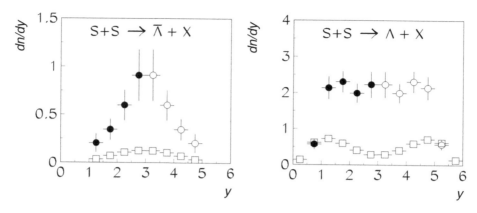

Fig. 8.4. On the left, the abundance of $\overline{\Lambda}$ in S–S collisions at $200A$ GeV, as a function of rapidity. The squares are the results for N–N collisions scaled up by the pion-multiplicity ratio. On the right, corresponding results for the abundance of Λ. Data produced by the NA35 collaboration [24].

become commonplace to study particle spectra as if either only parallel or only transverse flows were present: in a study of rapidity spectra, only $y_\parallel$ is considered, while $v_\perp$ is ignored; in a study of $m_\perp$ spectra, the longitudinal flow $y_\parallel$ is not considered.

8.3 Incomplete stopping

Considering that, at very high collision energies, the longitudinal scaling behavior is expected, see section 6.3, whereas in collisions of large nuclei at moderate energies a central fireball is more appropriate, it is natural that the real world is observed to be much more complex than these simple 'asymptotic' models.

A nice example of the case in which the baryon number just does not punch through is seen in the central $200A$-GeV S–S collisions at the SPS. We show, in Fig. 8.4, the production yields of $\overline{\Lambda}$ (left-hand side) and Λ (right-hand side) hyperons as functions of rapidity. The open circles in Fig. 8.4 are the directly measured data. The particle spectra must be symmetric around the CM rapidity, since this is a symmetric collision system. For this reason the open black circles are obtained by reflecting the measured data points (solid black circles) at the value $y = 2.96$.

The spectra arising from N–N interactions at the same energy are shown in Fig. 8.4 (open squares). These N–N-interaction comparison data have been multiplied by the rapidity-dependent pion-multiplicity-enhancement factor, which accounts for the increase in production of

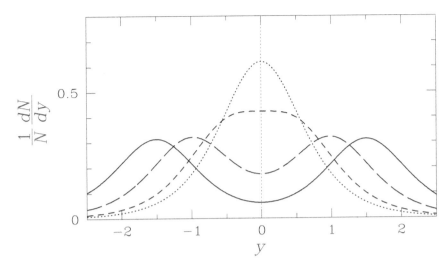

Fig. 8.5. Rapidity spectra of massless QGP quanta with flow: dotted, no flow $y_\| = 0$; short-dashed, $y_\| = 0.5$; long-dashed $y_\| = 1$; and solid, $y_\| = 1.5$ [178].

pions per nucleon observed on comparing S–S with N–N reactions. In the target ($y = 0.75 \pm 0.25$) and projectile ($y = 5.25 \pm 0.25$) fragmentation regions, this procedure gives a good agreement between yields of Λ particles in S–S and N–N scaled by the pion multiplicity. This suggests that, in the target/projectile fragmentation regions, the production of Λ has the same origin in both cases, presumably from individual N–N interactions.

However, in the central rapidity region in Fig. 8.4, new mechanisms of production of $\overline{\Lambda}$ and Λ are clearly visible. Inspecting the yield of $\overline{\Lambda}$, we see considerable localization at central rapidity of a particle made entirely from constituents not brought into reaction, $\overline{\Lambda}(\overline{u}\overline{d}\overline{s})$. Naturally, there must have been associated localization of the energy. The Λ rapidity spectrum is, in contrast, relatively flat. $\Lambda(uds)$ contains, aside from the strange quark made in the reaction, constituent quarks brought into the collision region by the projectile and target. Were the punch through of the light (u, d) quark content complete, we should see for Λ a distribution similar to $\overline{\Lambda}$, both in shape and in yield.

We consider now the rapidity spectra in the presence of a longitudinal flow $y_\|$, evaluating the $m_\perp$ integral in Eq. (8.22). The challenge is to describe a diversity of rapidity spectra of observed hadrons, which are very strongly varying between certain particles. The different behaviors are shown schematically in Fig. 8.5. We see how the thermal rapidity spectra of massless quanta in the deconfined phase, with $m = 0$, (and with $T = 145$ MeV, and $v_\perp = 0.52$, which values matter little), vary.

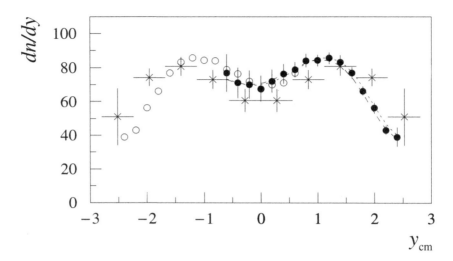

Fig. 8.6. Rapidity spectra of baryons $\langle b - \bar{b} \rangle$ observed by experiment NA49 [43] in central (5%) Pb–Pb interactions at $\sqrt{s_{NN}} = 17.2$ GeV (solid circles, direct measurement; open circles, reflection at y_{CM}). Stars are rapidity spectra of baryons for S–S interactions obtained by NA35 at $\sqrt{s_{NN}} = 18.4$ GeV, for the 3% most central events, scaled with participant number 352/52.

These gradually 'flow' apart as $y_\parallel$ is increased from $y_\parallel = 0$ (dotted line) to $y_\parallel = 1.5$ (solid line) in steps of 0.5.

Comparing with Fig. 8.4, we see that both limits (the central production of $\bar{\Lambda}$ and the flat distribution of Λ) are seen in Fig. 8.5. How can this be happening in the same reaction? In order to obtain different types of flow for different particles, we presume in the following illustrative example that hadrons arise from a mix of three quark fluids. The incoming valence quarks of colliding nuclei are retaining some ($v_\parallel^{\text{valence}} \neq 0$) memory of the original motion along the collision axis, and constitute the projectile and target fluids. However, all newly made pairs of quarks have practically no memory ($v_\parallel^{\text{pair}} \simeq 0$) of the initial condition of colliding matter, they are formed near y_{CM} and are constituents of the third fluid. In particular, pairs of strange quarks made in the plasma do not flow in the longitudinal direction.

For the protons produced, this model implies that all three quarks remember the incoming flow and their distribution should follow the solid or long-dashed line (depending on $y_\parallel$). For Λ, with one strange quark, we consider a mix of two thirds weight in the spectrum with flow and one thirds without. For particles like K^+ ($u\bar{s}$), we take a 50%–50% mix, and, for all newly made particles like $\bar{\Lambda}$ and Φ, we assume that only no-flow components contribute. To describe baryon rapidity spectra in Pb–Pb

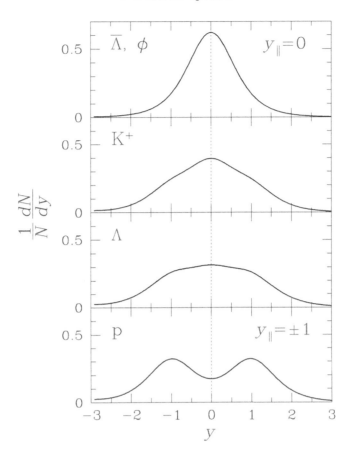

Fig. 8.7. Schematic representations of rapidity particle spectra within a thermal model with flow, parameters chosen for $\sqrt{s_{NN}} = 17.2$ GeV [178].

collisions at $\sqrt{s_{NN}} = 17.2$ GeV, reported by NA49 [43], see Fig. 8.6, we assume that the longitudinal flow is $y_\parallel \simeq \pm 1$, and choose this value without attempting to fit the spectra.

The proton rapidity spectrum is shown in the bottom panel in Fig. 8.7. We used $m_q = 0$, $T = 145$ MeV, and $v_\perp = 0.52c$, which parameters hardly matter and could be chosen very differently; these values were taken in view of the $m_\perp$ spectra we discuss below in section 8.4. We average over positive and negative flows $y_\parallel = \pm 1$, since the collision in the CM frame involves both. The strange-quark content of the Λ which contributes with relative strength 33% suffices to yield a flat distribution – see the second panel from the bottom in Fig. 8.7. The shape of the central rapidity plateau is in agreement with the results seen in NA49 data, as well as with those shown above in Fig. 8.4 for the NA35 S–S collisions.

In the third panel from the bottom, corresponding to the 50%–50% flow mix such as would be appropriate for $K^+(u\bar{s})$, the resulting rapidity shape is already peaked at the central rapidity. Finally, in the top panel, we show a prototype of the rapidity distribution arising for all hadrons made from completely newly made particles such as $\overline{\Lambda}$ and Φ. The measured Φ spectrum is again in qualitative agreement with this result [21].

A comparison of the baryon distributions between Pb–Pb and S–S collision, seen in Fig. 8.6, suggests that $y_\parallel$ is about 0.4 units of rapidity larger in the lighter collision system. Even though it is seemingly a small change, this opens by 50% the gap between the fluids, as we saw in Fig. 8.5, and a more pronounced central-rapidity reduction in abundances of certain particles is present for S–S compared with Pb–Pb collision systems.

8.4 Transverse-mass fireball spectra

The experimental study of the rapidity spectra is complemented by studies of particle-abundance distributions in the direction transverse to the collision axis. Under a Lorentz transformation along the collision axis, $p_\perp$ remains unchanged and thus

$$m_\perp = \sqrt{m^2 + \vec{p}_\perp^2}$$

is invariant. Transverse-mass $m_\perp$-particle spectra are therefore not directly distorted by flow motion of the fireball matter along the collision axis, and also no further consideration of the CM frame of reference is necessary, which in fixed target experiments is rapidly moving with respect to a laboratory observer.

There is also a great difference in the physics when we evaluate rapidity and transverse-mass spectra. As discussed in section 8.3, the rapidity spectra help us understand the degree of stopping and transparency of matter in collision, whereas the $m_\perp$ spectra offer insights into thermalization of matter after collision, and evolution of flow. In that sense $m_\perp$ spectra are often more interesting and also a greater challenge to describe in an *ab initio* study. Within the statistical model the focus in studying $m_\perp$ spectra is on determining the local temperature and transverse flow of the evolving fireball matter.

One could consider the particle spectra as functions of transverse-momentum $p_\perp$, but the regularities occurring for transverse-mass spectra for different particles suggest that the spectra have a thermal character. Therefore $m_\perp$ is a better variable to use in heavy-ion collisions, at least when $m_\perp$ is not too big: particles with high values of $m_\perp > 4$ GeV, at the temperatures we consider, are potentially produced in initial hard scattering of partons. These decay in yield as a power law, and hence dominate the exponentially decaying thermal particle yields at high $m_\perp$. Since hard parton scattering knows nothing about the mass of the final hadron observed, a better variable to look at to evaluate these processes is $p_\perp$. However, for $m_\perp \gg m$, there is little difference between $p_\perp$ and $m_\perp$, so we conclude that $m_\perp$ is overall the more suitable variable to consider in heavy-ion collisions. One of the surprising early results obtained at RHIC is the absence of high-$p_\perp$ particles in central interactions [17]. This suggests an effective parton thermalization mechanism.

In order to study the thermal properties in the fireball as 'reported' by the emitted particles, we analyze $m_\perp$ spectra of many different hadrons. The range of $m_\perp$, on the one hand, should not reach very large values, at which hadrons originating in hard parton scattering are relevant. On the other hand, we do need relatively small $m_\perp$, in order for the non-exponential structure associated with transverse flow and resonance decays to emerge.

The transverse-mass spectra of hyperons, which we have seen in Fig. 1.7 on page 20, are potentially very important in understanding and in modeling of the exploding QGP fireball. We have already in the S–Au 200A-GeV collisions the appearance of the exponential thermal spectra. The central-rapidity high-transverse-mass spectra of strange particles, K_s^0, $\overline{\Lambda}$, and Λ, given by the CERN–SPS WA85 collaboration, $m_\perp^{-3/2} \, dN_i/dm_\perp$, are shown in Fig. 8.8. The factor $m_\perp^{-3/2}$ is introduced in view of the form of Eq. (8.8). The resulting shape, shown in Fig. 8.8 on a semi-logarithmic display, can be fitted with a straight line. This exactly exponential behavior is initially surprising, considering that Eq. (8.8) required summation over the entire range of rapidity, given the rapidity acceptance range of WA85 limited to central $\Delta y < 1$ interval. However, effective summation over a wider range of y occurs, given the presence of collective longitudinal flow of matter. Similar results were also reported from the related work of the WA94 collaboration for S–S interactions [8].

We see, in Fig. 8.8, in the region of transverse masses presented, 1.5 GeV $< m_\perp <$ 2.6 GeV, not only that the particle spectra are exponential, $\propto \exp(-m_\perp/T_\perp)$, but also that the behaviors of all three different particles feature the same inverse slope, $T_\perp = 232 \pm 5$ MeV. This is not the actual temperature of the fireball, as noted earlier. The lower emission temperature of these particles, T_{tf}, is blue-shifted by the flow as is seen in Eq. (8.39), and can be approximately understood in terms of the Doppler factor in Eq. (5.36).

The same shape of $m_\perp$ spectrum appears in results from the WA80 collaboration results, for the neutral hadrons π^0 and η. In Fig. 8.9, we show the S–Au and S–S WA80 results at 200A GeV [28, 29, 234, 235], multiplying the invariant cross sections by the power $m_\perp^{-1/2}$ in order to establish

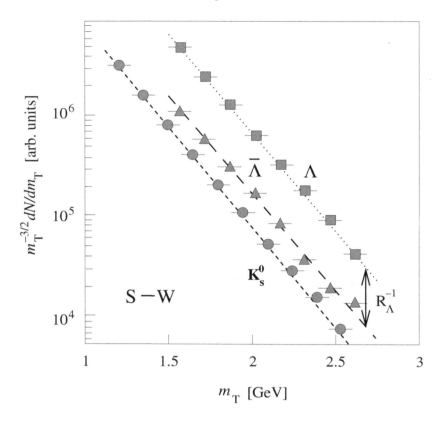

Fig. 8.8. Strange-particle spectra for Λ, $\overline{\Lambda}$, and K_S [225]. The line connecting the Λ and $\overline{\Lambda}$ spectra, denoted R_Λ^{-1}, shows how at fixed $m_\perp$ the ratio R_Λ of abundances of these particles can be extracted. Experimental WA85 results at $200A$ GeV [104, 116, 117].

a direct correspondence between the representations of the data of experiments WA85 and WA80. To determine the required multiplicative factor, we note that the particle-production cross section $d\sigma$ is controlled by the geometry of the collision, see section 5.2, and thus is the geometric interaction surface, σ_{inel}, multiplied by the yield of particles dN. Using Eq. (8.3) we obtain

$$m_\perp^{-1/2} E \frac{d^3\sigma}{d^3p} = \sigma_{\mathrm{inel}} \frac{dN}{2\pi \, m_\perp^{3/2} \, dm_\perp \, dy}. \tag{8.41}$$

Like WA85, the WA80 experiment also presented data for the central region in rapidity, $2.1 < y < 2.9$, and no further adjustment is needed in order to make the results exactly comparable.

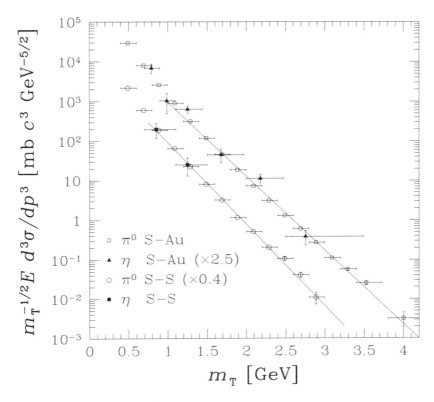

Fig. 8.9. Neutral-particle π^0 and η spectra (invariant cross sections divided by $m_\perp^{1/2}$) in the central-rapidity interval $2.1 < y < 2.9$ [225]. Upper solid line, S–Au thermal spectrum with temperature $T = 232$ MeV; lower solid line, S–S, $T = 210$ MeV. Experimental data at $200A$ GeV courtesy of the WA80 collaboration [28, 29, 234, 235].

The upper straight line (S–Au collisions) in Fig. 8.9 is the same exponential as we saw in Fig. 8.8, for the three different WA85 strange-particle spectra. While the WA85 data covered the interval 1.5 GeV $< m_\perp < 2.5$ GeV, the thermal exponential shape continues, in the WA80 data, through the highest data point at $m_\perp = 4$ GeV. The lower solid line in Fig. 8.9 is for S–S $200A$-GeV interactions and is drawn with $T = 210$ MeV. The choice of S–S temperature is based on the WA94 results obtained from their spectra of strange antibaryons [8]. It is noteworthy that the WA80 particle spectra shown in Fig. 8.9 span seven decades, and that over 5–6 decades the thermal spectral shape for neutral hadrons is in excellent agreement with the strange-particle spectral shape. We note that the rise in the yield of neutral mesons at low $m_\perp \simeq 0.5$ GeV is expected. It is due to secondary contributions to the yield of particles by decay of hadronic

Table 8.1.   Inverse (net) proton slopes $T_\perp$ for various reaction systems at $200A$ GeV ($158A$ GeV for Pb–Pb), increasing in size from left to right

Reaction	p–S	p–Au	d–Au	O–Au
$T_\perp$	154 ± 114	163 ± 5	172 ± 5	219 ± 5
y interval	$0.5 \le y \le 3.0$	$0.5 \le y \le 3.0$	$0.5 \le y \le 3.0$	$0.5 \le y \le 3.0$
Reaction	S–S	S–Ag	S–Au	Pb–Pb
$T_\perp$	235 ± 9	238 ± 2	276 ± 48	308 ± 15
y interval	$0.5 \le y \le 3.0$	$0.5 \le y \le 3.0$	$3.0 \le y \le 5.0$	$y \simeq 3$

resonances. While at high $m_\perp$ pion, kaon, and hyperon slopes agree, at small $m_\perp < 0.8$ GeV the pion spectrum is much steeper. Since most pions are produced at these $m_\perp$, a global fit to the pion data yields an inverse slope parameter which is much smaller than the value we can see in Fig. 8.9.

There is a clear difference between $T_\perp$ inverse slopes pertinent to different collision systems, and a systematic trend is visible: $T_\perp$ increases with the volume of the reaction zone. We show, in table 8.1, the $m_\perp$ inverse slopes of participating (net) protons for a number of collision systems studied by the NA35/NA49 collaboration [26, 43]. The shape of the $m_\perp$ spectra has been fitted to the simple form

$$\frac{dN}{dm_\perp \, dy} \sim m_\perp^\alpha \exp(-m_\perp/T). \tag{8.42}$$

The results presented in table 8.1 were obtained with $\alpha = 1$. Several effects contribute to an increase of $T_\perp$ with increasing size of the colliding system. With increasing number of participants in the collision, the fireball of dense matter becomes less transparent and thus colliding matter can be compressed more at a given collision energy. Moreover, larger systems have more time to develop the outward flow under the (higher) internal pressure, acquiring a greater collective velocity. Thus, what we see is that the initial fireball of the collision system is getting hotter and denser with increasing collision volume, which leads to a longer, and more violent explosion. This in turn enhances the transverse velocity at the time of production of particles. An in-depth analysis, which requires consideration of other particles apart from the (net) production of protons, confirms that the systematic increase of the inverse $m_\perp$ slope with increasing size of the system is associated with an increasing velocity of expansion of the source. The intrinsic temperature of emission from the fireball remains at the level of $T \lesssim 160$ MeV [59, 60, 176, 259].

Table 8.2.   Inverse slopes $T_\perp$ for various strange hadrons

	Λ	$\overline{\Lambda}$	Ξ^-	$\overline{\Xi}^+$	$\Omega^- + \overline{\Omega}^+$	ϕ
Pb–Pb	289 ± 2	287 ± 4	286 ± 9	284 ± 17	251 ± 19	305 ± 15
S–W	233 ± 3	232 ± 7	244 ± 12	238 ± 16		

To reach the most extreme conditions, collisions of the heaviest nuclei are required, and thus much of the experimental effort has gone into studying the Pb–Pb collision system. The highest inverse slopes are reported for several strange baryons and antibaryons by the WA97 collaboration [42]. Results presented in table 8.2 were obtained using, in Eq. (8.42), $\alpha = 1$ for Pb–Pb, and $\alpha = \frac{3}{2}$ for the S–W collision system. The corresponding spectra are shown in Fig. 1.7 on page 20. The most interesting result seen in table 8.2 is that there is practically the same inverse slope for baryons and antibaryons of the same type. This confirms the result reported by the WA85 collaboration for S–W interactions [6, 118], as is also shown in table 8.2.

The data point in table 8.2 for the $\phi(s\bar{s})$, in Pb–Pb collisions, is from the evaluation of the kaon-decay channel by the NA49 collaboration [21]. This data point disagrees with a preliminary result $T_{\mu\mu} = 227$ MeV, which was reported by the NA50 collaboration and obtained in the dimuon-decay channel [212].

If strange baryons and antibaryons were to be produced in an environment of baryon-rich confined matter, the difference in interactions of antibaryons, which have a large annihilation cross section at small momenta, should be visible as a baryon–antibaryon difference in the spectral shape, in particular at small $m_\perp$. The absence, to a very high precision, of any transverse-mass spectral asymmetry between strange baryons and antibaryons is a very important item of experimental evidence for a common mechanism of production of strange baryons and antibaryons by a source such as a QGP fireball which treats matter and antimatter in the same way. In order to suppress interactions within a hadronic-matter phase possibly formed after the QGP state hadronizes, either a sudden breakup of the fireball, arising after considerable super-cooling, or sequential evaporation of hadrons in time, without formation of a hadron phase, is required. This symmetry between matter and antimatter has not been reproduced in transport models, in which confined hadron degrees of freedom appear.

In Fig. 8.10, we see, for Pb–Pb collisions at $158A$ GeV, results shown in table 8.2 along with other inverse slopes $T_\perp$ ordered as functions of particle mass. Several different results are shown for pions, which arise

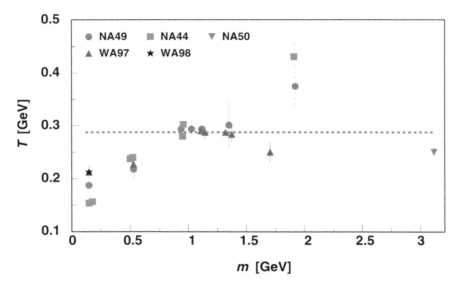

Fig. 8.10. Mass dependences of inverse slopes observed in Pb–Pb interactions at 158A GeV; symbols indicate the experiment from which data is drawn, as coded in the figure.

for different (low) $m_\perp$, y windows considered in different experiments. Even so, there is some unresolved variance between different pion results. Ten different hadronic particles with 0.9 GeV $< m <$ 1.5 GeV exhibit a common inverse slope indicated by the horizontal dashed line. There is general agreement that the increase in the slope seen on comparing pions, kaons, and baryons (obtained within an overlapping range of $p_\perp$, not of $m_\perp$) is due to the presence of a strong transverse flow of matter from which these particles originate [59]. The observation of a thermal charmonium spectrum (the point at $m =$ 3.1 GeV) both in S- and Pb-induced reactions [11, 14], with $m_\perp$ slopes similar to those for the other heavy hadrons, suggests that thermalization of hadrons is a universal phenomenon in heavy-ion collisions. The thermal shape of the observed charmonium spectra is somewhat surprising considering the 'standard-model' reaction picture of suppression of charmonium, see section 1.6.

The highest value of T in Fig. 8.10, at $m =$ 1.9 GeV, for the deuteron, confirms that these particles are not produced thermally. Production of deuterons is believed to arise predominantly from the final-state interaction between nearly free-streaming nucleons. The inverse slope for Ω and $\overline{\Omega}$ (at $m =$ 1.6 GeV) seems to be about two standard deviations below expectation. It is understood to be due to excess production of Ω and $\overline{\Omega}$ at low $p_\perp$; see Fig. 8.11.

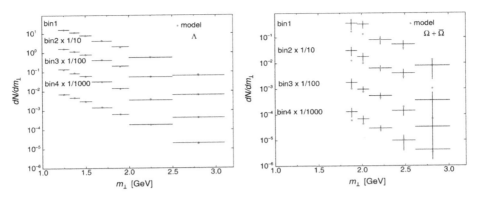

Fig. 8.11. Thermal analysis of Λ (left) and $\Omega+\overline{\Omega}$ (right) $m_\perp$ spectra for various centralities of collision [259].

8.5 Centrality dependence of $m_\perp$-spectra

A study of transverse-mass spectra [259] has been performed for the precisely known strange-hadron spectra of the experiment WA97 [42], reported for several centrality bins; this data, with all centrality bins combined, is shown in Fig. 1.7 on page 20. The shapes of the various particle spectra depend in a complex, nonlinear, but unique, way, on the temperature used, and on the velocity of transverse flow, and these parameters are determined universally for all particles considered in each collision centrality.

In an early study of hadron spectra it was suggested that spectra alone could not separate $T_{\rm tf}$ and $v_{\rm tf}$, as these quantities are highly correlated; see the Doppler formula Eq. (5.36). As we will see, these two parameters can be determined without any need for other experimental input, when precise experimental data are available for $m_\perp$ spectra reaching down to relatively low values of $p_\perp$. This is possible for the following reason: it is assumed that, after hadronic resonances have decayed, their decay products do not rescatter from surrounding matter, thus the non-thermal spectrum is combined with the primary thermal spectrum to form the final observed spectrum. By choosing the yield of resonances to be determined by the temperature seen in the spectrum, the shape of the computed spectrum becomes a highly nonlinear function of temperature and velocity. Since there is a minimum transverse momentum required in order to observe a particle, the yields above $p_\perp^{\rm min}$ depend also on the transverse-flow velocity. This method assumes that the chemical (particle-production) freeze-out temperature $T_{\rm f}$ is assumed to be nearly equal to the thermal (spectrum-shaping) freeze-out temperature $T_{\rm tf}$. The results obtained are consistent with this assumption.

The final $m_\perp$ distribution for particles is composed of directly produced particles and decay products originating in the 'root' particle R decaying to the observed particle X and any Z, with variables $R(M, M_T, Y) \to X(m, m_\perp, y) + Z$ [238]:

$$\frac{dN_X}{dm_\perp} = \frac{dN_X}{dm_\perp}\Big|_{\text{direct}} + \sum_{\forall R \to X+Z} \frac{dN_X}{dm_\perp}\Big|_{R \to X+2+\cdots}. \tag{8.43}$$

Only first-generation, and only two-body, decays were considered, as is appropriate for the hyperons and kaons. The decay contribution to the yield of X is

$$\frac{dN_X}{dm_\perp^2\, dy}\Big|_R = \frac{g_R b_{RX}}{4\pi p^*} \int_{Y_-}^{Y_+} dY \int_{M_{T_-}}^{M_{T_+}} dM_T^2\, \mathcal{J}\, \frac{d^2 N_R}{dM_T^2 dY}. \tag{8.44}$$

Here, g_R and b_{RX} are the R-degeneracy and branching into X, and $p^* = \sqrt{E^{*2} - m^2}$ with $E^* = (M^2 - m^2 - m_2^2)/(2M)$, are the energy and momentum of the decay product X in the restframe of its parent R. The limits on the integration are the maximum values accessible to the decay product X:

$$Y_\pm = y \pm \sinh^{-1}\left(\frac{p^*}{m_\perp}\right),$$

$$M_{T_\pm} = M\frac{E^* m_\perp \cosh \Delta Y \pm p_\perp \sqrt{p^{*2} - m_\perp^2 \sinh^2 \Delta Y}}{m_\perp^2 \sinh^2 \Delta Y + m^2},$$

and

$$\mathcal{J} = \frac{M}{\sqrt{P_T^2 p_\perp^2 - (ME^* - M_T m_\perp \cosh \Delta Y)^2}},$$

where $\Delta Y = Y - y$.

The primary particle spectra (both those directly produced and parents of decay products) are derived from the thermal Boltzmann distribution. As discussed earlier in this chapter, in general the longitudinal flow does not significantly influence $m_\perp$ spectra. Thus it is possible, in order to simplify the evaluation of integrals, to disregard longitudinal flow and to allow spherical symmetry of the transverse flow. A second hadronization-surface 'velocity' seen in Eq. (8.39), $v_f^{-1} \equiv dt_f/dx_f$, was considered. Thus the thermal distribution of directly produced and parent particles R had the form

$$\frac{d^2 N}{dm_\perp\, dy} \propto \left(1 - \frac{\vec{v}_f^{-1} \cdot \vec{p}}{E}\right) \gamma m_\perp \cosh y \exp\left[-\gamma \frac{E}{T}\left(1 - \frac{\vec{v} \cdot \vec{p}}{E}\right)\right], \tag{8.45}$$

where $\gamma = 1/\sqrt{1 - v^2}$.

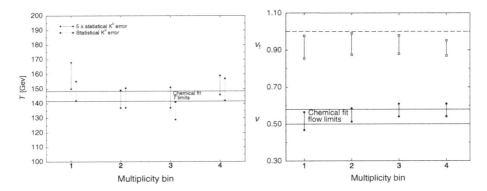

Fig. 8.12. The thermal freeze-out temperature T (left), flow velocity v (bottom right), and breakup (hadronization hyper-surface-propagation) velocity v_f (top right) for various collision-centrality bins. The upper limit $v_f = 1$ (dashed line) and chemical-freeze-out-analysis limits for v (solid lines) are also shown. For the temperature, results obtained with increased error for kaon spectra are also shown.

Simultaneous analysis of the spectra of Λ, $\overline{\Lambda}$, Ξ, $\overline{\Xi}$, $\Omega + \overline{\Omega}$, and $K_S = (K^0 + \overline{K^0})/2$ in four centrality bins was performed. In each centrality bin pronounced minima in T, v_f and v plane are observed for the total statistical error:

$$\chi^2 = \sum_i \left(\frac{F_i^{\text{theory}} - F_i}{\Delta F_i} \right)^2,$$

evaluated relative to the experimental precision of measurement ΔF_i of the result F_i. The chemical parameters, which are not well determined by a momentum-distribution fit, are not varied. Since the statistics of kaons was very high, and thus the statistical precision of data potentially was significantly greater than systematic error, also a global fit with a five-fold-increased kaon error was performed [259].

Some of the resulting $m_\perp$ spectra for particles are shown, in Fig. 8.11, in each part for the four bins separately. On the left-hand side, we see as an example the Λ spectrum. The description of the shape, in all four centrality cases, is very satisfactory, also for all other particles considered, except for $\Omega + \overline{\Omega}$ in the right-hand panel in Fig. 8.11. In all four centrality bins for the sum $\Omega + \overline{\Omega}$, the two lowest $m_\perp$ data points are underpredicted. This low-$m_\perp$ excess explains why the inverse $m_\perp$ slopes for Ω and $\overline{\Omega}$ are reported to be smaller than the values seen for all other strange (anti)hyperons in Fig. 8.10. This behavior suggests that soft Ω and $\overline{\Omega}$ are produced in a significant manner by mechanisms beyond the statistical model, which we discuss further at the end of section 19.3.

The parameters shaping the spectral form, which arise in this description of hyperon $m_\perp$ spectra, are shown in Fig. 8.12, on the left-hand side the thermal freeze-out temperature $T_{\rm tf}$, and on the right-hand side the transverse velocity v (bottom) and the breakup (hadronization) speed parameter $v_{\rm f}$ (top). The value of $v_{\rm f}$ is near to the velocity of light, which is consistent with the picture of a sudden breakup of the fireball. The horizontal lines delineate the ranges of the result of chemical particle-yield analysis, and are similar to those presented in table 19.3 on page 360. The range of values of T seen in this table is slightly different as the results presented were updated. The $m_\perp$-spectral-shape analysis is found to be consistent with the purely chemical analysis of strange and non-strange hadron production.

An important objective of this complex analysis is to see whether different centrality bins yield results consistent with the same physics. There is no indication, in the left-hand panel of Fig. 8.12, of a significant or systematic change of T with centrality, or dominance of the result by the kaon spectrum alone. The resulting temperature, here dominated by the thermal shape, agrees with the temperature obtained from analysis of particle yields alone, table 19.3, which is dominated by the chemical freeze-out temperature.

The flow (expansion) velocity v (lower part of the right-hand panel of Fig. 8.12), even though it is flat to within the experimental error, reveals a slight but systematic increase with centrality, and thus size of the system. This is expected, since the more central events involve a greater volume of matter, which allows more time for the development of the flow.

For all four centralities these results show that there is no need to introduce a two-staged freeze-out; in fact, we can conclude that $T_{\rm tf} \simeq T_{\rm f}$. The myth of unequal thermal and chemical freeze-out temperatures is rooted in the high temperature obtained in chemical-*equilibrium* analysis of hadron yields. However, results of such an analysis of the experimental data lack the required statistical confidence, even though the systematic behavior of the particle production data is well reproduced, as we shall discuss at the end of section 9.2. Specifically, once a fully descriptive set of parameters is introduced, allowing for precise data description, thermal and chemical freeze-out conditions are found to be the same.

The results of the analysis described in this section are consistent with strange hadrons being produced by the new state of matter at CERN in all centrality bins explored by the experiment WA97, i.e., for numbers of participants greater than $\simeq 100$. The low-centrality fifth bin, now being studied by experiment WA57, see Fig. 9.5 [87], exhibits different characteristics, with less enhancement of the production of multistrange hadrons.

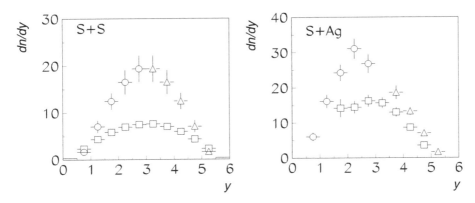

Fig. 9.1. The abundance of $1.6\Lambda + 4K_S + 1.6\overline{\Lambda}$ as a function of rapidity. On the left, S–S; on the right, S–Ag (open circles are the directly measured data). The triangles are reflected data points for S–S and reflected interpolated data employing S–S and S–Ag. The squares in the S–S case are the results for N–N collisions scaled up by the pion-multiplicity ratio; for S–Ag these are the scaled-up p–S results. Data courtesy of the NA35 collaboration [128].

9 Highlights of hadron production

9.1 The production of strangeness

Strangeness is a valuable tool for understanding the reaction mechanism, since it has to be made during the collision. The question is that of how it is produced. In terms of experimental information, the first thing we would like to establish is whether the mechanism producing strangeness involves a hot fireball at central rapidity, or whether perhaps a lot of strangeness originates from the projectile/target-fragmentation region.

Results of the experiment NA35 [128] are shown in Fig. 9.1 as functions of rapidity for the case of S–S 200A-GeV collisions. We consider the overall abundance of $\langle s + \bar{s} \rangle$. The open circles are the measured data points, the open triangles are the symmetrically reflected data points, and squares on the left-hand side are the results of N–N (isospin-symmetric nucleon–nucleon) collisions scaled up by the ratio in pion multiplicity, whereas on the right-hand side the p–S results are scaled up. We show the rapidity yield obtained by integrating the transverse-mass $m_\perp$ distribution for the total yield of strangeness:

$$\frac{d\langle s + \bar{s} \rangle}{dy} = 1.6 \frac{d\Lambda}{dy} + 4 \frac{dK_S}{dy} + 1.6 \frac{d\overline{\Lambda}}{dy}. \tag{9.1}$$

We note that, on doubling the K_S yield, we include K_L, and, on doubling again, we add both K^+ and K^-, which explains the factor 4.

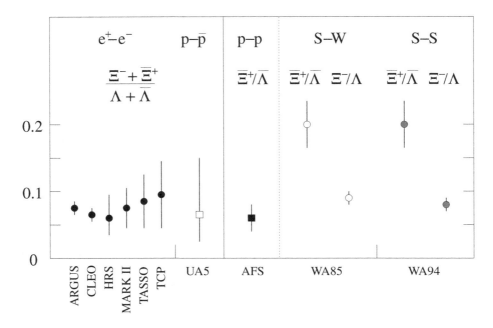

Fig. 9.2. The ratio (at fixed $p_\perp$) of (multi)strange baryon–antibaryon particle abundances, measured in the central rapidity region for $200A$-GeV S–S/W collisions, compared with ratios obtained in lepton- and nucleon-induced reactions. Data assembled by the WA85/94 collaboration [1–5, 7].

In Fig. 2.6 on page 32, we saw that a factor 1.5 allows one to extrapolate the yields of Λ and $\overline{\Lambda}$ to include all singly strange hyperons. However, there is also strong production of multistrange Ξ and $\overline{\Xi}$, with $\Xi^-/\Lambda \simeq$ 0.1 and $\overline{\Xi}^+/\overline{\Lambda} \simeq 0.2$, as is seen in Fig. 9.2. Assuming similar rapidity distributions for Ξ^- and Λ and $\overline{\Xi}^+$ and $\overline{\Lambda}$, and remembering that, for each charged Ξ, there is its neutral isospin partner, this implies that the coefficient of Λ should have been 1.9, and that of $\overline{\Lambda}$ 2.3, in order to account for the production of multistrange hyperons. Thus the enhancement in production of strange particles seen in Fig. 9.1 is slightly (by $\simeq 6\%$) understated.

The difference between scaled N–N and S–S data is most pronounced at central rapidity ($y_{\rm CM} = 2.97$), and it disappears within one unit of the projectile- and target-fragmentation regions. At central rapidity $y \simeq 3$, a new source of strangeness not present in the N–N-collision system contributes. The S–S system is relatively small, thus stopping is small, and it is quite impressive that the enhancement in production of strangeness is observed only at central rapidity. Since other SPS experiments involved heavier nuclei and/or lower energy, and thus certainly involved more stopping, we can be sure that the excess of strangeness at the SPS originates at central rapidity.

We show, in the right-hand panel of Fig. 9.1, similar results for S–Ag collisions: the open circles are the measured points, open triangles are estimates based on S–S and the 'reflected' S–Ag results, and the open squares are pion-multiplicity-scaled p–S results. An enhancement in production of strangeness is also seen here, though the asymmetry of the collision system makes it more difficult to understand the effect quantitatively.

The abundant strangeness, in the central rapidity region, is at the origin of the effective production of multistrange antibaryons. The WA85 and WA94 collaborations [1–5, 7] explored the relative abundances at central rapidity of the various strange baryons and antibaryons produced in S–W and S–S reactions. The central-rapidity particle ratios have been obtained at $p_\perp \geq 1$ GeV. The results for relative abundances are reported when yields of particles of unequal masses are compared, both with $p_\perp \geq 1$ GeV and using as cutoff a fixed value of $m_\perp \geq 1.7$ GeV. The results at fixed $p_\perp$ are shown in Fig. 9.2. In the left-hand panel, we see the annihilation and production results from e^+–e^- and p–$\bar{\text{p}}$ reactions, in the middle panel, we see the ISR–AFS p–p measurement of $\overline{\Xi}/\overline{\Lambda}$, which is a factor of five below the S–W and S–S result, even though the ISR energy $\sqrt{s_{NN}}$ was nearly four times higher than is available at the SPS. In S–S and S–W interactions a clear enhancement of the $\overline{\Xi}^+/\overline{\Lambda}$ ratio is observed, which has been predicted as a signature of the QGP [215, 226].

A more extreme picture of the enhancement is found when we compare the yields of various hyperons at fixed $m_\perp$, as would be done in a thermal model considering coalescence of quarks to give hadrons. This means that, on comparing, e.g., $\overline{\Xi}$ with $\overline{\Lambda}$, we are looking at particles at different $p_\perp$. The experimental results reported by the WA85 collaboration, for S–W interactions at 200A GeV, are at 'fixed $m_\perp$':

$$\left.\frac{\overline{\Xi^-}}{\overline{\Lambda + \Sigma^0}}\right|_{m_\perp} = 0.4 \pm 0.04\,, \qquad \left.\frac{\Xi^-}{\Lambda + \Sigma^0}\right|_{m_\perp} = 0.19 \pm 0.01. \tag{9.2}$$

We introduce the average singly strange hyperon yield Y(qqs), which at fixed $m_\perp$ is the same for all components:

$$Y = \Lambda = \Sigma^0 = \Sigma^+ = \Sigma^-. \tag{9.3}$$

Thus, the actual antihadron ratio is indeed twice as large as that measured:

$$\left.\frac{\overline{\Xi^-}}{\overline{Y}}\right|_{m_\perp} \simeq 0.8 \pm 0.1\,. \tag{9.4}$$

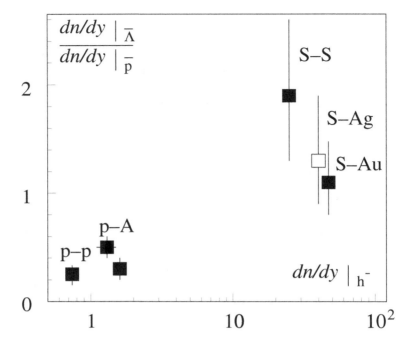

Fig. 9.3. The ratio of the rapidity density dn/dy for $\overline{\Lambda}/\overline{p}$, measured at central y, as a function of the negative-hadron central-rapidity density $dn/dy|_{h^-}$. NA35 collaboration [46, 47, 138].

We now consider the quark content:

$$\frac{\overline{\overline{\Xi}}}{\overline{\overline{Y}}}\bigg|_{m_\perp} = \frac{\overline{ss}\overline{q}}{\overline{s}\overline{q}\overline{q}}\bigg|_{m_\perp} = \frac{\overline{s}}{\overline{q}}\bigg|_{m_\perp} \simeq 0.8. \tag{9.5}$$

We see that, at the time of production of antihyperons, there has been comparable availability of antiquarks at high momentum, $\bar{u} = \bar{d} = 1.2\bar{s}$. This result is hard to explain other than in terms of QGP, for which, at the values of statistical parameters applicable here, near to chemical equilibrium all three antiquark flavors are at nearly equal abundances. Since the abundance of light quarks comprises valence quarks, it is twice as large, as can be seen in Fig. 10.3 on page 203, $u = d \simeq 2.5s$, consistent with the half as large baryon ratio seen on the right-hand side of Eq. (9.2). These results involving the abundances of multistrange antibaryons have not been explained in terms of hadron-cascade models.

Consistent with this result is the observation of the NA35 collaboration [24, 25, 54] regarding the $\overline{\Lambda}/\overline{p}$ ratio. In Fig. 9.3, we show this ratio as a function of the negative-hadron rapidity density $dn/dy|_{h^-}$ at central y. The p–p and p–A reactions are at small values of $dn/dy|_{h^-}$, whereas

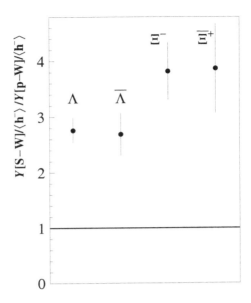

Fig. 9.4. Ratios of abundances of particles (Λ, $\overline{\Lambda}$, Ξ and $\overline{\Xi}$) normalized with respect to the abundance of h⁻: S–W results divided by p–W results at 200A GeV in the rapidity window $2.5 < y < 3$ for 1.4 GeV/$c < p_\perp < 3$ GeV/c. WA85 collaboration [41].

the S–S, S–Ag, and S–Au reactions are accompanied by a relatively high $dn/dy|_{h^-}$. We observe that there is an increase in this ratio by nearly a factor five, and, even more significantly, the abundance of the *heavier* and *strange* $\overline{\Lambda}$ is similar to if not greater than the abundance of $\bar{p}$.

The enhancement in production of strange hyperons and antihyperons can be studied by comparing it directly with the yield seen in p–A interactions. For this purpose, one obtains specific yields of strange particles 'sp', normalized with respect to the yield of negative hadrons h⁻. These can be compared with such yields in proton-induced interactions, i.e., we look at the enhancement E_s^i in production of a strange particle i defined as

$$E_s^i \equiv \frac{Y_{sp}^i(\text{S–A})}{Y_{sp}^i(\text{p–A})}, \tag{9.6}$$

where $i = (\Lambda, \overline{\Lambda}, \Xi, \overline{\Xi})$. The results are presented in Fig. 9.4 [41]. We see an enhancement of this specific yield in nuclear interactions, compared with p–A collisions, and this enhancement increases with increasing strangeness content. The stronger enhancement in production of multistrange hadrons is expected for hadronization involving enhancements in yield and density of strangeness compared with p–A interactions.

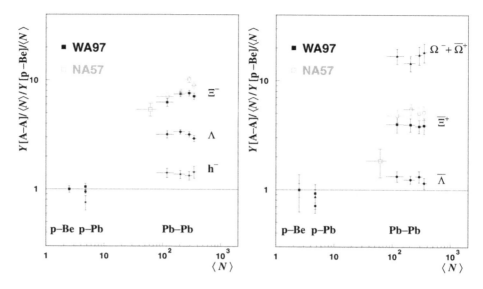

Fig. 9.5. Yields Y per wounded nucleon $\langle N \rangle$ in Pb–Pb relative to p–Be collisions from WA97 [41] (solid data points), and from NA57 for Ξ^- and $\overline{\Xi}^-$ [108] (open data points).

This presentation of enhancement in production of strange hadrons understates the magnitude of the effect since it is diluted by the overall enhancement of the yield of h$^-$, which is also expected to arise due to the enhancement of entropy in QGP, see section 7.1. Therefore a method to extrapolate the yields from p–A interactions to A–A interactions was devised, by scaling with the number of participating (wounded) nucleons $\langle N \rangle$ [41]. The Pb–Pb 158A GeV results of the WA97 collaboration are shown in Fig. 9.5, using the p–Be interaction results as reference. Solid points are the four centrality bins considered. We have presented some of these results in Fig. 1.6 on page 19. The absolute enhancement in production of strange particles compared to the p–Be extrapolated yields increases strongly with the strangeness content.

In order to understand whether there is a threshold for the enhancement to occur, the experiment NA57 has repeated the measurement of the experiment WA97, and has extended the reach by studying more peripheral collisions. In the most peripheral fifth data bin the number of participants is 60. At this time, only the Ξ^- and $\overline{\Xi}^-$ NA57 data for Pb–Pb are available, as is seen in Fig. 9.5. The rapid drop in enhancement of the production of $\overline{\Xi}^-$ is most remarkable, and, if it is confirmed in, e.g., $\overline{\Omega}$ results, this can be seen as definitive evidence for a rather sudden onset of the formation of QGP as a function of the size of the system.

We do not yet have a similar measurement for the onset of the enhancement as a function of energy. The specific production of strangeness per hadron is shown in Fig. 1.5 on page 17. However, there is a rapid change in hadron yield with energy, thus this result, as we have discussed in section 1.6, is not fully representative of the energy dependence of strangeness production. Another complication with the study of the energy dependence is the presence of the energy threshold for antibaryon production. Thus the energy dependence of the enhancement in production of strange antibaryons has to be evaluated, not with reference to the p–A reaction system, but entirely within the A–A system. The low-energy measurement, presumably under pre-QGP conditions, can be used to establish a basis against which the production at higher energy can be studied. In such a measurement, we can hope and expect to see a sudden onset with energy of the yield of multistrange (anti)baryons, if indeed a new state of matter is being created.

Another important topic in the production of strange hadrons, is the symmetry of particle and antiparticle spectra seen in Fig. 1.7 on page 20. As discussed, see table 8.2, the fitted inverse-slope parameters agree at the level of 1%, when statistics is good enough: $T^\Lambda = 289 \pm 3$ MeV, to be compared with $T^{\overline{\Lambda}} = 287 \pm 4$ MeV [42]. There is no evidence for any difference at small momenta, for which the annihilation reaction would be most significant. Thus we can literally 'see', in Fig. 1.7, that these particles are escaping from the central fireball without further interaction with hadronic gas.

9.2 Hadron abundances

Despite the relative smallness of the S–S collision system, and the highest available fixed-target heavy-ion energy, the remarkable difference between Λ and $\overline{\Lambda}$ rapidity distributions, shown in Fig. 8.4, proves that neither is the baryon-punch-through Bjørken reaction picture applicable, nor do we see stopping of the valence quarks in the central region, as we discussed in section 8.3. How does the situation look for nonstrange hadrons? In Fig. 9.6, we see the rapidity distribution of negative particles h⁻, which comprise π^-, K^-, and $\overline{p}$, shown per collision event for the 5% most central Pb–Pb 158A-GeV collisions (full triangles), 3% most central S–S 200A-GeV collisions (stars, multiplied by factor 6.5), and the N–N interactions (full dots, multiplied by a factor 176) [43]. Detailed study of the chemical composition of hadrons produced in these experiments, section 19.3, leads to an estimate that the π^- make up 92% of these particles, K^- make up 6.6%, and $\overline{p}$ contribute 1.1% in Pb–Pb fixed target interactions at 159A GeV.

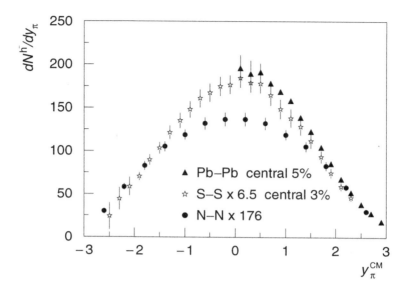

Fig. 9.6. The pseudorapidity distribution of negative hadrons (sum of π^-, K^- and $\bar{p}$) in Pb–Pb, scaled S–S, and N–N collisions [43].

The abundance of h^- produced in the Pb–Pb reactions can be obtained from Fig. 9.6. On fitting the h^- distribution to a Gaussian shape,

$$\frac{dn}{d\eta} = n_0 e^{-(y-y_{\rm CM})^2/\Gamma}, \qquad \frac{\sigma^2}{2} = \Gamma, \tag{9.7}$$

one finds $\sigma \simeq 1.4$ (half width at half maximum). The integral of the distribution gives the total multiplicity of all negative hadrons, after an additional minor correction [43]: $\langle h^-\rangle = 695 \pm 30$. Thus, in geometrically central Pb–Pb 158A-GeV interactions, the total hadron multiplicity is nearly 2400 per event, when we allow for positively charged and neutral hadrons. Since the maximum negative-hadron abundance at $y = 0$ is $dh^-/dy|_{\rm max} \simeq 200$, we also have $dn_{\rm h}/dy|_{\rm max} \simeq 680$ in these events.

To compare with Pb–Pb data, the S–S yields have been scaled up by a factor 6.5, this factor arising from the ratio of participants, which are in Pb–Pb measured to be 352 ± 12, and in S–S 52 ± 3. A further factor 0.96 is introduced by the NA49 collaboration to account for the difference in collision energy, which is somewhat higher in the S–S system. However, no correction for the fact that S–S is proton–neutron symmetric and Pb–Pb asymmetric was made. This correction may be significant since π^- is the carrier of the excess in valence d quarks, and there are 80 more d–$\bar{\rm d}$ present than u–$\bar{\rm u}$. Given that $\langle h^-\rangle \simeq 700$, an enrichment of π^- by valence quarks of projectile and target could reach a non-negligible level of 10%.

Were this 'isospin' correction introduced, it would perhaps be difficult to distinguish the Pb–Pb from S–S results in Fig. 9.6, and both are notably greater than the N–N-based expectations: the observed yields have been scaled by the factor 176 which is the ratio of Pb–Pb to N–N participants. We conclude that, both in Pb–Pb and in S–S collisions, we observe a similar per participant central rapidity excess in hadron multiplicity, and thus also an excess in production of entropy; see chapter 7.

The relevance of the 40% excess in production of h⁻ in A–A collisions at SPS energies is amplified by the opposite behavior seen at the lower AGS energies, for which there is a 20% suppression compared with the scaled nucleon–nucleon results. One can qualitatively argue about the AGS result obtained at much lower energies as follows: whereas in N–N reactions all pions produced reach the detectors, pions produced in a series of N–N collisions at the AGS are deposited in dense baryonic matter, where some can be absorbed. Moreover, a notable amount of the available collision energy is used to do work to compress colliding nuclear matter, and this energy is not available to produce pions, a point already noted in early work with relativistic heavy ions [233]. We will not pursue the implications of the suppression of pion production at fixed-target collision energies below $15A$ GeV, for the study of the nuclear-matter equations of state.

Since there is a change in pattern of behavior of pion production as a function of the collision energy, we consider in a more systematic way whether this can be understood in terms of a general change in pattern of behavior as a function of collision energy. Gaździcki [126] proposed to explore this effect as a function of the Fermi-energy variable [121, 172, 174]

$$F \equiv \frac{(\sqrt{s_{NN}} - 2m_N)^{3/4}}{(\sqrt{s_{NN}})^{1/4}}, \tag{9.8}$$

where $\sqrt{s_{NN}}$ is the CM energy for a nucleon–nucleon pair and m_N is the nucleon mass. There are several advantages in using F as an energy variable. The measured mean multiplicity of pions in N–N interactions is approximately proportional to F [127]. In the Landau model [172, 174], both the entropy and the initial temperature of the matter (for $\sqrt{s_{NN}} \gg 2m_N$) are also proportional to F.

In Fig. 9.7, we see the average yield of all pions $\langle \pi \rangle$ per average participating nucleon $\langle N_p \rangle$. The data is from [127], with the most recent results presented in [254]. The lower straight line follows open diamonds, which are results from N-N interactions, whereas the upper line follows the high-energy SPS (squares) and RHIC (open crosses) results. The cross over from the one behavior to the other is seen at the lower range of SPS energies, whereas the AGS results (triangles) indeed fall below the

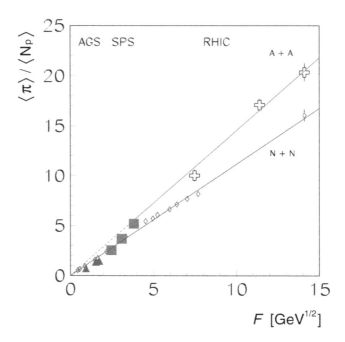

Fig. 9.7.   The per-participant average yield of pions $\langle\pi\rangle/\langle N_p\rangle$ as a function of the Fermi-energy variable F, results from A–A and N–N interactions.

N–N results. Both high-energy SPS and RHIC results are seen to follow the 'high-entropy' branch (see chapter 7), which differs clearly from the low-entropy reactions at the AGS and in N–N collisions.

Not only the yield of pions harbors a mystery. The shape of the SPS rapidity distribution for hadrons, see Fig. 9.6, is not fully understood today, and we can not convincingly explain why there is so little difference in shape among the three reactions shown. Generally, one would expect the h^- yield in S–S reactions to be 'wider' in rapidity than that for Pb–Pb collisions. Instead, what we see in Fig. 9.6 is that the rapidity shape of h^- produced in N–N reactions is the same as that observed in Pb–Pb reactions, apart from an additional central-rapidity contribution. However, we recall the qualitative study seen in Fig. 8.5, along with the observation that about half of all pions observed are actually decay products of hadronic resonances. The dilemma in understanding this distribution is in fact one of the reasons that encourages us to focus on the study of particle spectra that are fully made of newly created matter, such as $\overline{\Lambda}$, see Fig. 8.4. These are clearly a more sensitive, and less model-dependent, probe of novel physics occurring in the central rapidity region.

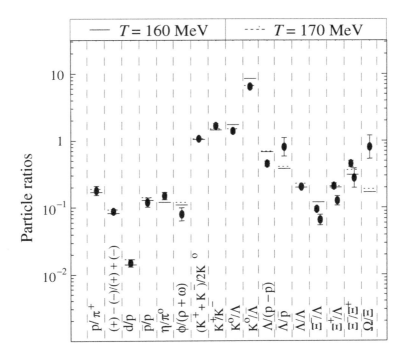

Fig. 9.8.   Particle ratios (experimental dots) seen in A–A 200*A*-GeV reactions for various particle species (shown in horizontal order), compared with the prediction of the thermal model for two different freeze-out temperatures and quark fugacity $\lambda_q = 1.42$ [82].

Further below, we will return to consider, in Fig. 9.19, the rapidity distribution of all charged hadrons observed at the nearly eight-fold higher RHIC energy. There is some spreading of the distribution, which need not be entirely due to the rapidity-versus-pseudorapidity effect we discussed in Fig. 8.3, originating most probably from the expected onset of transparency and outflow of baryon number from the central region.

The spectra of many identified hadronic particles have been measured over sufficiently large ranges of rapidity and transverse mass to allow extrapolation to cover all of the relevant kinematic domain, and the total particle-production yield can be established. The total yields of particles are not dependent on the deformation of the spectra arising from the collective flow motion within the source. Consideration of relative abundance ratios eliminates biases from the various experimental set-ups, in particular the event trigger bias cancels out.

We show the compilation of CERN (200*A*-GeV) and AGS (14*A*-GeV) data in Figs. 9.8 and 9.9, in a procedure in which chemical-yield equilibrium of hadron abundances is assumed in a statistical model. In some

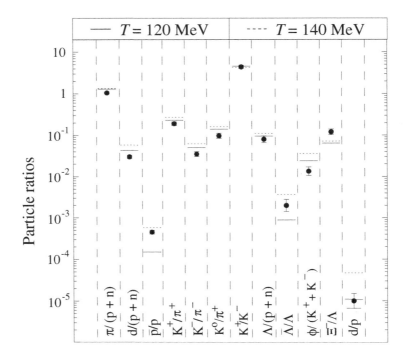

Fig. 9.9. Particle ratios seen in A–A 14A-GeV reactions (solid lines) for various particle species compared with the prediction of the thermal model for two different freeze-out temperatures and quark fugacity $\lambda_q = 4$ [82].

cases seen in these figures, the experimental errors are smaller than the size of the experimental 'dot'.

We see that the particle ratios span typically several decades, yet the systematic pattern of theoretical and experimental results coincides. Thus, the production of hadronic particles occurs, without doubt, near to the chemical equilibrium. We note that the two chemical freeze-out (particle-production) source temperatures used were $T = 160$ and 170 MeV for 200A-GeV data and $T = 120$ and 140 MeV for the AGS 14A-GeV results. Other parameters used in Figs. 9.8 and 9.9 include, for 200-GeV data, the quark fugacity $\lambda_q = 1.42$ and, for 14-GeV data, $\lambda_q = 4$. Conservation of strangeness is imposed as a constraint, i.e., the number of strange quarks and antistrange quarks, in different hadrons, balances exactly. We will discuss how to perform this calculation in chapter 11.

The first impression we have is that we see a rather good systematic agreement in the behavior of the particle yields with this statistical equilibrium-abundance model: almost all gross features of the data, for both sets, are well reproduced. Before we proceed, let us therefore pause to wonder if we should abandon the kinetic, i.e., collisional theory of particle production, and focus solely on the experimental fact that the observed hadronic multiplicities are the result of a preestablished statistical distribution, which works so well. In a sense, this finding confirms a 15 year-old prediction that such a result can be naturally explained in terms of a dynamic theory of a transient deconfined state hadronizing in a coalescence model [166]. Only a detailed study of the subtle deviations in hadron yields from precise statistical equilibrium yields allows one to understand the hadronization mechanism [69], and therefore ultimately also to explore the properties of the hadronizing QGP state.

Indeed, looking closer at Figures 9.8 and 9.9, we see systematic deviations involving, in particular, (multi)strange particles: in the $200A$-GeV data the yield anomalies mostly involve strange antibaryons. The net deviations in the total hadron yields are in fact greater – for example, were the chemical freeze-out condition set to reproduce, in the $200A$-GeV case, the ratio $\Lambda/(p - \bar{p})$ exactly, we would have enhanced the disagreement in the ratio $\overline{\Lambda}/\bar{p}$ further. There is clear evidence, in these two figures, that yields of strange particles require greater attention, beyond chemical-equilibrium mode, and we devote much of our effort in this book to understanding the physics behind this phenomenon.

Figures 9.8 and 9.9 demonstrate that the yields of strange antibaryons compared to non-strange hadrons in general vary between 50%–150% of the chemical-equilibrium yield. This strangeness 'fine structure' yield variation is one of the reasons that the measurement of abundances of rarely produced (strange)antibaryons is an excellent diagnostic tool in the study of the properties of the dense hadronic matter, as we have discussed above and in section 2.2. For this reason, we will discuss the production of strangeness and strange antibaryons as signature of QGP in considerably more detail in part VI.

9.3 Measurement of the size of a dense-matter fireball

An important aspect of hadron-production studies is the measurement of two-particle Bose–Einstein correlations, which permits one to evaluate the size of the space–time region. Also Fermi–Dirac correlations can be considered, but practical considerations have favored the measurement of the positive-boson interference. The two-particle intensity interferometry originates from the ambiguity in the path between the source and the detector for indistinguishable quantum particles. The two-particle intensity method was developed by Hanburry-Brown and Twiss as means of determining the dimensions of distant astronomical objects, and is referred to in short hand as 'HBT'; see [75] and references therein.

HBT analysis is today a wide subject of specialization, which could fill this book. We will briefly introduce the method of analysis and present

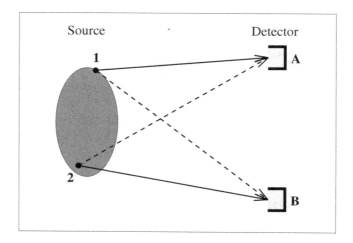

Fig. 9.10. HBT interference: detectors A and B see quantum waves emitted from different source locations 1 and 2 differently.

some recent results. In Fig. 9.10, we illustrate an uncorrelated source of particles π_1 and π_2, with momentum p_1 and p_2, emitted from some points 1 and 2, respectively, within an emission region (the shadowed region in Fig. 9.10). Particles are counted at points A and B. When two observed quantum particles are identical, two indistinguishable histories are possible, drawn with full and dashed lines, respectively, in Fig. 9.10.

The intensity-interference pattern is observed as an enhancement in the number of like-particle (boson) pairs originating from a single source, normalized with respect to a random sample of particles from two different interactions. This enhancement is studied by means of the two-particle correlation function,

$$C_2(p_A, p_B) = \frac{\rho(A, B)}{\rho(A) * \rho(B)}, \tag{9.9}$$

where the numerator represents events with particles registering in both detectors, and the denominator the number of pairs of uncorrelated particles. If no correlation in particle intensity exists, the counts in both detectors are independent, which means that $\rho(A, B) \rightarrow \rho(A) * \rho(B)$.

The correlation function is in principle dependent on the momenta of both particles observed. A first simple measure of the size of the source is obtained by considering C_2 for similar transverse momenta of both particles. Summing over all other variables ('projecting'), one finds that, as a function of the difference in transverse momenta $\vec{q}_\perp = \vec{p}_{A,\perp} - \vec{p}_{B,\perp}$, $C_2(p_A, p_B)$ exhibits a clear correlation peak near $q_\perp = 0$.

More generally, the shape of the enhancement as a function of the available momentum variables contains information on the geometric source parameters and thus both the size and the shape of the source, and, when models are considered, also its dynamic evolution. The difference in momentum for pions $\vec{q} = \vec{p}_A - \vec{p}_B$, can be decomposed using as a reference vector the sum of the pion momenta (the pair momentum) $\vec{p} = \vec{p}_A + \vec{p}_B$, as well as the collision axis of nuclei. The 'longitudinal' direction 'l' with corresponding difference in momentum q_l is also referred to as the 'beam' direction, as is appropriate for fixed-target experiments.

The 'out' component of the transverse-momentum-difference vector $\vec{q}_\perp$ (as before, transverse with respect to the beam axis) is the projection onto the pion-momentum axis of $\vec{q}_\perp$:

$$q_o = \frac{\vec{q}_\perp \cdot \vec{p}_\perp}{|\vec{p}_\perp|}. \tag{9.10}$$

The 'side' component q_s is the remaining second component of $\vec{q}_\perp$ and its magnitude is

$$q_s = \sqrt{\vec{q}_\perp^2 - q_o^2}. \tag{9.11}$$

In the fits of the correlation functions, one likes to sharpen the definition of the longitudinal (beam) component, considering that $\vec{p}$ is in general not normal to the axis,

$$q_l^2 = q_z^2 - q_o^2 + \frac{p_o q_o - p_z q_z}{p_o^2 - p_z^2}, \tag{9.12}$$

where q_z is the magnitude of the difference in momenta for the pair along the longitudinal (beam) axis, and p_z is the same component of the sum of momenta of the pair.

The correlation C_2 is fitted to the form comprising three source-shape parameters R_i:

$$C_2 = D\left(1 + \lambda e^{-(q_o^2 R_o^2 + q_s^2 R_s^2 + q_l^2 R_l^2)}\right). \tag{9.13}$$

Here*, $0 \le \lambda \le 1$, and, for the ideal HBT situation, $\lambda = 1$. Other geometric parametrizations have been considered, and also further interference terms between the geometric parameters in Eq. (9.13) have been introduced [272].

The interpretational situation in heavy-ion collisions is complicated by the finite lifetime, and the strong dynamic evolution of the particle-emitting source. Thus detailed interpretation of the observed correlations between the particles produced requires development of model-dependent understanding, and a considerable amount of effort continues to be devoted to the interpretation of the data. Generally, the following hypotheses are made regarding the source of particles:

---

* In this section λ is not a fugacity.

1. emission of particles is chaotic;
2. correlated particles do not arise primarily from decay of resonances, though a strong resonance input is expected for pion correlations;
3. particles do not interact subsequent to strong-interaction freeze-out – corrections for Coulomb effects are often applied; and
4. kinematic correlations, e.g., conservation of energy–momentum, are of no relevance.

A considerable wealth of available experimental results leads to a few conclusions of relevance to the understanding of the reaction mechanisms operating in relativistic nuclear collisions.

• No evidence is found for a major expansion of the hadronic fireball, which would be required, e.g., for a (long-lived) mixed (HG/QGP) intermediate phase. The nuclear-collision geometry determines the size of the source for pions and kaons.
• The size of the particle source is similar though a bit smaller for strange (kaons) than it is for non-strange particles (pions). Thus the conditions for production of these rather different particles are surprisingly similar.
• There is a proportionality of the central hadron multiplicity yield to the geometric volume of the source.
• Evidence for the occurrence of transverse flow of the particle source is seen.

These results are consistent with a reaction picture in which the (deconfined) fireball expands and then rather suddenly disintegrates and hadronizes. In such a process even the momentum freeze-out of final-state particles occurs at a relatively early stage of the evolution of strongly interacting matter.

Figure 9.11, compiled by the STAR collaboration [18], shows the experimental results, i.e., parameters introduced in Eq. (9.13), as functions of the collision energy $\sqrt{s_{NN}}$ for pion-intensity interferometry. These results of diverse experimental groups (see the top of the figure) are only compatible with a compact pion source being present at all reaction energies. While this is expected for the lower AGS energies, at which the fireball of nuclear matter is expected to have nuclear size, the actual slight decrease in size seen at the RHIC with $\sqrt{s_{NN}} = 130$ GeV and CERN–SPS with $\sqrt{s_{NN}} = 17$ GeV implies that, despite a rapid observed expansion of the fireball, there is even more sudden production of hadrons without an extended period of hadronization.

As a function of collision energy, we see in Fig. 9.11 that the parameter $\lambda \leq 1$ falls smoothly and rapidly from unity (the ideal expected value for an incoherent source) to about 0.5 at the RHIC; this decrease is attributed partially to an increase in the fraction of pions arising from hadron resonances at higher energies. λ is also affected by several experiment-specific

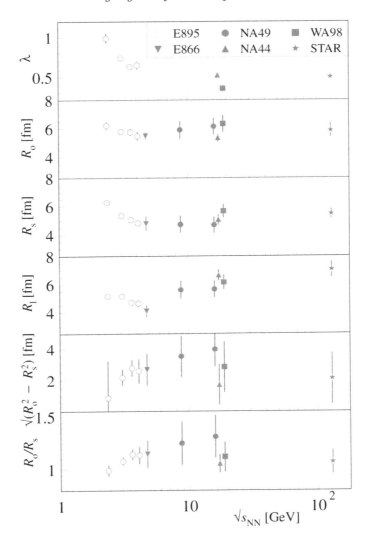

Fig. 9.11. Sizes of fireballs of excited nuclear matter derived from pion-correlation analysis, compilation and RHIC result by the STAR collaboration [18].

background effects and thus the physics of this behavior is not explored in depth.

The two parameters R_s and R_o correlate most directly to the geometry of the emitting source. This is illustrated in Fig. 9.12. The source of pions is here presumed to be a shallow surface structure; the 'out' direction is toward the eye of the observer. If the source is longitudinally deformed, and the observer is at a more transverse location, the effect is amplified

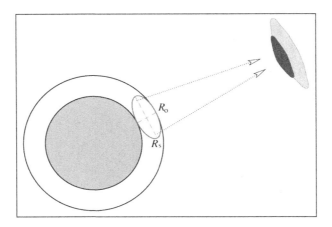

Fig. 9.12.   Surface hadronization offers a possible explanation for why the HBT 'side' radius R_s can be larger than the radius measured out toward the eye of the observer, R_o.

by the geometry of the source. In contrast to this situation, simulations involving a long-lived phase of pion matter with pions originating from the volume of the fireball of dilute matter lead to $R_\mathrm{o}/R_\mathrm{s} > 1$. Experimental results, seen at the bottom of Fig. 9.11, deviate from this expectation most clearly at RHIC energies.

$R_\mathrm{o}/R_\mathrm{s} \simeq 1$ signals a rather short duration of pion production: one can show that, when spatio-temporal correlations vanish, $\delta \equiv \sqrt{R_\mathrm{o}^2 - R_\mathrm{s}^2}$ is a measure of the life span of the emitting source, which, as can also be seen in Fig. 9.11, is not as large as the equilibrium hadronization models require. The source volume $R_\mathrm{s}^2 R_\mathrm{l}$ is found to increase along with the total produced multiplicity of particles, as the centrality of the collision is varied at RHIC energies.

In addition to the overall behavior shown in Fig. 9.11, the STAR collaboration notes that the size parameters decrease significantly with increasing $m_\perp$ of particles. In that regard, the $m_\perp$ dependence at the RHIC is similar to, but stronger than that observed in central Pb–Pb collisions at the CERN–SPS facility. This suggests that the emission of hadrons at the RHIC is occurring from a more rapidly expanding surface source than is the emission at the SPS.

9.4 Production of transverse energy

So far, we have been describing production of hadronic particles and particle spectra. However, it is almost always simpler to measure the total 'flow' of energy contained in the various particles, rather than abundances of the many kinds of particles. The energy distribution can be used to determine the extent to which energy from the longitudinal motion of the colliding nuclei participates in the nuclear interaction. This is most easily done by considering the energy emitted in the transverse direction.

Since fragments of projectile and target contaminate the longitudinal flow, it is the transverse to the beam axis component of energy which is considered as a suitable measure of the amount of CM energy, Eq. (5.2), made available in the reaction for production of particles. Therefore, one studies the distribution of the energies of the particles, weighted by the sine of their angle θ_i with the beam axis (see Fig. 5.6), which is called the transverse energy,

$$E_\perp = \sum_i E_i \sin \theta_i. \tag{9.14}$$

The resulting distribution of transverse energy produced, $dE_\perp/d\theta$, as a function of the angle θ, can be converted into a distribution in pseudorapidity η by employing Eq. (5.25).

Experimentally, $dE_\perp/d\eta$ is determined with the help of a segmented calorimeter: particles entering a segment, covering a range of θ, deposit their energy, which is determined by exploiting various mechanisms of interaction of particles in matter – hence the name 'calorimeter' which derives from the name of a common heat measuring device. In fixed-target experiments, the laboratory-frame angle is not very large, see Fig. 5.7, and thus the calorimeter is typically located relatively far away, in front of the beam axis.

We show the transverse-energy-distribution data reported by the experiments WA98 [191] and NA49 [27], for 158A-GeV Pb–Pb (fixed-target) CERN experiments in Fig. 9.13. The key feature of this result is that there is a pronounced peak in the transverse energy distribution, slightly forward of the rapidity value $y = 2.9$. The shift in pseudorapidity distribution is a result of the definition of pseudorapidity; see Fig. 5.8 on page 88. This well-peaked distribution is consistent with the expectations based on the observed negative-particle distribution shown in Fig. 9.6.

The study of the transverse energy spectra has systematically been carried out for many systems and collision impact parameters, by numerous groups. It is worth noting that, for relatively small projectiles, doubling the mass of the projectile increases the geometric number of participants, in small-impact-parameter collisions, by approximately a factor $2^{2/3} \simeq 1.59$, corresponding to the increase in area of the impact surface on a large target. For example, an increase in $E_\perp$ by about a factor 1.6 was observed for the S–Au reactions relative to the O–Au reactions [201].

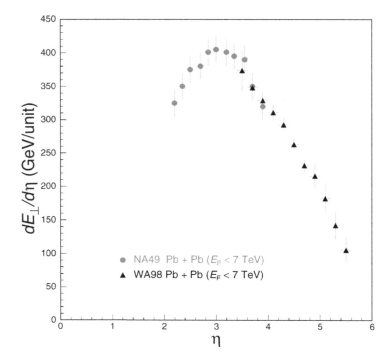

Fig. 9.13. The transverse energy distribution as a function of pseudorapidity for Pb–Pb fixed-target collisions at $158A$ GeV, with a central collision trigger. Combination of NA49 results [27] and WA98 results [191].

The three main trends observed on the SPS energy scale are

- the increase in the transverse energy with increasing mass of the colliding system,
- the increase in transverse energy with increasing energy, and
- the increase in the transverse energy with the number of participating nucleons, derived from the geometric centrality of the colliding nuclei; see section 5.2.

The measurement of transverse energy at the RHIC has produced a rather unexpected result, which we will discuss next.

9.5 RHIC results

With the first physics run at the RHIC, in 2000, a new domain of collision energy has been reached. These results were obtained at $\sqrt{s_{\mathrm{NN}}} = 130$ GeV and have produced some surprises when one compares them with SPS results. One is the discovery that the size of the fireball is barely different from that at the SPS, section 9.3, the other addresses the suppression

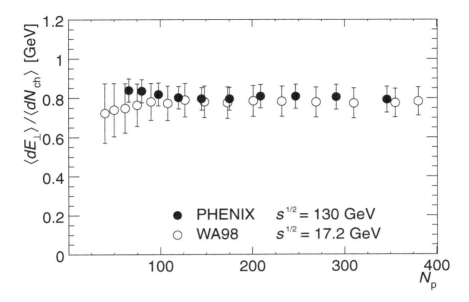

Fig. 9.14. The transverse energy per charged particle (solid dots) as a function of the number of participants at RHIC (PHENIX, 130 GeV) and at SPS (WA98, Pb–Pb fixed-target collisions at $158A$ GeV).

of hard parton production noted in section 8.4. In Fig. 9.14 we see the constancy of the transverse-energy yield per charged hadron produced.

Once the number of participants exceeds 100, there is no difference from the results we presented in Fig. 9.13 for the most central collisions, when results are expressed per participant, both PHENIX and WA98 results are shown in Fig. 9.14. This agreement between two different energy regimes is natural should the hadron-production mechanism at the RHIC and SPS be the same, as would be expected if a new state of matter were formed, hadronizing in both cases under similar conditions. The difference between the SPS and the RHIC is in the hadron-multiplicity yield, which is related to the total entropy available to hadronize.

There is more total transverse energy produced at the RHIC at central rapidity, than there is at the SPS, and this is seen on considering the pseudorapidity density of transverse energy per pair of participants, shown in Fig. 9.15. We note that the number of colliding pairs is half of all participants, i.e., in case of p–p reaction, there are two participants and one pair, and thus in this case the experimental data can be shown as measured.

We thus conclude that the extra deposition of energy per unit of rapidity at the RHIC is converted into extra hadronic particles, which explains the remarkable result we saw in Fig. 9.14. It will be most interesting to

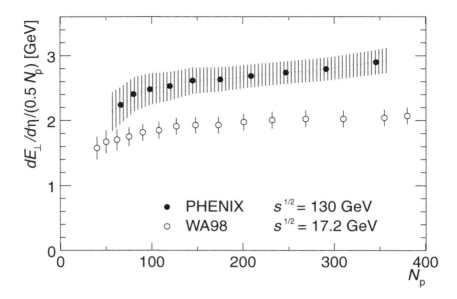

Fig. 9.15. The transverse-energy pseudorapidity density per pair of participants as a function of the number of participants, obtained at RHIC (PHENIX) and at SPS (WA98) [16].

see whether this trend continues in the near future; that is, whether, at the highest RHIC energy, $\sqrt{s_{NN}} = 200$ GeV, the transverse energy per hadron produced will remain constant and only an increase in production of hadronic particles will be observed.

We now turn to the excitation function of hadron production (dependence on $\sqrt{s_{NN}}$), and we include the first results at 200 GeV from RHIC. The central-rapidity charged-hadron yield per pair of participants is shown as a function of the collision energy $\sqrt{s_{NN}}$ (on a logarithmic scale) in Fig. 9.16. We see three experimental heavy-ion-multiplicity yields at the RHIC: Au–Au results at $\sqrt{s_{NN}} = 56$, 130, and 200 GeV (filled black data points) [50], CERN–SPS NA49 Pb–Pb results at $\sqrt{s_{NN}} = 17.2 = 2 \times 8.6$ GeV and 4.3 GeV (open circles), and the low-energy AGS results. This is compared with p–$\bar{\text{p}}$ inelastic-collision results of UA5 (CERN) and CDF (Fermilab). The interpolation line for the p–$\bar{\text{p}}$ results defines reference yields used in Fig. 9.17. The importance of the RHIC results is clear, since without these one could argue that the top-energy SPS point is in agreement with the p–$\bar{\text{p}}$ line, which, given RHIC results, we recognize to be near a crossing point of two very different types of behavior. We recall that some of these data are also shown in Fig. 9.7.

The maximum-energy result from the RHIC ($\sqrt{s_{NN}} = 200$ GeV) falls on a nearly straight line, which begins near the intercept $\sqrt{s_{NN}} = 1$ GeV,

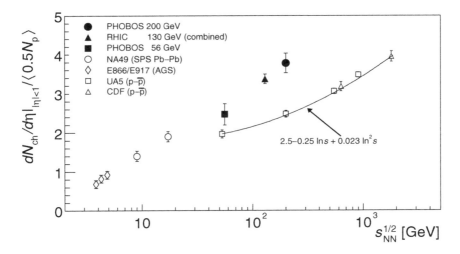

Fig. 9.16. The charged-hadron multiplicity per pair of participants at central rapidity as a function of $\sqrt{s_{NN}}$, on a logarithmic display. Shown are results for the 6% most central Au–Au collisions at the RHIC (PHOBOS) at $\sqrt{s_{NN}} = 56$, 130, and 200 GeV (filled black data), SPS Pb–Pb results (open circles), AGS results (open diamonds), along with the high-energy p–p̄ data (open squares and triangles), which are fitted to an empirical formula [50].

and follows within error all other experimental heavy-ion points. Thus, to a remarkable accuracy, the central-rapidity multiplicity in heavy-ion collisions is described by the empirical relation

$$\boxed{\frac{dN_{\text{ch}}}{d\eta} = (1.6 \pm 0.1)\frac{1}{2}N_{\text{part}} \, \log\left(\frac{\sqrt{s_{NN}}}{\text{GeV}}\right).} \tag{9.15}$$

The overall yield of particles produced is increasing faster than linearly with $\log \sqrt{s_{NN}}$. However, the decrease in stopping just compensates for the increase in rapidity density, distributing the increase in particle yield over a wider range of (pseudo)rapidity. If this simple scaling, Eq. (9.15), were to continue to the LHC energy range, the rapidity density per participant would be 'only' 6–7 per pair of participants. For the 6% most central events, corresponding to $N_{\text{part}} = 365$ in Pb–Pb interactions at the LHC, a relatively low $dN_{\text{ch}}/d\eta \lesssim 2500$ charged particles yield per unit of pseudorapidity is thus expected, based on this simple extrapolation.

We now consider the charged particle yield at central rapidity, per participant pair. The dependence on the number of participants N_{part}, shown in Fig. 9.17, (from the PHENIX collaboration) agrees with the results obtained by the PHOBOS collaborations [48, 281]. In Fig. 9.17, we also see to the left the UA5 p–p̄ ($\sqrt{s} = 130$ GeV) interpolated value. The periph-

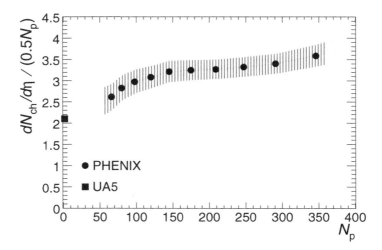

Fig. 9.17.   Production of charged hadrons at central rapidity in Au–Au collisions at $\sqrt{s_{\mathrm{NN}}} = 130A$ GeV, obtained at various collision centralities, and presented per pair of participants as a function of the number of participants (solid circles, the uncertainty is shaded) [281]. The solid square is the interpolated result from UA5 p–p̄.

eral yield in Au–Au interactions for 50–100 participants extrapolates well to this point. A slight increase in the specific yield of charged hadrons is noted for the most central collision. Overall, an increase of 50% in specific yield of hadrons per participant is observed on comparing N–N with p–p̄ reactions. This very characteristic behavior allows discrimination between models of hadron production. This is a topic in rapid evolution which we will not further pursue at this time.

The primary reason to move to the highest accessible nuclear collision energies is the desire to create a matter–antimatter-symmetric state of dense matter akin to the conditions present in the early Universe. A baryon-free QGP state should be accessible in LHC experiments. However, at $\sqrt{s} = 130A$ GeV at the RHIC, considerable matter–antimatter asymmetry is still observed. A measure of the baryon content is obtained by inspecting the central-rapidity antiproton-to-proton ratio p̄/p. In Fig. 9.18, devised by the STAR collaboration [19], to the right, we see that, in the mid-rapidity region, this ratio is appreciably different from unity. In view of the systematic behavior seen in the p–p interactions (open symbols), this is not unexpected, though there was some hope that a rapid onset of longitudinal expansion of matter could precipitate the creation of the baryon-free region at the RHIC.

The low-energy 'AGS' point is showing the production threshold, the observed small ratio is not visible on the scale of the figure. We note

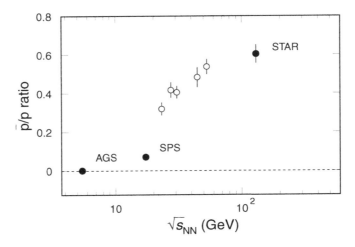

Fig. 9.18.   The mid-rapidity antiproton-to-proton ratio ($\bar{p}/p$) measured in central heavy-ion collisions (filled symbols) and p–p collisions (open symbols).

a marked increase of the antimatter-to-matter ratio on going from SPS to RHIC energies. As expected, the conversion of kinetic energy available in the interaction into hadron multiplicity, and here specifically antibaryons, is more effective in A–A reactions at RHIC energies than it is in p–p and even matter–antimatter p–$\bar{p}$ interactions, as can be seen in Fig. 9.16.

The (pseudo)rapidity shape of the charged-particle distribution, see Fig. 9.19, as measured by the PHOBOS collaboration, displays a flat top, as could be expected in the punch-through case, see Fig. 5.2 on page 74. The presence of a slight central dip could be in part due to pseudorapidity being used as a variable, see section 8.1. For the most central 3% collisions one finds a charged-hadron multiplicity of $\langle h^+ + h^- \rangle = 4100 \pm 100\,(\text{statistical}) \pm 400\,(\text{systematic})$, within the interval $|\eta| \leq 5.4$ [230]. This is nearly a 3-fold increase compared with the SPS yield (for h^- see Fig. 9.6), while the collision CM energy is 7.5-fold higher. This implies that a high fraction of the collision energy is available for production of particles at RHIC energies. This fraction is less than for the SPS due to greater transparency at higher energy.

The charged-hadron rapidity distributions, shown in Fig. 9.19, are seen to fall within the rapidity gap between projectile and target rapidities. We see again the physics motivation to desire a rapidity separation, which is available at the RHIC collider: particles produced at central rapidity cannot be confounded with contributions from fragmentation of the projectile and target.

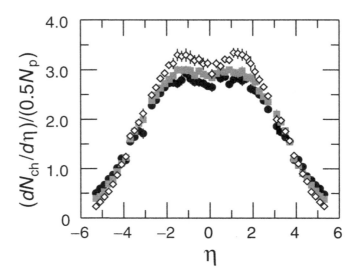

Fig. 9.19. The rapidity distribution of charged hadrons in Au–Au collisions at $\sqrt{s_{NN}} = 130A$ GeV obtained at various collision centralities implying numbers of participants $\langle N \rangle = 102$, 216, and 354 for open circles, squares, and diamonds, respectively. Phobos collaboration [230].

Turning briefly to strangeness we note that the STAR experiment reported the result

$$\frac{dN_{K^+}}{dy}\bigg|_{y=0} = 35 \pm 3.5, \quad \frac{dN_{K^-}}{dy}\bigg|_{y=0} = 30 \pm 3.$$

Allowing for strangeness in neutral kaons and hyperons

$$\frac{d\bar{s}}{dy}\bigg|_{y=0} = \frac{ds}{dy}\bigg|_{y=0} > 100.$$

This very large abundance of strangeness has to be compared with the yield of non-strange hadrons:

$$\frac{d\pi^+}{dy} \simeq \frac{d\pi^-}{dy} \simeq 235.$$

The primary number of mesons is $\simeq 175$, considering resonance cascading, see section 7.3. Strangeness is thus reaching near symmetry with light flavors. We will return to a full analysis of this interesting subject in section 19.4.

It is quite clear, given the RHIC results, that the nucleus–nucleus collisions differ substantially from elementary hadronic collisions such as p–p and p–p̄ in their hadron-production efficiency. The conditions reached at the RHIC are clearly more extreme than at SPS. The $\sqrt{s_{NN}} = 130$ GeV, RHIC central-rapidity hadron yield is about 40% higher at 7.5-fold higher CM collision energy than that seen at SPS, as can be seen in Fig. 9.16. Inspecting Eq. (7.27) and noting that τ_0, the initial thermal equilibration time, is most likely shorter at the RHIC than it is at the SPS, we conclude that the initial entropy density is at least 40%, and probably more than that, greater than that at the SPS. Along with the entropy density, given that the energy per particle seen at the RHIC is similar to results observed at CERN, see Fig. 9.14, we expect a similar enhancement of initial energy density reached during the 2000 RHIC run at $130A$ GeV, compared with CERN top-energy results.

Can we estimate more precisely the initial energy density produced at the RHIC? In order to convert Eq. (7.27) into a relation for the initial energy density, we can use $E/S = \epsilon/\sigma$, the energy per unit entropy available. For a relativistic (massless) gas with $P = \epsilon/3$ at negligible baryon density, the Gibbs–Duham relation, Eq. (10.30), implies that $\epsilon_0 = \frac{3}{4}T_0\sigma_0$. Thus, Eq. (7.27) also means that

$$\epsilon_0 = \frac{A^{1/3}}{\pi(1.2\,\text{fm})^2} \frac{T_0}{\tau_0} \frac{9}{4} \frac{dN_{\text{ch}}/(0.5A)}{dy}. \tag{9.16}$$

This slightly unusual form Eq. (9.16), as stated here, shows that the initial energy density, apart from an increase due to an enhancement in multiplicity density, also increases at RHIC compared to SPS due to two likely changes in the initial condition: a shortening of the initial thermalization time τ_0 for the more dense initial state formed at the higher RHIC energy, and also an associated increase of the initial temperature T_0 at which thermalization has occurred. Of course, it is very difficult to pin down quantitatively these two contributions to the initial increase in energy density. On the other hand, combined they could be as important as the increase in the energy density due to an increase in the final state particle multiplicity.

In order to arrive at an estimate for ϵ_0, we take $T_0 = 300$ MeV and $\tau_0 = 1$ fm, and, for $A = 350$, we use the result seen in Fig. 9.17, $dN_{\text{ch}}/(0.5A) = 3.5$. We obtain $\epsilon_0 = 16$ GeV fm^{-3}. Lattice calculations seen in Fig. 15.3 on page 300 suggest [159]: $\epsilon_0 \simeq 11T_0^4$, which yields for $\epsilon_0 = 16$ GeV fm^{-3} a temperature $T_0 = 325$ MeV. Given our ignorance of the value of τ_0 and remaining uncertainties in lattice studies of QGP equations of state, we estimate that the energy density and temperature reached at the RHIC are $\epsilon_0 \simeq 15$–20 GeV fm^{-3} and $T_0 \simeq 320$–330 MeV $\geq 2T_c$, where T_c, the critical temperature for deconfinement, is estimated to be about 160 MeV [159]; section 15.5.

The 'Bjørken energy formula' used often in such estimates arises from Eq. (9.16) by the substitution $3T_0N_{\text{ch}}/(0.5A) \rightarrow \langle E \rangle$. $\langle E \rangle$ is the mean energy per pair of participants,

$$\varepsilon_0 \simeq \frac{A^{1/3}}{\pi(1.2\,\text{fm})^2\tau_0}\frac{d\langle E_\perp\rangle}{d\eta}. \tag{9.17}$$

This expression leads to a lower energy-density estimate, since it does not account for the factor T_0/T_f implicitly present in Eq. (9.16) and omitted in Eq. (9.17).

IV

Hot hadronic matter

10 Relativistic gas

10.1 Relation of statistical and thermodynamic quantities

The first law of thermodynamics describes the change in energy dE of a system in terms of a change in volume dV and entropy dS:

$$dE(V, S) = -P\,dV + T\,dS, \tag{10.1}$$

$$T = \left(\frac{\partial E}{\partial S}\right)_V, \qquad P = -\left(\frac{\partial E}{\partial V}\right)_S. \tag{10.2}$$

The coefficients of the first law are the temperature T and the pressure P. Both can be introduced as the partial derivatives of the energy $E(V, S)$. E is a function of the extensive variables V and S, i.e., variables that increase with the size of the system. Below, we include into this consideration the baryon number, see Eq. (10.12), which is also an extensive variable.

The free energy,

$$F(V, T) \equiv E - TS, \tag{10.3}$$

is the quantity in which, as indicated, the dependence on the entropy is replaced by the dependence on temperature, an intensive variable that does not change with the size of the system. Namely,

$$dF(V, T) = dE - T\,dS - S\,dT = -P\,dV - S\,dT, \tag{10.4}$$

and, as a consequence of the transformation Eq. (10.3), we obtain in analogy to Eq. (10.2)

$$S = -\left(\frac{\partial F}{\partial T}\right)_V, \qquad P = -\left(\frac{\partial F}{\partial V}\right)_T. \tag{10.5}$$

For an extensive system with $F \propto V$, a very useful relation for the entropy density σ follows from Eq. (10.5):

$$\sigma \equiv \frac{S}{V} = \frac{\partial P}{\partial T}. \tag{10.6}$$

Since the free energy F depends on $T = 1/\beta$, we relate it to the statistical functions introduced in chapter 4. To establish the connection in a more quantitative manner, we obtain the energy E in terms of the free energy F. We substitute S into Eq. (10.4), in order to arrive at a relation for $E(V, T)$:[*]

$$E = F(V, T) + TS(V, T),$$
$$= F(V, T) - T \frac{\partial}{\partial T} F(V, T) \equiv -\frac{\partial}{\partial(1/T)}\left(\frac{-F}{T}\right). \tag{10.7}$$

This relation has the same form as Eq. (4.13), and we deduce that

$$F(V, T) = -\beta^{-1}[\ln Z(V, \beta) + f]. \tag{10.8}$$

The integration constant f could be a function of V. However, by definition of P, Eq. (10.5), and using Eq. (10.8),

$$P = \frac{\partial}{\partial V}\left(\beta^{-1} \ln Z\right) + T \frac{\partial f}{\partial V}. \tag{10.9}$$

In the definition of the partition function Eq. (4.14), the individual energies E_i depend on the volume V. We obtain

$$P = \frac{-\sum_i (\partial E_i/\partial V)e^{-\beta E_i}}{\sum_i e^{-\beta E_i}} + T \frac{\partial f}{\partial V} = -\left\langle \frac{\partial E_i}{\partial V} \right\rangle + T \frac{\partial f}{\partial V}. \tag{10.10}$$

Only if $\partial f/\partial V$ vanishes does the proper relation between work and pressure arise: the work done on the system, when the volume V is decreased by dV (dV is negative), is equivalent to the mean value of the change of all the energy levels brought about by a change of the volume. A constant f is a physically irrelevant ambiguity in the relationship Eq. (10.8) between the free energy F and the canonical partition function Z, and can be discarded. Thus, Eqs. (10.8) and (10.9) read

$$F(V, T) = -\beta^{-1} \ln Z(V, \beta), \quad \beta P = \frac{\partial \ln Z}{\partial V}, \quad \beta^{-1} = T. \tag{10.11}$$

This well-known equation establishes the bridge between the thermal (T, F, and P) and statistical (β and $\ln Z$) quantities. The volume V is present in both formulations, but, in fact, since $\ln Z$ and F are extensive in V for infinite volumes, V in general disappears from many further considerations.

---

[*] Clearly $E(V, S)$ is not the same function as $E(V, T)$, which is indicated by stating the variables on which E depends, rather than introducing a new symbol.

We now allow for the presence of a conserved (baryon) number b, i.e., we consider $E(V, S, b)$. This necessitates the introduction of the (baryo)chemical potential μ. μ is the incremental energy cost required to change the baryon number at fixed pressure and entropy:

$$dE(V, S, b) = -P\, dV + T\, dS + \mu\, db, \tag{10.12}$$

$$P = -\left(\frac{\partial E}{\partial V}\right)_{S,b}, \quad T = \left(\frac{\partial E}{\partial S}\right)_{V,b}, \quad \mu = \left(\frac{\partial E}{\partial b}\right)_{V,S}. \tag{10.13}$$

The coefficients P, T, and μ are, as before, fixed in terms of the partial differentials of $E(V, S, b)$ with respect to its three variables.

The familiar generalization of the free energy Eq. (10.4), often called the thermodynamic potential, $\mathcal{F}(V, T, \mu)$, is defined by the transformation

$$\mathcal{F}(V, T, \mu) \equiv E(V, S, b) - ST - \mu b. \tag{10.14}$$

On evaluating the differentials as in Eq. (10.4), we indeed see that, as suggested on the left-hand side of Eq. (10.14), the thermodynamic potential is a function of V, T, and μ,

$$d\mathcal{F} = -P\, dV - S\, dT - b\, d\mu, \tag{10.15}$$

where

$$P = -\left(\frac{\partial \mathcal{F}}{\partial V}\right)_{T,\mu}, \quad S = -\left(\frac{\partial \mathcal{F}}{\partial T}\right)_{V,\mu}, \quad b = -\left(\frac{\partial \mathcal{F}}{\partial \mu}\right)_{V,T}. \tag{10.16}$$

A series of arguments that has allowed us to establish Eq. (10.11) fixes a relation between $\mathcal{F}(V, T, \mu)$ and the grand partition function $\mathcal{Z}(V, T, \lambda)$:

$$\boxed{\mathcal{F}(V, T, \mu) = -\beta^{-1} \ln \mathcal{Z}(V, \beta, \lambda), \quad \beta = 1/T, \quad \lambda = e^{\mu/T}.} \tag{10.17}$$

The thermodynamic pair of variables (T, μ) is often used for describing the properties of $\mathcal{F}$, instead of the grand-canonical statistical quantities (β and λ). To do this, it is quite important that appropriate attention be paid to the simple relation

$$\mu = \mu(\lambda, \beta) = \beta^{-1} \ln \lambda. \tag{10.18}$$

Consider, for example, the expression for the energy. With

$$\mathcal{Z}(V, \beta, \lambda) = \tilde{\mathcal{Z}}(V, T, \mu), \tag{10.19}$$

we obtain

$$E = -\frac{d}{d\beta} \ln \mathcal{Z}(V, \beta, \lambda), \tag{10.20}$$

$$= T^2 \frac{d}{dT} \ln \tilde{\mathcal{Z}}(V, T, \mu) + \frac{d}{d\mu} \ln \tilde{\mathcal{Z}}(V, T, \mu) \left. \frac{d\mu(\lambda, \beta)}{d\beta} \right|_{\beta=T^{-1}}. \tag{10.21}$$

The second form is clearly much different from the simple statistical relation Eq. (10.20). However, given Eq. (10.18), we have

$$\left.\frac{d\mu(\lambda, \beta)}{d\beta}\right|_{\beta=T^{-1}} = -T\mu, \tag{10.22}$$

and thus the last term in Eq. (10.21) is $\mu \, d\mathcal{F}/d\mu$. We hence obtain a form of the the important Gibbs–Duham relation, see Eq. (10.26) below,

$$\boxed{E(V,T,\mu) = \mathcal{F}(V,T,\mu) + TS(V,T,\mu) + \mu b(V,T,\mu),} \tag{10.23}$$

where the baryon number and entropy are

$$\mathcal{F}(V,T,\mu) = -P(T,\mu)V, \tag{10.24a}$$

$$b = -\frac{d}{d\mu}\mathcal{F}(V,T,\mu) = \lambda \frac{d}{d\lambda}\ln \mathcal{Z}(V,\beta,\lambda), \tag{10.24b}$$

$$S = -\frac{d}{dT}\mathcal{F}(V,T,\mu) = \frac{d}{dT}T\ln \tilde{\mathcal{Z}}(V,T,\mu). \tag{10.24c}$$

The expression for the entropy, Eq. (10.24c), takes a much more complex form in terms of statistical variables. Namely, Eq. (10.23) implies that

$$S = \frac{1}{T}(E - \mathcal{F} - \mu b) = \ln \mathcal{Z} - \beta \frac{\partial \ln \mathcal{Z}}{\partial \beta} - (\ln \lambda)\lambda \frac{\partial \ln \mathcal{Z}}{\partial \lambda}. \tag{10.25}$$

In an extensive system, we can greatly simplify Eq. (10.23). We replace $\mathcal{F}$ by $-PV$, Eq. (10.16), and obtain the usual form of the Gibbs–Duham relation:

$$\boxed{P = T\sigma + \mu\nu - \epsilon, \qquad \sigma = \frac{S}{V}, \quad \nu = \frac{b}{V}, \quad \epsilon = \frac{E}{V}.} \tag{10.26}$$

For completeness of the discussion, we mention now two more quantities, the enthalpy $H(P,S,b)$ and the Gibbs free energy $G(P,T,b)$. To obtain these two quantities with a new mix of variables, we continue the process of replacement of variables. We recall that, at first, we moved from $E(V,S) \rightarrow F(V,T)$ and subsequently from $E(V,S,b) \rightarrow \mathcal{F}(V,T,\mu)$, i.e., we replaced the extensive variables by the intensive variables. The one extensive variable left is the volume itself. Since we address in this book an isolated system (a fireball) that can expand its volume with entropy and baryon number remaining nearly constant, elimination of V in favor of P would seem a logical step and indeed the statistical (partition-function) analog to $G(P,T,b)$ is the recently proposed generalization of the (grand) canonical partition function to the pressure partition function $\Pi(P,\beta,b)$ [142]. We will not pursue this interesting subject further in this book.

However, we note that, for a given pressure, entropy, and baryon number, it should be convenient to introduce the enthalpy,

$$\boxed{H(P, S, b) \equiv E(V, S, b) + PV.}$$ (10.27)

The volume occupied by the system is obtained as the change of H with respect to pressure at fixed S and b:

$$dH = V\, dP + T\, dS + \mu\, db.$$ (10.28)

In thermal physics, a quantity often considered in the study of a freely expanding isolated system is the specific enthalpy per particle, the so-called heat function h,

$$\frac{H}{b} \equiv h = \frac{\epsilon}{\nu} + \frac{P}{\nu}.$$ (10.29)

The Gibbs free energy G is introduced to facilitate consideration of evolution, not at a constant entropy, but in a 'heat bath', i.e., at a given temperature, though at a fixed baryon number (not in a 'baryon bath'):

$$\boxed{G(P, T, b) \equiv E + PV - TS} = \mu(P, T, b)\, b.$$ (10.30)

Both the enthalpy $H(P, S, b)$ and the Gibbs free energy $G(P, T, b)$ have not yet been used much in the study of heavy-ion-fireball dynamics.

10.2 Statistical ensembles and fireballs of hadronic matter

We extend the discussion of physical ensembles introduced in chapter 4. The concept of an ensemble consisting of weakly coupled physical systems, $\mathcal{M} = \{M_i,\ i = 1, \ldots, N\}$, was introduced by Gibbs and Boltzmann. It helped to establish a conceptual foundation of statistical physics. A large number, $N \to \infty$, of such systems is normally considered. The otherwise negligible interactions between individual systems M_i are such that both energy and (conserved) quantum numbers (such as, e.g., the baryon number) can be exchanged between the systems. This establishes a 'bath' of energy and baryon number, in which each individual system is immersed, and with which it can equilibrate its properties.

When we examine the microscopic properties in the ensemble, such as energies of individual members, we speak of a *micro-canonical* ensemble. Furthermore, we distinguish between the *canonical* ensemble and the *grand-canonical* ensemble: in both cases, we have adopted a statistical distribution in energy. However, in the canonical ensemble, we still treat discrete quantum numbers (particle number, baryon number, etc.) microscopically, whereas in the grand-canonical ensemble, we have also adapted the statistical-ensemble distribution for the discrete properties

such as baryon number.   Of course, when we have more than one conserved discrete quantum number, any of these properties can be treated
in 'canonical' or 'grand-canonical' way.

A colloquial way to explain the difference between the ensembles is
to say that, in the micro-canonical ensemble, we consider each individual
system M_i as being decoupled from the others. In the canonical ensemble,
we allow only for coupling of energy, and in the grand-canonical ensemble,
we allow for exchange of energy and quantum number (particle number).
Practically, we maintain the picture of an ensemble of many weakly coupled systems M in place, but, for describing physical properties within
micro-canonical, canonical, or grand-canonical ensembles, we use different
physical variables.

Said differently, in a theoretical description, it is our choice how we
characterize the properties of the system as long as there is a precise
mathematical transformation we can use to make a transition between
the different descriptions of the same physical situation, and the description of choice is what is most convenient. There is an exception to this
'convenience principle': the color-confining nature of strong interactions
imposes color neutrality on all 'drops' of QGP we consider, thus, in principle, we may not use a color-grand-canonical ensemble, and, if it is used,
the question to consider is that of whether results obtained in this way
make good physical sense.

What is, in our context, the individual ensemble element M_i? Can
we view it as a single hadron, or do we have to take the entire drop of
highly excited hadronic matter formed in the nuclear collision as being the
element in the Gibbs ensemble? In chapter 4, we have wondered if a single
particle can be seen as the element of the ensemble. This is motivated
by the fact that, in our physical environment, the number of particles is
not fixed, and their variable number is an expression of the sharing of the
total energy and baryon number (or other conserved number). In support
of this point of view, we will next show in section 10.3 that, allowing
for a change in numbers of particles, the state of maximum entropy at
fixed energy and baryon number is the conventional statistical-equilibrium
distribution.

We know that there are physical processes of particle production that
allow conversion of energy into particles, such that their yields reach
(chemical) equilibrium. The well-known Boltzmann collision dynamics
assures that the momentum distributions are equilibrated in (binary) collisions. Microscopic processes of particle production and interaction can
establish, in a particle ensemble M_i, a distribution that is normally associated with ensemble elements consisting of larger drops of matter.

Each high-energy heavy-ion reaction forms a many-body system, a fireball, which evolves into a final state with thousands of particles. The

study of the average rather than individual microscopic properties of such a large system makes sense, if the distribution of individual properties of the sub-components has a 'peaked' shape. What this means is that most individual objects considered should be found near to their common average, just a few may be far from it. On intuitive grounds, it seems that otherwise many-body systems equilibrate exceedingly slowly, if at all. Since there are many examples of dynamic systems that do not satisfy this criterion, we see that in general there is *a priori* no guarantee that strongly interacting confined matter will ever equilibrate. However, experimental results suggest that strongly interacting particle systems of practically any size approach statistical equilibrium very rapidly. Why this is the case remains today an open issue, see section 5.6.

In the following, we develop further the physical properties of ideal relativistic gases introduced in section 4.4. We assemble useful formulas, including in the discussion chemically nonequilibrated gases, which have been treated only sparingly before.

10.3 The ideal gas revisited

The additivity of different gas fractions (i.e., flavors) 'f' originates from the additive property of the logarithm of the partition functions Eq. (4.19):

$$\mathcal{Z} = \prod_f \mathcal{Z}_f, \qquad \ln \mathcal{Z} = \sum_f \ln \mathcal{Z}_f. \tag{10.31}$$

For an ideal Fermi gas, such as a quark gas in the deconfined phase, we have for each flavor, as seen in Eqs. (4.38) and (4.39),

$$\ln \mathcal{Z}_F = g_F V \int \frac{d^3 p}{(2\pi)^3} [\ln(1 + \gamma \lambda e^{-\beta \varepsilon}) + \ln(1 + \gamma \lambda^{-1} e^{-\beta \varepsilon})], \tag{10.32}$$

where the degeneracy factor is, e.g., $g_F = g_s g_c$ and comprises $g_s = 2$ for spin-$\frac{1}{2}$ degeneracy and $g_c = 3$ for color. For bosons with degeneracy g_B, in principle, we must allow for the possibility of condensation (macroscopic occupancy) in the lowest energy state ε_0:

$$\ln \mathcal{Z}_B = -g_B V \int \frac{d^3 p}{(2\pi)^3} [\ln(1 - \gamma \lambda e^{-\beta \varepsilon}) + \ln(1 - \gamma \lambda^{-1} e^{-\beta \varepsilon})]$$
$$- g_B [\ln(1 - \gamma \lambda e^{-\beta \varepsilon_0}) + \ln(1 - \gamma \lambda^{-1} e^{-\beta \varepsilon_0})]. \tag{10.33}$$

We will not address further in this book the condensation phenomena and will not pursue further the last term in Eq. (10.33).

Differentiating with respect to the energy of the particle, see Eq. (4.41), we obtain the single-particle distribution functions. For the fermions and antifermions seen in Eq. (4.42), respectively, we have

$$f_F(\varepsilon, \mu) = \frac{1}{\gamma^{-1}e^{\beta(\varepsilon-\mu)} + 1}, \tag{10.34a}$$

$$\bar{f}_F(\varepsilon, \mu) = \frac{1}{\gamma^{-1}e^{\beta(\varepsilon+\mu)} + 1}, \tag{10.34b}$$

and similarly for bosons and antibosons:

$$f_B(\varepsilon, \mu) = \frac{1}{\gamma^{-1}e^{\beta(\varepsilon-\mu)} - 1}, \tag{10.35a}$$

$$\bar{f}_B(\varepsilon, \mu) = \frac{1}{\gamma^{-1}e^{\beta(\varepsilon+\mu)} - 1}. \tag{10.35b}$$

We will also use the short-hand notation

$$f_{F,B}^{\pm} = f_{F,B} \pm \bar{f}_{F,B}, \tag{10.36}$$

since these combinations occur in evaluations of statistical properties of gases.

The particle densities are

$$\rho_F \equiv \frac{N_F}{V} = \frac{1}{V}\lambda\frac{d}{d\lambda}\ln \mathcal{Z}_F = g_F\int\frac{d^3p}{(2\pi)^3}f_F^-, \tag{10.37a}$$

$$\rho_B \equiv \frac{N_B}{V} = \frac{1}{V}\lambda\frac{d}{d\lambda}\ln \mathcal{Z}_B = g_B\int\frac{d^3p}{(2\pi)^3}f_B^-. \tag{10.37b}$$

These distributions determine the local equilibrium particle densities, for example, the local density of quarks and antiquarks given by the integral of the Fermi distribution, Eqs. (10.34a) and (10.34b):

$$n_q = \int\frac{d^3p}{(2\pi)^3}\frac{1}{1 + \gamma_i^{-1}\lambda_i^{-1}e^{\varepsilon(p)/T}} \rightarrow \gamma_i\lambda_i\int\frac{d^3p}{(2\pi)^3}e^{\varepsilon(p)/T}, \tag{10.38a}$$

$$n_{\bar{q}} = \int\frac{d^3p}{(2\pi)^3}\frac{1}{1 + \gamma_i^{-1}\lambda_i e^{\varepsilon(p)/T}} \rightarrow \gamma_i\lambda_i^{-1}\int\frac{d^3p}{(2\pi)^3}e^{\varepsilon(p)/T}. \tag{10.38b}$$

The Boltzmann limit, which is applicable when the phase-space cells have small overall occupancy, is also indicated in Eqs. (10.38a) and (10.38b). In this limit, the chemical-abundance factors enter as coefficients of the distributions. We note that, while the chemical potential enhances the abundance of particles, it suppresses the abundance of antiparticles.

In order to obtain other statistical properties, such as, e.g., the energy content of the system, one can also apply the rule that the occupation functions Eqs. (10.34a)–(10.35b) and Eq. (4.46) can be folded with the quantity of interest. To obtain the energy density, we fold with the single-particle energy ε the sum of particle and antiparticle spectra. The correctness of this prescription is seen on evaluating the derivative of $\ln \mathcal{Z}$ with respect to β, Eq. (10.20):

$$\epsilon_B = g_B \int \frac{d^3p}{(2\pi)^3} \varepsilon f_B^+, \tag{10.39a}$$

$$\epsilon_F = g_F \int \frac{d^3p}{(2\pi)^3} \varepsilon f_F^+, \tag{10.39b}$$

$$\epsilon_g = g_g \int \frac{d^3p}{(2\pi)^3} \varepsilon f_g. \tag{10.39c}$$

The gluon distribution f_g is seen in Eq. (4.46). The total energy density is the sum of all contributing terms:

$$\epsilon = \sum_i \epsilon_i. \tag{10.40}$$

10.4 The relativistic phase-space integral

To evaluate the properties of ideal relativistic gases, we need to evaluate the relativistic momentum integral, which appears in all phase-space integrals in a similar form. To do this we consider the definition of the Bessel function K_ν,

$$K_\nu(z) = \frac{\sqrt{\pi}(z/2)^\nu}{\Gamma(\nu + \frac{1}{2})} \int_1^\infty e^{-zt}(t^2 - 1)^{\nu - \frac{1}{2}} \, dt, \quad \Re\nu > -\frac{1}{2}, \tag{10.41}$$

valid for $|\arg z| < \pi/2$. We used before the case $\nu = 1$, Eq. (8.7), which arises on substituting in Eq. (10.41) $t \to \cosh t$. The connection to the class of integrals which we now require is obtained by recognizing that $z = \beta m$ and substituting into Eq. (10.41):

$$t \to \sqrt{p^2 + m^2}/m. \tag{10.42}$$

With $\varepsilon = \sqrt{p^2 + m^2}$, we obtain

$$K_\nu(\beta m) = \frac{\sqrt{\pi}}{\Gamma(\nu + \frac{1}{2})} \left(\frac{\beta}{2m}\right)^\nu \int_0^\infty \frac{p^{2\nu}}{\varepsilon} e^{-\beta\varepsilon} \, dp. \tag{10.43}$$

On integrating by parts with the relation

$$\frac{\partial}{\partial p} e^{-\beta\varepsilon} = -\beta \frac{p}{\varepsilon} e^{-\beta\varepsilon},$$

we obtain $(\nu > \frac{1}{2})$

$$K_\nu(\beta m) = \frac{\sqrt{\pi}}{\Gamma(\nu - 1/2)} \frac{1}{m} \left(\frac{\beta}{2m}\right)^{\nu-1} \int_0^\infty p^{2\nu-2} e^{-\beta\varepsilon} \, dp. \tag{10.44}$$

We recall that

$$\Gamma(\tfrac{1}{2}) = \sqrt{\pi}; \quad \Gamma(\tfrac{3}{2}) = \sqrt{\pi}/2; \quad \Gamma(\tfrac{5}{2}) = \tfrac{3}{2}\Gamma(\tfrac{3}{2}); \quad \ldots.$$

Two interesting limits arise from the well known series expansion of the Bessel function Eq. (10.41).

- The non-relativistic limit in which we use p/m as the small parameter:

$$K_\nu(z) \to \sqrt{\frac{\pi}{2z}} e^{-z} \left(1 + \frac{4\nu^2 - 1}{8z} + \frac{(4\nu^2 - 1)(4\nu^2 - 9)}{2!(8z)^2}\right.$$

$$\left. + \frac{(4\nu^2 - 1)(4\nu^2 - 9)(4\nu^2 - 25)}{3!(8z)^3} + \cdots\right). \tag{10.45}$$

We note that this expansion is rather slowly convergent. The special case of interest to us is

$$\frac{K_1(z)}{K_2(z)} = 1 - \frac{3}{2}\frac{1}{z} + \frac{15}{8}\frac{1}{z^2} - \frac{15}{8}\frac{1}{z^3} + \frac{135}{128}\frac{1}{z^4} + \mathcal{O}(z^{-5}). \tag{10.46}$$

- The relativistic limit, in which the mass is negligible relative to the typical energies, and thus effectively $m \simeq 0$. For the relevant two cases, we have, for $z \to 0$,

$$K_1(z) = \frac{1}{z} + \left[\ln\left(\frac{z}{2}\right) + \gamma_E\right]\frac{z}{2} + \left[\ln\left(\frac{z}{2}\right) + \gamma_E - \frac{5}{4}\right]\frac{z^3}{16} + \cdots,$$

$$K_2(z) = \frac{2}{z^2} - \frac{1}{2} - \left[\ln\left(\frac{z}{2}\right) + \gamma_E - \frac{3}{4}\right]\frac{z^2}{8}$$

$$- \left[\ln\left(\frac{z}{2}\right) + \gamma_E - \frac{17}{12}\right]\frac{z^4}{96} + \cdots,$$

leading to

$$\frac{K_1(z)}{K_2(z)} = \frac{z}{2} + \left[\ln\left(\frac{z}{2}\right) + \gamma_E\right]\frac{z^3}{4} + \cdots. \tag{10.47}$$

We recall that

$$\gamma_E = \lim_{n\to\infty} \sum_{k=1}^n \frac{1}{k} - \ln n = 0.577\,215\,664\,9 \ldots$$

is the Euler constant.

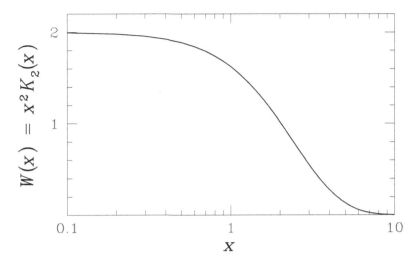

Fig. 10.1. The relativistic distribution function $W(x) = x^2 K_2$.

With this preparation, we are now in a position to study properties of relativistic gases. Let us first look at the classical limit, the first, 'classical' term in the expansion of the partition function Eq. (4.51). For $n = 1$, we obtain, from Eq. (4.53),

$$\ln \mathcal{Z}_{\text{cl}} = Z^{(1)} = \sum_{\text{f}} \gamma_{\text{f}}(\lambda_{\text{f}} + \lambda_{\text{f}}^{-1}) Z_{\text{f}}^{(1)}, \tag{10.48}$$

with

$$Z_{\text{f}}^{(1)} = g_{\text{f}} V \int \frac{d^3 p}{(2\pi)^3} e^{-\beta \varepsilon(\boldsymbol{p})} \equiv g_{\text{f}} \frac{\beta^{-3} V}{2\pi^2} W(\beta m_{\text{f}}). \tag{10.49}$$

The sum, in Eq. (10.48), includes all Fermi and Bose particles.

We encounter in Eq. (10.49) the function Eq. (10.44), with $\nu = 2$,

$$W(\beta m) \equiv \beta^3 \int e^{-\beta \varepsilon} p^2 \, dp = (\beta m)^2 K_2(\beta m), \tag{10.50a}$$

$$\rightarrow 2, \quad \text{for } m \rightarrow 0, \tag{10.50b}$$

$$\rightarrow \sqrt{\frac{\pi m^3}{2T^3}} e^{-m/T}, \quad \text{for } m \gg T, \tag{10.50c}$$

where, in the last limit, we exploited the large-argument limit shown in Eq. (10.45). As can be seen from Fig. 10.1, the change between the two asymptotic limits occurs near $\beta m = 1$.

In the classical (Boltzmann) limit indicated by superscript 'cl' the number of particles of each species, Eq. (4.56), is given by

$$N|^{\mathrm{cl}} = \lambda \frac{\partial}{\partial \lambda} \ln \mathcal{Z}_{\mathrm{cl}} = \lambda Z^{(1)} = g\lambda \frac{\beta^{-3} V}{2\pi^2} W(\beta m),$$ (10.51)

where we have combined the factors $\gamma^{\pm 1}\lambda \to \lambda$, since, when only one particle species is considered, one fugacity suffices. The most useful and often quoted property of a relativistic gas is the average energy per particle:

$$\left.\frac{E}{N}\right|^{\mathrm{cl}} = \frac{-(\partial/\partial\beta)\ln \mathcal{Z}_{\mathrm{cl}}}{\lambda(\partial/\partial\lambda)\partial\ln \mathcal{Z}_{\mathrm{cl}}} = 3T + m\frac{K_1(\beta m)}{K_2(\beta m)}.$$ (10.52)

The fugacity coefficients λ for the particles cancel out. To obtain Eq. (10.52), we exploited the property of the function $W = x^2 K_2$,

$$\frac{d}{dx}W(x) = -x^2 K_1(x),$$ (10.53)

arising from the recursion relation of the K-functions,

$$\frac{d}{dx}K_\nu(x) = -K_{\nu-1}(x) - \frac{\nu}{x}K_\nu(x),$$ (10.54a)

written in the form

$$\frac{d}{dx}(x^\nu K_\nu(x)) = -x^\nu K_{\nu-1}(x).$$ (10.54b)

• In the relativistic limit $\beta m \to 0$, we can use Eq. (10.47) and obtain

$$\left.\frac{E}{N}\right|^{\mathrm{cl}}_{m=0} = 3T.$$ (10.55)

Equation (10.55) can be improved; see Eq. (10.68). However, further refinement in the limit $m/T \to 0$ requires that quantum statistics be considered.

• In the non-relativistic limit $\beta m \gg 1$, the ratio appearing in Eq. (10.52) is as given in Eq. (10.46), and we obtain

$$\left.\frac{E}{N}\right|^{\mathrm{cl}}_{\mathrm{nr}} = m + \frac{3}{2}T\left(1 + \frac{5}{4}\frac{T}{m} - \frac{5}{4}\frac{T^2}{m^2} + \frac{45}{64}\frac{T^3}{m^3}\cdots\right), \quad \frac{m}{T} > 1.$$ (10.56)

Note that, to obtain the correct first $3T/2$ term in the non-relativistic limit, the next-to-leading term in Eq. (10.46) needs to be considered. The slow convergence of the series, Eq. (10.56), is also to be remembered, i.e., the non-relativistic limit requires a truly a non-relativistic $m \gg T$ condition. For $m \simeq T$, the relativistic limit offers a better approximation.

We continue with a more thorough discussion of the energy per baryon in the next subsection, addressing there, in particular, the differences arising for bosons and fermions.

10.5 Quark and gluon quantum gases

In the deconfined QGP phase, we have to consider the quantum nature of the effectively massless, relativistic quark and gluon gases. Many of the results for quantum gases that we require arise in terms of a series expansion of which the Boltzmann approximation is the first term. We are, in particular, interested in the properties of the equation of state, i.e., the relation between the energy density ϵ, Eq. (10.40), and the pressure P, Eq. (10.11).

Integrating by parts Eqs. (10.32) and (10.33),

$$\pm \int \frac{d^3 p}{(2\pi)^3} \ln(1 \pm \gamma \lambda e^{-\beta \varepsilon}) = \frac{\beta}{3} \int \frac{d^3 p}{(2\pi)^3} |\vec{p}| \frac{\partial \varepsilon}{\partial |\vec{p}|} f_{F,B} \qquad (10.57)$$

$$= \frac{\beta}{3} \int \frac{d^3 p}{(2\pi)^3} \frac{\vec{p}^2}{\varepsilon} f_{F,B},$$

where the factor $\frac{1}{3}$ arises from the $p^2\, dp$ momentum integral. We have used the (relativistic-dispersion) relation Eq. (4.31) in the last equality. We obtain for the pressure Eq. (10.11), noting Eqs. (10.32), (10.33), and (10.36), and using Eq. (4.31) to eliminate the momentum,

$$3P = g_F \int \frac{d^3 p}{(2\pi)^3} \left(\varepsilon - \frac{m^2}{\varepsilon} \right) f_F^+ + g_B \int \frac{d^3 p}{(2\pi)^3} \left(\varepsilon - \frac{m^2}{\varepsilon} \right) f_B^+ \leq \epsilon. \qquad (10.58)$$

Since the particle-occupation probabilities $f_{B,F}^+$ are always positive, the terms proportional to m^2 in Eq. (10.58) always reduce the pressure. For this reason the maximum absolute value of the ideal-gas pressure, for given thermal parameters, is subject to the relativistic bound

$$\boxed{\epsilon - 3P = g_F m \int \frac{d^3 p}{(2\pi)^3} \left(\frac{m}{\varepsilon} \right) f_F^+ + g_B m \int \frac{d^3 p}{(2\pi)^3} \left(\frac{m}{\varepsilon} \right) f_B^+ \geq 0.} \qquad (10.59)$$

The right-hand side of Eq. (10.59) is cast into the form which is natural considering the trace of the energy–momentum tensor of quantum fields. It can be evaluated using Eq. (10.43), when the series expansion of quantum distributions exists. The leading (Boltzmann) term is

$$\epsilon - 3P = \frac{g T^4}{2\pi^2} x^3 K_1(x), \quad x = m/T. \qquad (10.60)$$

For high temperatures relative to (vanishingly small) mass, we find the relativistic equation of state,

$$\boxed{3P \to \epsilon, \qquad \text{for} \quad \beta m \to 0,} \qquad (10.61)$$

corresponding to the maximum mobility of particles. Massive particles move slowly relative to the velocity of light and are far away from this limit. In fact, the pressure in the normal world around us is vanishingly small, seen on the scale of energy density comprising the rest mass. This is expressed in the power of 30 arising in Eq. (1.2), which separates the pressure on Earth from that in the QGP.

We turn now to consider the energy per particle in relativistic quantum gases in more detail. We employ the series expansion appearing in Eq. (4.51), and obtain, by expanding Eqs. (10.32) and (10.33) for each particle species,

$$\ln \mathcal{Z} = \frac{\beta^{-3}V}{2\pi^2} \sum_{n=1}^{\infty} g_n \frac{\lambda^n}{n^4} (n\beta m)^2 K_2(n\beta m). \qquad (10.62)$$

We have combined the factor $(-)^{n+1}$ for fermions with the degeneracy g to form the factor g_n. This expansion Eq. (10.62) can not be used if the condition $m - \mu < 0$ arises. This happens, in particular, for massless quarks at finite baryon density. We will be able to deal with this interesting case exactly for $m \to 0$; see Eq. (10.74). In the HG phase, for a very narrow parameter range, allowing in particular condensation of kaons, the expression Eq. (10.62) is also not valid. Apart from these exceptions, Eq. (10.62) can be used as the basis for the evaluation of the properties of hot, strongly interacting matter.

Using the series expansion Eq. (10.62), the quantum generalization of the classical particle number, Eq. (10.51), is

$$\boxed{N = \lambda \frac{\partial}{\partial \lambda} \ln \mathcal{Z} = \frac{\beta^{-3}V}{2\pi^2} \sum_{n=1}^{\infty} g_n \frac{\lambda^n}{n^3} (n\beta m)^2 K_2(n\beta m).} \qquad (10.63)$$

As noted, the masses and fugacities are such that $m - \mu > 0$, so that the series expansion exists. The relativistic limit $m \to 0$ is now

$$\left. \frac{N}{V} \right|_{m=0}^{\mathrm{B}} = \frac{gT^3}{\pi^2} \zeta(3), \qquad (10.64)$$

$$\left. \frac{N}{V} \right|_{m=0}^{\mathrm{F}} = \frac{gT^3}{\pi^2} \eta(3). \qquad (10.65)$$

We have introduced the Riemann zeta function

$$\zeta(k) = \sum_{n=1}^{\infty} \frac{1}{n^k}. \qquad (10.66\mathrm{a})$$

We note that

$$\zeta(2) = \frac{\pi^2}{6}, \qquad \zeta(3) \simeq 1.202, \qquad \zeta(4) = \frac{\pi^4}{90}. \qquad (10.66\mathrm{b})$$

For a Fermi occupation function, the signs of the terms in the sums in Eq. (10.63) are alternating, which leads to the eta function

$$\eta(k) = \sum_{n=1}^{\infty} (-1)^{n-1} \frac{1}{n^k} = (1 - 2^{1-k})\zeta(k), \tag{10.67a}$$

and thus

$$\eta(3) = \frac{3}{4}\zeta(3) = 0.901\,5, \quad \eta(4) = \frac{7}{8}\zeta(4) = \frac{7}{720}\pi^4. \tag{10.67b}$$

The generalization of the energy per particle, Eq. (10.52), to quantum statistics yields

$$\frac{E}{N} = 3T \frac{\displaystyle\sum_{n=1}^{\infty} g_n \frac{\lambda^n}{n^4}\left((nx)^2 K_2(nx) + \frac{1}{3}(nx)^3 K_1(nx)\right)}{\displaystyle\sum_{n=1}^{\infty} g_n \frac{\lambda^n}{n^3}(nx)^2 K_2(nx)}, \tag{10.68}$$

where $x = m\beta$. For the non-relativistic limit, the Boltzmann approximation Eq. (10.56) is quite appropriate, resulting in Eq. (10.56). The ultra-relativistic limit with $m \to 0$ yields

$$\left.\frac{E}{N}\right|_{m=0}^{\text{B}} = 3T\frac{\zeta(4)}{\zeta(3)} = 2.70\,T, \tag{10.69}$$

$$\left.\frac{E}{N}\right|_{m=0}^{\text{F}} = 3T\frac{\eta(4)}{\eta(3)} = 3.15\,T. \tag{10.70}$$

The factor $\frac{7}{8}$ seen in Eq. (10.67b) enters Eq. (10.70) and is the source of the reduction of the number of fermionic degrees of freedom in a Fermi gas compared with that in a Bose gas. However, we have to allow for the presence both of quarks and of antiquarks; thus the radiation term in the quark gas is actually $\frac{7}{4}$ times as large as that in the gluon gas, apart from the other statistical flavor and color multiplication factors.

In Fig. 10.2, we show the variation of the energy per particle in units of m, as a function of T/m. The solid line depicts the Boltzmann limit; the long-dashed line, fermions with $\lambda = 1$; and the short-dashed line, bosons. Asymptotic conditions are indicated by dotted lines corresponding to the limits Eqs. (10.55), (10.69), and (10.70). For finite mass, we see at $T \to 0$ the non-relativistic linear rise common to all three cases, Eq. (10.56). The transition from non-relativistic to relativistic behavior occurs within the entire temperature domain shown in Fig. 10.2.

Our discussion of the properties of gases of relativistic particles cannot be complete without a review of the particularly interesting case of a free

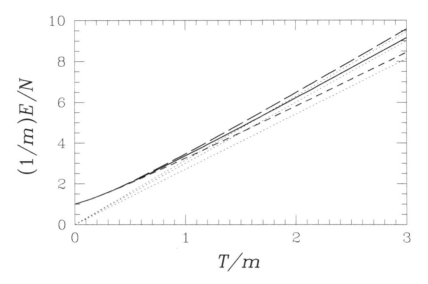

Fig. 10.2.   The energy per particle in units of m, as a function of T/m. Solid line: Boltzmann limit; long-dashed line: fermions; short-dashed lines: bosons. Asymptotic conditions are indicated by dotted lines. Quantum gases are evaluated with fugacities $\gamma, \lambda = 1$.

gas of massless quarks. In particular, at finite baryon density, for which we cannot expand the Fermi distribution function in the presence of strong quantum degeneracy, this analytically soluble case offers the only practical method for studying the behavior of an ideal gas of quarks. To see this, let us assume that we are at a finite positive chemical potential, which means that there is a net number of quarks present. Fermi distributions as functions of ε/T both for particles, Eq. (10.34a), and for antiparticles (dashed), Eq. (10.34b), are shown in Fig. 10.3, for a typical situation of $\mu/T = 0.5$ (that is, $\lambda = 1.65$).

Let us restate the mathematical problem more precisely. The grand partition function of the Fermi system, Eq. (10.32), can be written, using Eq. (10.57), in the form

$$3\frac{T}{V} \ln \mathcal{Z}_{\mathrm{F}} = g_{\mathrm{F}} \int \frac{d^3 p}{(2\pi)^3} \frac{\vec{p}^2}{\varepsilon} \left(\frac{1}{e^{\beta(\varepsilon-\mu)} + 1} + \frac{1}{e^{\beta(\varepsilon+\mu)} + 1} \right). \qquad (10.71)$$

Our usual series expansion would work for the momentum range such that $\varepsilon > \mu$. For the massless case, we would have to split the integral into two, and expand in a slightly different fashion. On approaching this problem in this way, one actually discovers that, for $m = 0$, the partition function can be exactly integrated.

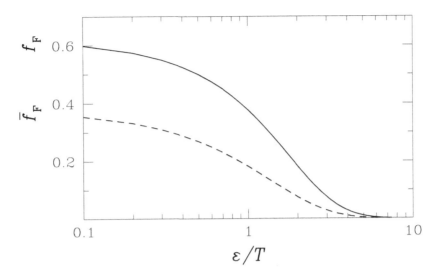

Fig. 10.3. A comparison of particle (solid) and antiparticle (dashed) Fermi distribution functions, as functions of ε/T for $\mu/T = 0.5$.

We substitute the arguments of f_F and $\bar{f}_F$ with $x = \beta(\varepsilon \pm \mu)$:

$$3\frac{T}{V}\ln\mathcal{Z}_F = \frac{g_F}{2\pi^2}T^4\bigg(\int_{\beta(m-\mu)}^{\infty}dx\,\frac{[(x+\mu/T)^2 - (m/T)^2]^{3/2}}{e^x + 1}$$

$$+ \int_{\beta(m+\mu)}^{\infty}dx\,\frac{[(x-(\mu/T))^2 - (m/T)^2]^{3/2}}{e^x + 1}\bigg). \quad (10.72)$$

A systematic expansion in m/T was carried out in [113]. We consider the leading term for $m = 0$. For what follows it is important to note that, in Eq. (10.72), one of the factors under the integral is, for $m \to 0$,

$$[(x \pm \mu/T)^2 - (m/T)^2]^{3/2} \to (|x \pm \beta\mu|)^3.$$

The range of the integrals is now split to be from $\pm\beta\mu \to 0$ and from $0 \to \infty$. The final-range integrals can be recombined to give an elementary polynomial integral,

$$\int_{-\beta\mu}^{0}dx\,\frac{|x+\beta\mu|^3}{1+e^x} - \int_{0}^{\beta\mu}dx\,\frac{(x-\beta\mu)^3}{1+e^x}$$

$$= \int_{0}^{\beta\mu}dx\,\frac{(\beta\mu-x)^3}{1+e^{-x}} + \int_{0}^{\beta\mu}dx\,\frac{(\beta\mu-x)^3}{1+e^x},$$

$$= \int_{0}^{\beta\mu}dx\,(\beta\mu-x)^3 = \frac{(\beta\mu)^4}{4}, \quad (10.73)$$

where we have changed variable from x to $-x$ in the first integrand. This term is the usual Fermi-integral contribution remaining in the limit $T \to 0$. The remaining infinite-range integral is evaluated by expansion in power series along the lines of the method shown in section 10.5, and it leads, in a straightforward way, to the first two terms in large parentheses in the following final result:

$$\ln \mathcal{Z}_{\mathrm{F}}|_{m=0} = \frac{g_{\mathrm{F}} V \beta^{-3}}{6\pi^2} \left(\frac{7\pi^4}{60} + \frac{\pi^2}{2} \ln^2 \lambda + \frac{1}{4} \ln^4 \lambda \right). \tag{10.74}$$

The net quark density follows immediately from Eq. (10.74):

$$\rho_{\mathrm{q}} \equiv n_{\mathrm{q}} - n_{\bar{\mathrm{q}}} = 3\rho_{\mathrm{q}} \frac{1}{V} \lambda \frac{\partial}{\partial \lambda} \ln \mathcal{Z}_{\mathrm{F}}|_{m=0},$$

$$\rho_{\mathrm{q}} = \frac{g_{\mathrm{F}} \beta^{-3}}{6\pi^2} \left(\pi^2 \ln \lambda + \ln^3 \lambda \right) = \frac{g_{\mathrm{F}}}{6} \left(\mu T^2 + \frac{\mu^3}{\pi^2} \right). \tag{10.75}$$

At zero temperature, the second term is the well-known expression for the degenerate Fermi gas. However, already at a modestly high temperature $T > \mu/\pi$, the first term dominates. In the range of parameters of interest to us, when $\lambda_{\mathrm{q}} \simeq 1.2$–$2.5$ and $T > 140$ MeV this is always the case. The resulting proportionality of the quark (i.e., baryon) density to the chemical potential, and the accompanying quadratic temperature dependence, offer a very counterintuitive environment for a reader used to working with cold Fermi gases.

We refer to section 4.6 for the energy and pressure of the quark and gluon gases, and a more thorough discussion the properties of a QGP is given in chapter 16.

10.6 Entropy of classical and quantum gases

We consider next the single-particle entropy associated with hadrons. We recall the expressions for entropy presented in section 7.1, Eq. (7.1) for a Fermi–Bose gases and Eq. (7.2) for a Boltzmann gas. We use the Gibbs–Duhem relation Eq. (10.26) as well as the statistical-physics analog, Eq. (10.25) and obtain

$$\frac{S}{N} = \frac{PV + E}{TN} - \frac{\mu}{T} = \frac{\ln \mathcal{Z} - \beta \dfrac{\partial}{\partial \beta} \ln \mathcal{Z}}{\lambda \dfrac{\partial}{\partial \lambda} \ln \mathcal{Z}} - \ln \lambda. \tag{10.76}$$

We will consider several cases of physical interest and note that, for pions, even at a temperature $m/T \simeq 1$, the relativistic Boltzmann limit is of interest, while the non-relativistic limit is of interest primarily for understanding the entropy of the baryon contribution.

• The classical-gas case: the partition function is, in the Boltzmann approximation, proportional to the fugacity; see, e.g., Eq. (4.40). Thus Eq. (10.76) simplifies to

$$\frac{S}{N}\bigg|^{\text{cl}} = 1 + \frac{E}{TN} - \frac{\mu}{T} = 1 - \frac{\beta\dfrac{\partial}{\partial\beta}\ln\mathcal{Z}}{\lambda\dfrac{\partial}{\partial\lambda}\ln\mathcal{Z}} - \ln\lambda. \tag{10.77}$$

Here, $\ln\lambda = \mu/T$, with the understanding that λ, in the present context, is synonymous with the abundance fugacity γ, and in what follows for the pion gas $\mu \equiv T\ln\gamma$.

By inserting Eq. (10.52) into Eq. (10.77), we obtain

$$\frac{S}{N}\bigg|^{\text{cl}} = 4 + \beta m\frac{K_1}{K_2} - \frac{\mu}{T}. \tag{10.78}$$

We consider first the limit $m/T \to 0$, Eq. (10.47):

$$\frac{S}{N}\bigg|^{\text{cl}} \simeq 4 + \frac{m^2}{2T^2} + \left[\ln\left(\frac{m}{2T}\right) + \gamma_{\text{E}}\right]\frac{m^4}{4T^4} - \frac{\mu}{T} + \cdots, \quad \frac{m}{T} \to 0. \tag{10.79}$$

An expansion suitable for the non-relativistic case, $m/T \gg 1$, can also be obtained using Eq. (10.46):

$$\frac{S}{N}\bigg|^{\text{cl}} \simeq \frac{5}{2} + \frac{m}{T} + \frac{15}{8}\frac{T}{m} - \frac{15}{8}\frac{T^2}{m^2} + \frac{135}{128}\frac{T^3}{m^3} - \frac{\mu}{T} + \cdots, \quad \frac{m}{T} \gg 1. \tag{10.80}$$

Numerical calculation shows that the Boltzmann specific entropy is monotonically falling, as shown in Fig. 10.4 by the solid line, toward the asymptotic value $S/N = 4$ (dashed line), Eq. (10.78). The two approximants, Eqs. (10.79) and (10.80), are depicted as dotted lines. Both fourth-order approximants describe the exact result well in general, except near to the physically interesting case $m/T \simeq 1$, for which the entropy per particle is $S/N = 4.4$, in the absence of a chemical potential, i.e., for a chemically equilibrated classical (Boltzmann) gas.

• The low-density nucleon gas case: we consider the non-relativistic expansion, $m/T > 1$, but we need to retain in our consideration the baryon number fugacity. Moreover, it is the entropy per baryon rather than the entropy per particle which is of interest. Recalling that in the Boltzmann approximation $\ln\mathcal{Z}_{\text{cl}} \propto (\lambda + \lambda^{-1}) = 2\cosh(\mu/T)$, and $b = \lambda(d/d\lambda)\ln\mathcal{Z}_{\text{cl}} \propto (\lambda - \lambda^{-1}) = 2\sinh(\mu/T)$, Eq. (10.77) now implies that

$$\frac{S}{b}\bigg|_{N}^{\text{cl}} = \left(4 + \beta m\frac{K_1(\beta m)}{K_2(\beta m)}\right)\coth\left(\frac{\mu}{T}\right) - \frac{\mu}{T}. \tag{10.81}$$

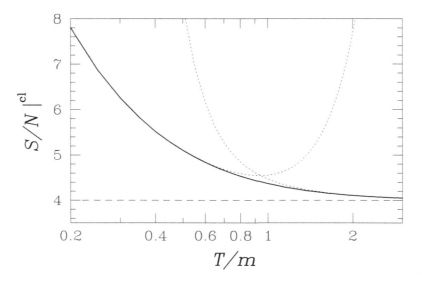

Fig. 10.4.   The entropy per particle of a classical (Boltzmann) gas. Dashed line, asymptotic value $S/N = 4$; dotted line, fourth-order approximants, Eqs. (10.79) and (10.80).

Using the asymptotic expansion Eq. (10.46), we obtain

$$\frac{S}{b}\Big|_N^{\text{cl}} = \frac{5}{2}\left(1 + \frac{3}{4}\frac{T}{m} - \frac{3}{4}\frac{T^2}{m^2} + \frac{27}{64}\frac{T^3}{m^3} + \cdots\right)\coth\left(\frac{\mu}{T}\right)$$
$$- \frac{\mu - m\coth(\mu/T)}{T}. \qquad (10.82)$$

• We find the entropy for quantum quark and gluon gases using the Gibbs–Duham relation in the form Eq. (10.76) and the relativistic equation of state Eq. (10.61). The entropy is

$$S|_{m=0} = \frac{4PV}{T} - \sum_f \frac{\mu_f}{T} N_f = \frac{4E}{3T} - \sum_f \frac{\mu_f}{T} N_f, \qquad (10.83)$$

where the sum over different kinds of component f is implied, $E = \sum_f E_f$, etc. For each component, we obtain in the relativistic limit $m/T \ll 1$, and, for $\mu = 0$, dividing by N, and using Eqs. (10.69) and (10.70),

$$\frac{S}{N}\Big|_{m=0}^{\text{B}} = 4\frac{\zeta(4)}{\zeta(3)} = 3.61, \qquad (10.84)$$

$$\frac{S}{N}\Big|_{m=0}^{\text{F}} = 4\frac{\eta(4)}{\eta(3)} = 4.20. \qquad (10.85)$$

To obtain the complete dependence on m/T, we use Eq. (10.68). For each particle species, subject to the existence of the series representation of the integral, as addressed earlier, the result is

$$\frac{S}{N} = 4 \frac{\displaystyle\sum_{n=1}^{\infty} \frac{(u\lambda)^n}{n^4}\left((nx)^2 K_2(nx) + \frac{1}{4}(nx)^3 K_1(nx)\right)}{\displaystyle\sum_{n=1}^{\infty} \frac{(u\lambda)^n}{n^3}(nx)^2 K_2(nx),} - \ln\lambda, \qquad (10.86)$$

where $x = m\beta$. $u = -1$ for fermions and $u = 1$ otherwise. For the non-relativistic limit $x > 1$, one can use Eq. (10.46) in Eq. (10.86) to obtain

$$\frac{S}{N} = \frac{\displaystyle\sum_{n=1}^{\infty} \frac{(u\lambda)^n}{n^4}(nx)^2 K_2(nx) I(nx)}{\displaystyle\sum_{n=1}^{\infty} \frac{(u\lambda)^n}{n^3}(nx)^2 K_2(nx)} - \ln\lambda, \qquad (10.87)$$

$$I(nx) = \frac{5}{2} + nx + \frac{15}{8}\frac{1}{nx} - \frac{15}{8}\frac{1}{(nx)^2} + \frac{135}{128}\frac{1}{(nx)^3} + \cdots, \qquad (10.88)$$

which, for $n = 1$, yields the result Eq. (10.82), once we rearrange terms of two components to include particles and antiparticles and divide by the baryon number (particle–antiparticle difference).

For the case of a vanishing chemical potential, the non-relativistic Boltzmann approximation, Eq. (10.80), is quite appropriate. In Fig. 10.5, we compare the entropy per particle, evaluated at zero chemical potential ($\lambda = 1$), for the Fermi (long-dashed line), Bose (short-dashed line) and Boltzmann (solid line, see Fig. 10.4) particles.

11 Hadronic gas

11.1 Pressure and energy density in a hadronic resonance gas

We now consider the physical properties of a hadronic, confined phase, such as energy density, pressure, and abundances of various particles, assuming that we have a locally thermally and chemically equilibrated phase. Although full chemical equilibrium is most certainly not attainable in the short time of the nuclear-collision interaction, see chapter 5, this study provides very useful guidance and a reference point for understanding the properties of hadronic matter out of chemical equilibrium.

There are two ways to look at a hadronic gas: the first is that we can study its properties using the known hadronic states. This approach will

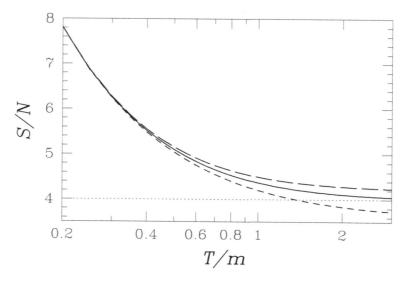

Fig. 10.5. Comparison of entropy per particle for Fermi and Bose gases, their classical Boltzmann limit, as a function of T/m: long-dashed Fermi gas, short-dashed Bose gas, solid line Boltzmann classical limit.

lead to difficulties when and if the temperature is high, since the contribution of high-mass resonances is apparently not convergent. Even though the population of each such state is suppressed exponentially by the Boltzmann factor $e^{-m/T}$, the number of states rises exponentially with mass, and compensates for this effect. This phenomenon was noticed almost 40 years ago. This led to the development of the statistical-bootstrap model (SBM) and the Hagedorn-gas model, which we will address in chapter 12.

In the physically most relevant hadron-gas domain, 70 MeV $> T > 170$ MeV, each distinguishable hadron distribution is far from quantum degeneracy. Therefore, we can use the Boltzmann approximation. The only exceptional case is the pion, which, when necessary and appropriate, will be considered as a Bose particle. Each of the hadronic states is considered as a separate contributing fraction in the thermal and chemically equilibrated gas phase, with all fugacities set at $\lambda = 1$ (no net quantum numbers, e.g., $b = 0$ etc.). The result is shown in Fig. 11.1. We included 4627 (counting spin and isospin degeneracy) hadronic states listed by the particle data group (PDG) [136]. No doubt many more hadronic resonances exist. However, as the mass of the new resonances increases, they become more difficult to characterize, given the dense background of the neighboring resonances, and normally increasing decay width, both of which effects are reducing the signal-to-noise ratio in the experiment.

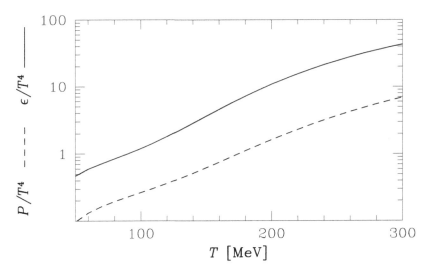

Fig. 11.1. The energy density (solid line) and pressure (dashed line) in units of T^4 for all known hadrons (on a logarithmic scale) as functions of temperature T. All fugacities are set to unity.

The worrying fact is that the energy and pressure seem to grow well beyond the values spanned by the lattice calculations; see section 15.5. This happens since we have allowed very many hadrons to be present in the same volume. Even though each kind is relatively rare, the large number of resonances implies a considerable total particle density. However, hadrons are not point-like, and in some sense the presence of particles fills the space available. In the context of the statistical bootstrap, we will argue in section 12.3 that each hadron occupies a fraction of the spacial volume. This qualitative argument leads to a correction that relates the physically observable P and ϵ to the point-particle result (subscript 'pt') we have so far studied [144]:

$$P = \frac{P_{\mathrm{pt}}}{1 + \epsilon_{\mathrm{pt}}/(4B)}, \qquad \epsilon = \frac{\epsilon_{\mathrm{pt}}}{1 + \epsilon_{\mathrm{pt}}/(4B)}. \qquad (11.1)$$

The energy density of a hadron is assumed to be $4B$, where, as before, B is the bag constant, and we recall the benchmark value, $B^{1/4} = 171$ MeV, corresponding to $4B = 0.45$ GeV fm^{-3}. This excluded volume modifies and limits the growth both of ϵ and of P with temperature. The magnitude of the effect depends on details of the implementation and on the parameters used. However, ϵ/P is little influenced by this phenomenological uncertainty.

The dynamics of HG matter described in, e.g., the hydrodynamic approach in section 6.2 depends in a critical way on the ratio of the inertia

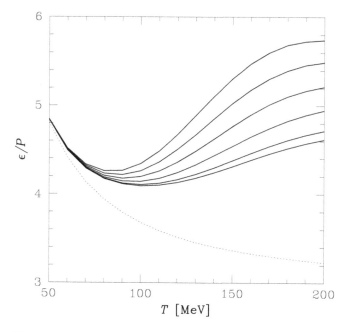

Fig. 11.2. The ratio of energy density and pressure for a hadronic gas as a function of the temperature T. Dotted line: pure pion gas; solid lines: gas comprising pions, nucleons, kaons, and $\Delta(1232)$, from bottom to top for $\lambda_q = 1, 1.2, 1.4, 1.6, 1.8$, and 2.

(energy density) to force (pressure) . In Fig. 11.2 for several simple cases, we show the HG ratio ϵ/P, as a function of temperature. The dotted curve is for the pure pion gas, and we see how the relativistic equation of state is approached for $T > 100$ MeV. Remarkably, a very different result is seen once heavy hadrons are introduced. The solid lines include, apart from pions, a few more massive states: nucleons, kaons, and $\Delta(1232)$. The solid lines from bottom to top are for $\lambda_q = 1, 1.2, 1.4, 1.6, 1.8$, and 2. We recognize that increasing λ_q (i.e. increasing the massive-baryon component) leads to a greater ratio of inertia to force. This result is clearly independent of the (schematic) finite-volume correction we introduced in Eq. (11.1). A fully realistic calculation of this situation is presented in Fig. 11.3, for the case $\lambda_s = 1.1$ and $\gamma_s/\gamma_q = 0.8$ for $\lambda_q = 1$ to 2 in steps of 0.2 from bottom to top, and $\gamma_q = 1$ (dashed lines), or $\gamma_q = e^{m_\pi/(2T)}$ (full lines). Imagine that a hadron phase is formed from a deconfined QGP at some temperature $T > 140$ MeV. In view of these results, we then expect an accelerating flow of matter as the ratio of inertia to force decreases, until a minimum is reached at $T = 90$ MeV. At this point, the HG phase most likely ceased to exist, in the sense that the distance particles travel

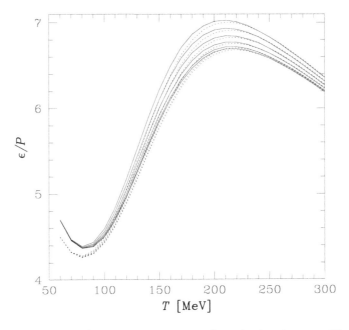

Fig. 11.3. The energy density over pressure for a hadronic gas with statistical parameters $\lambda_s = 1.1$ and $\gamma_s/\gamma_q = 0.8$, with $\lambda_q = 1$ to 2 in steps of 0.2 from bottom to top, and $\gamma_q = 1$ (dashed lines), or $\gamma_q = e^{m_\pi/(2T)}$ (full lines).

between scattering had become too large. At the time of writing this book, it is not clear whether the hadronic phase is present at all, since in the sudden breakup of a QGP a direct transition to free-streaming hadrons produced sequentially in time can be imagined. However, these results apply certainly to the case in which no QGP is formed, namely at sufficiently low collision energy.

11.2 Counting hadronic particles

There are several discrete quantum numbers of a hadron gas that are conserved and require introduction of independent chemical potentials. The chemical potentials for conservation of baryon number and strangeness, μ_B and μ_S, are the best known. Alternatively, and more conveniently for our purposes, one can use the quark chemical potentials μ_q and μ_s for light and strange quarks, respectively. We will often differentiate between the u and d quarks, and use μ_u and μ_d.

This choice of quark chemical potentials is a matter of convenience and is made in order to facilitate the translation of QGP-phase variables into HG-phase variables; in no way does it assume deconfinement of quarks.

The relationship between the two sets of chemical potentials, quark-based and traditional hadron-conserved-quantum-number based, is given by the natural relations

$$\mu_b = 3\mu_q, \qquad \mu_b = 3T \ln \lambda_q, \qquad (11.2a)$$
$$\mu_S = \mu_q - \mu_s, \qquad \mu_s = \mu_b/3 - \mu_S, \qquad (11.2b)$$

where the minus signs are due to the conventional assignment of strangeness -1 to the strange quark. Expressed in term of the fugacities we have:

$$\lambda_b = \lambda_q^3, \qquad \lambda_S = \lambda_q/\lambda_s. \qquad (11.3)$$

B_h and S_h are the baryon number and strangeness of hadron 'h', and its chemical potential can be written either in terms of μ_b and μ_S, or in terms of μ_q and μ_s:

$$\mu_h = B_h \mu_b + S_h \mu_S, \qquad (11.4a)$$
$$\mu_h = \nu_h^q \mu_q + \nu_h^s \mu_s, \qquad (11.4b)$$

where ν_h^q and ν_h^s count the numbers of light and strange valence quarks inside the hadron, respectively, with antiquarks counted with a minus sign. By adapting the quark-based chemical potentials for hadrons, we recognize the fact that, in the quark model, the quantum numbers of hadrons are obtained by adding the quantum numbers of their constituent quarks.

The particle numbers are more directly addressed in the partition function in terms of fugacities. Since the fugacities are obtained by exponentiating the chemical potentials, Eq. (4.18), the fugacity of each hadronic species is simply the product of the fugacities of the valence quarks. We view a hadron as simply a carrier of the valence quarks, which determine the fugacity and chemical potential of each particle. For example, we have

$$
\begin{aligned}
\text{p}: \quad & \mu_p = 2\mu_u + \mu_d, & \lambda_p = \lambda_u^2 \lambda_d \,; \\
\text{n}: \quad & \mu_n = \mu_u + 2\mu_d, & \lambda_n = \lambda_u \lambda_d^2 \,; \\
\Lambda: \quad & \mu_\Lambda = \mu_u + \mu_d + \mu_s, & \lambda_\Lambda = \lambda_u \lambda_d \lambda_s \,; \qquad \text{etc.}
\end{aligned}
$$

We distinguish between the up and down quarks by introducing separate chemical potentials μ_u and μ_d, which is tantamount to introduction of the chemical potential μ_Q related to the conservation of electrical charge. In view of the quark baryon number $\frac{1}{3}$ and the quark charges $-\frac{1}{3}$ and $+\frac{2}{3}$, the relations between the chemical potentials are

$$\mu_u \equiv \tfrac{1}{3}\mu_b + \tfrac{2}{3}\mu_Q, \qquad \mu_d \equiv \tfrac{1}{3}\mu_b - \tfrac{1}{3}\mu_Q. \qquad (11.5)$$

The average of μ_u and μ_d is the quark chemical potential μ_q:

$$\mu_q \equiv \frac{\mu_u + \mu_d}{2}. \tag{11.6}$$

The definitions Eqs. (11.5) and (11.6) imply a modification of Eq. (11.2a),

$$\mu_q \rightarrow \mu_q = \tfrac{1}{3}\mu_b + \tfrac{1}{6}\mu_Q, \tag{11.7}$$

which is rarely considered. It arises from the fact that a quark system containing (nearly) equal numbers of u and d quarks would still have a net (positive) charge of a sixth the total number of u and d quarks, arising from the electrical charge of the proton in the initial state formed by the colliding nuclei.

The asymmetry in the number of u and d quarks is best described by the quantity

$$\delta\mu = \mu_d - \mu_u = -\mu_Q, \tag{11.8}$$

where the negative sign in the last equality reminds us that the d quark has negative charge. Inverting Eq. (11.7), we obtain

$$\mu_b = 3\mu_q\left(1 + \frac{1}{6}\frac{\delta\mu}{\mu_q}\right). \tag{11.9}$$

In a free-quark gas with $\mu_q < \pi T$, we have, in view of Eq. (10.75),

$$\mu_d \propto \langle d - \bar{d}\rangle, \qquad \mu_u \propto \langle u - \bar{u}\rangle, \tag{11.10}$$

where the net number (number of quarks minus that of antiquarks) of light quarks enters. In a QGP, we find the remarkably simple relation [216]

$$\frac{1}{6}\frac{\delta\mu}{\mu_q} = \frac{1}{3}\frac{\mu_d - \mu_u}{\mu_d + \mu_u} = \frac{1}{3}\frac{\langle d - \bar{d}\rangle - \langle u - \bar{u}\rangle}{\langle d - \bar{d}\rangle + \langle u - \bar{u}\rangle} = \frac{n - p}{A}, \tag{11.11}$$

with $A = n + p$, and n and p are the neutron and proton contents of the matter which formed the QGP phase.

For the case of greatest asymmetry available, in Pb–Pb collisions, we have $\delta\mu/(6\mu_q) = 0.21$. In the HG phase a similarly sized effect for $\delta\mu/(6\mu_q)$ to that in a QGP is found, considering this issue numerically; see figure 1 in [183]. Especially in studying yields of individual particles, the specific quark u and d content can play a noticeable role. To see this, let us compare the u and d fugacities:

$$\frac{\lambda_d}{\lambda_u} = e^{\delta\mu/T} = \lambda_q^{\delta\mu/\mu_q}. \tag{11.12}$$

For Pb–Pb interactions under baryon-rich conditions, a λ_d/λ_u ratio significantly different from unity results. Some dilution of this phenomenon will occur if a QGP is formed due to the contribution of hadronizing gluons, which do not differentiate between the two light u and d flavor states. We conclude that the u–d asymmetry can not be completely ignored when one is considering the abundances of hadronic particles in baryon-rich fireballs.

Flavor-changing weak interactions are too slow to matter on the time scale of heavy-ion collisions. The strong and electro-magnetic interactions do not mix the quark flavors u, d, and s. These are separately conserved on the time scale of hadronic collisions. Only the number of quark–antiquark pairs of the same flavor changes; that is, pairs can be produced or annihilated. The fugacities we call γ_q and γ_s serve to count the number of pairs of light and strange quark, respectively, at any given time. In general, these pair-abundance fugacities are rapidly evolving in time, in contrast to the fugacities λ_q and λ_s. In fact, for entropy-conserving evolution of a fireball of QGP, the fugacity λ_q is nearly constant, and, as we now shall address, as long as local conservation of strangeness is maintained, $\lambda_s \simeq 1$.

Comparing the QGP with the HG phase, the value of the strangeness fugacity λ_s is in a subtle and important way different. Given the mobility of individual quarks in the QGP phase, and ignoring the influence of electrical charge in this qualitative discussion, the phase space of both s and $\bar{s}$ quarks must be the same, irrespective of the baryon content. To balance the s and $\bar{s}$ distributions, we have $\lambda_s = 1$, irrespective of the value of λ_q, see, e.g., Eq. (4.42). It is instructive to check the phase-space integral describing the density of strangeness in order to appreciate these remarks, and to recall the precise physical difference between the fugacities λ_s and γ_s:

$$
\langle n_s \rangle - \langle n_{\bar{s}} \rangle = \int \frac{d^3 p}{(2\pi)^3} \left(\frac{1}{\gamma_s^{-1} \lambda_s^{-1} \exp\left(\frac{\sqrt{p^2 + m_s^2}}{T} \right) + 1} \right.
$$

$$
\left. - \frac{1}{\gamma_s^{-1} \lambda_s \exp\left(\frac{\sqrt{p^2 + m_s^2}}{T} \right) + 1} \right). \tag{11.13}
$$

We note the change in the power of λ_s between these two terms, and recognize that this integral can vanish only for $\lambda_s \to 1$. We discuss in the following section the small but significant asymmetry in λ_s due to the Coulomb charge present in baryon-rich quark matter: long-range electro-magnetic interactions influence strange and antistrange particles differently, and a slight deviation $\lambda_s > 1$ is needed in order to compensate for this effect in the QGP phase.

Now, let us look at the HG phase. Strange quarks are bound in states comprising also light quarks. The presence of a net baryon number assures that there is an asymmetry in abundance of light quarks and antiquarks, and thus also, e.g., of strange baryons and antibaryons, with more hyperons than antihyperons being present. Owing to this asymmetry, strangeness cannot be balanced in the HG with the value $\lambda_s = 1$, unless the baryon density vanishes locally. We will address this important issue in a more quantitative manner in section 11.4. We have learned that a determination of $\lambda_s = 1$ in the hadron-abundance analysis is indicating production of these hadrons directly in a breakup of a QGP phase, since a value different from unity is expected when a HG phase breaks up.

11.3 Distortion by the Coulomb force

It has been recognized for a long time that the Coulomb force can be of considerable importance in the study of relativistic heavy-ion collisions. It plays an important role in the HBT interferometry method of analysis of the structure of the particle source [57, 209]; section 9.3. The analysis of chemical properties is also subject to this perturbing force, and in consideration of the precision reached experimentally in the study of particle ratios, one has to keep this effect in mind.

We consider a Fermi gas of strange and antistrange quarks allowing that the Coulomb potential V_C is established by the excess charge of the colliding nuclei. Within a relativistic Thomas–Fermi phase-space occupancy model [193], and for finite temperature in a QGP, we have as generalization of Eq. (11.13) [177]

$$\langle N_s \rangle - \langle N_{\bar{s}} \rangle = \int_{R_f} g_s \frac{d^3r \, d^3p}{(2\pi)^3} \left(\frac{1}{1 + \gamma_s^{-1} \lambda_s^{-1} e^{(E(p) - \frac{1}{3} V_C(r))/T}} \right.$$

$$\left. - \frac{1}{1 + \gamma_s^{-1} \lambda_s e^{(E(p) + \frac{1}{3} V_C(r))/T}} \right), \quad (11.14)$$

which clearly cannot vanish for $V_C \neq 0$, in the limit $\lambda_s \to 1$.

In Eq. (11.14), the subscript R_f on the spatial integral reminds us that only the classically allowed region within the fireball is covered in the integration over the level density; $E = \sqrt{m^2 + \vec{p}^2}$, and, for a uniform charge distribution within a radius R_f of charge Z_f,

$$V_C = \begin{cases} -\dfrac{3}{2} \dfrac{Z_f e^2}{R_f} \left[1 - \dfrac{1}{3} \left(\dfrac{r}{R_f} \right)^2 \right], & \text{for} \quad r < R_f ; \\ -\dfrac{Z_f e^2}{r}, & \text{for} \quad r > R_f. \end{cases} \quad (11.15)$$

One obtains a rather precise result, for the range of parameters of interest to us, using the Boltzmann approximation:

$$\langle N_{\mathrm{s}}^{\mathrm{B}} \rangle - \langle N_{\bar{\mathrm{s}}}^{\mathrm{B}} \rangle = \gamma_{\mathrm{s}} \left(\int g_s \frac{d^3 p}{(2\pi)^3} e^{-E/T} \right)$$
$$\times \int_{R_{\mathrm{f}}} d^3 r \left(\lambda_{\mathrm{s}} e^{V_{\mathrm{C}}/3T} - \lambda_{\mathrm{s}}^{-1} e^{-V_{\mathrm{C}}/3T} \right). \qquad (11.16)$$

The Boltzmann limit allows us also to verify and confirm the signs: the Coulomb potential is negative for the negatively charged s quarks with charge $\frac{1}{3}$, which is made explicit in the potential terms in all expressions above. We have

$$\tilde{\lambda}_{\mathrm{s}} \equiv \lambda_{\mathrm{s}} \ell_{\mathrm{C}}^{1/3} = 1, \qquad \ell_{\mathrm{C}} \equiv \frac{\int_{R_{\mathrm{f}}} d^3 r \, e^{V/T}}{\int_{R_{\mathrm{f}}} d^3 r}. \qquad (11.17)$$

$\ell_{\mathrm{C}} < 1$ expresses the Coulomb deformation of strange quark phase space. ℓ_{C} is not a fugacity that can be adjusted to satisfy a chemical condition, since consideration of λ_i, $i = $ u, d, s, exhausts all available chemical balance conditions for the abundances of hadronic particles, and allows introduction of the fugacity associated with the Coulomb charge of quarks and hadrons; see section 11.2. Instead, ℓ_{C} characterizes the distortion of the phase space by the long-range Coulomb interaction. This Coulomb distortion of the quark phase space is naturally also present for u and d quarks, but appears less significant given that λ_{u} and λ_{q} are empirically determined. On the other hand this effect compensates in part the u–d abundance asymmetry effect we have discussed in Eqs. (11.5)–(11.12).

Choosing $T = 140$ MeV and $m_{\mathrm{s}} = 200$ MeV, and noting that the value of γ_{s} is practically irrelevant since this factor cancels out in the Boltzmann approximation, see Eq. (11.16), we find for $Z_{\mathrm{f}} = 150$ that the value $\lambda_{\mathrm{s}} = 1.10$ is needed for $R_{\mathrm{f}} = 7.9$ fm in order to balance the Coulomb distortion. One should remember that the dimensionless quantities m_{s}/T and $R_{\mathrm{f}}T$ determine the magnitude of the effect we study. Chemical freeze-out at higher temperature, e.g., $T = 170$ MeV, leads for $\lambda_{\mathrm{s}} = 1.10$ to somewhat smaller radii, which is consistent with the higher temperature used.

The influence of the Coulomb force on chemical freeze-out is relevant in central Pb–Pb interactions, wheras for S–Au/W/Pb reactions, a similar analysis leads to a value $\lambda_{\mathrm{s}} = 1.01$, which is little different from the value $\lambda_{\mathrm{s}} = 1$ expected in the absence of the Coulomb deformation of phase space. Another way to understand the varying importance of the Coulomb effect is to note that, while the Coulomb potential acquires in the Pb–Pb case a magnitude comparable to the quark chemical potential, it remains small on this scale for S–Au/W/Pb reactions.

11.4 Strangeness in hadronic gas

We now describe the abundance of strange particles in the hadronic-gas phase. This is, compared with the QGP, a very complicated case, since there are many particles which are carriers of 'open' strangeness. Moreover, strong interactions result in the presence of numerous hadronic resonances with open strangeness. The postulate of the dominance by hadron resonance formation of hadron–hadron interactions [140] allows a vast simplification of the theoretical treatment. Regarding the hadronic-gas phase as a mixture of various non-interacting hadronic-resonance gases, all information about the interaction is contained in the mass spectrum $\tau(m^2, b)$ which describes the number of hadrons of baryon number b in a mass interval dm^2. We will address this postulate in more detail in chapter 12. Within this approach to strong interactions, the logarithm of the total partition function is additive in its strange and not strange sectors, so long as the various gas fractions interact mainly via formation of hadronic resonances. We then have

$$\ln Z = \ln Z^{\text{non-strange}} + \ln Z^{\text{strange}}. \tag{11.18}$$

In the grand-canonical description, one finds that the non-strange hadrons influence the strange ones by providing a background value of statistical parameters, such as the baryochemical potential μ_{b}, which are accessible to direct measurement. We conclude that, in order to understand abundances of strange particles, it is sufficient to consider $\ln Z^{\text{strange}}$. In the Boltzmann approximation, it is easy to write down the partition function for the strange-particle fraction of the hadronic gas, $\mathcal{Z}_{\text{s}}$. Including the possibility of an only partially saturated strange phase space through the factor γ_{s}, and similarly γ_{q} for light quarks, but suppressing for simplicity the isospin asymmetry $\delta\mu$, Eq. (11.8), we have

$$\ln \mathcal{Z}_{\text{s}}^{\text{HG}} = \frac{VT^3}{2\pi^2}\Big[(\lambda_{\text{s}}\lambda_{\text{q}}^{-1} + \lambda_{\text{s}}^{-1}\lambda_{\text{q}})\gamma_{\text{s}}\gamma_{\text{q}}F_{\text{K}} + (\lambda_{\text{s}}\lambda_{\text{q}}^2 + \lambda_{\text{s}}^{-1}\lambda_{\text{q}}^{-2})\gamma_{\text{s}}\gamma_{\text{q}}^2F_{\text{Y}}$$
$$+(\lambda_{\text{s}}^2\lambda_{\text{q}} + \lambda_{\text{s}}^{-2}\lambda_{\text{q}}^{-1})\gamma_{\text{s}}^2\gamma_{\text{q}}F_{\Xi} + (\lambda_{\text{s}}^3 + \lambda_{\text{s}}^{-3})\gamma_{\text{s}}^3F_{\Omega}\Big]. \tag{11.19}$$

In the phase-space function F_i all kaon (K), hyperon (Y), cascade (Ξ), and omega (Ω) resonances plus their antiparticles are taken into account:

$$F_{\text{K}} = \sum_j g_{\text{K}_j}W(m_{\text{K}_j}/T); \quad \text{K}_j = \text{K}, \text{K}^*, \text{K}_2^*, \ldots, \quad m \leq 1780 \text{ MeV},$$

$$F_{\text{Y}} = \sum_j g_{\text{Y}_j}W(m_{\text{Y}_j}/T); \quad \text{Y}_j = \Lambda, \Sigma, \Sigma(1385), \ldots, \quad m \leq 1940 \text{ MeV},$$

$$F_{\Xi} = \sum_j g_{\Xi_j}W(m_{\Xi_j}/T); \quad \Xi_j = \Xi, \Xi(1530), \ldots, \quad m \leq 1950 \text{ MeV},$$

$$F_\Omega = \sum_j g_{\Omega_j} W(m_{\Omega_j}/T); \quad \Omega_j = \Omega, \Omega(2250). \tag{11.20}$$

The g_i are the spin–isospin degeneracy factors, $W(x) = x^2 K_2(x)$, see Eq. (10.50a) and Fig. 10.1, where K_2 is the modified Bessel function, Eq. (10.44).

We need to understand, in terms of experimental observables, the chemical properties of the fireball at the time of hadron production. The method of choice is the study of particle ratios [167, 216]; section 9.1. In order to obtain the mean abundances of various strange particles, we introduce for each species its own dummy fugacity (which we subsequently will set equal to unity). The explicit expressions for these ratios turn out to be very simple, and one quickly deduces from the following examples the principles which allow one to construct any ratio:

$$\frac{\langle n_{\bar\Lambda} \rangle}{\langle n_\Lambda \rangle} = \lambda_q^{-4} \lambda_s^{-2}; \tag{11.21a}$$

$$\frac{\langle n_{\bar\Xi} \rangle}{\langle n_\Xi \rangle} = \lambda_q^{-2} \lambda_s^{-4}; \tag{11.21b}$$

$$\frac{\langle n_{\bar\Omega} \rangle}{\langle n_\Omega \rangle} = \lambda_s^{-6}; \tag{11.21c}$$

$$\frac{\langle n_{K+} \rangle}{\langle n_{K-} \rangle} = \lambda_s^{-2} \lambda_q^2; \tag{11.21d}$$

$$\frac{\langle n_K \rangle}{\langle n_\Lambda \rangle} = \lambda_s^{-2} \lambda_q^{-1} \gamma_q^{-1} \frac{F_K}{F_\Lambda}. \tag{11.21e}$$

In a more colloquial notation found also in this book, one omits $\langle n \rangle$, using as the symbol for the particle considered the subscript only.

The baryochemical potential, or more simply, the quark fugacity λ_q, can be deduced from the above stated ratios. Best for this purpose is to consider the ratios not involving quark-pair fugacities γ_q and γ_s. Any two ratios containing only λ_q and λ_s can be combined to evaluate these quantities. Since many more than two ratios are available, a check of the procedure is possible. This, in fact, constitutes a strong confirmation of the validity of phase-space characterization of particle yields. Postponing detailed discussion to chapter 19, we note that all groups that applied this method to study the chemical properties have found extremely good consistency. This implies that the production of particles as different as kaons K and anticascades $\overline\Xi$ occurs by a similar mechanism, and nearly at the same instance in the evolution of the fireball; these particles know of each other, either due to processes of rescattering in the HG phase, or simply because they have been produced directly with yields corresponding to the relative size of the phase space.

An example of the consistency relation can be obtained by combining the ratios of cascades, lambdas, and kaons,

$$\frac{\overline{\Xi}/\Xi}{\overline{\Lambda}/\Lambda} = \frac{K^+}{K^-},$$ (11.22)

which is very well satisfied in all measurements of which we are aware. It is important to note that Eq. (11.22) applies a full '4π' yield. For the central-rapidity particle yield ratio, a correction containing the influence of the velocity of expansion of the fireball has to be applied.

Although the proper determination of the chemical properties is best achieved in a global fit of hadron yields, it is important that we see how the physics of this determination works. It can be seen that the multitude of strange hadrons allows us to determine the value of λ_s in many different ways, for example,

$$\frac{\overline{\Lambda}/\Lambda}{(\overline{\Xi}/\Xi)^2} = \lambda_s^6$$ (11.23)

and similarly

$$\frac{\overline{\Xi}/\Xi}{(\overline{\Lambda}/\Lambda)^2} = \lambda_q^6.$$ (11.24)

This estimate produces an answer for the value of these parameters that very accurately agrees with results of global fits.

It is equally easy to fix the ratio γ_s/γ_q since comparison of hyperons of unequal strangeness content always yields this pair fugacity ratio. The difficulty is that we have to understand the ratio of phase spaces of the various baryons, which is controlled by the temperature, when we consider the full yield. A first estimate is obtained by comparing in the same $m_\perp$ range, e.g., Λ and Ξ. How this is done is shown in Fig. 8.8 on page 150. Even then, the feed from higher resonances is important and temperature remains an input into the determination of γ_i.

11.5 The grand-canonical conservation of strangeness

Using the partition function Eq. (11.19), we can calculate the net strangeness by evaluating

$$\langle N_s \rangle - \langle N_{\bar{s}} \rangle = \lambda_s \frac{\partial}{\partial \lambda_s} \ln \mathcal{Z}_s^{HG}.$$ (11.25)

We find

$$\langle n_s \rangle - \langle n_{\bar{s}} \rangle = \frac{T^3}{2\pi^2} \left[(\lambda_s \lambda_q^{-1} - \lambda_s^{-1} \lambda_q) \gamma_s \gamma_q F_K \right.$$

$$+ (\lambda_s \lambda_q^2 - \lambda_s^{-1} \lambda_q^{-2}) \gamma_s \gamma_q^2 F_Y$$
$$+ 2(\lambda_s^2 \lambda_q - \lambda_s^{-2} \lambda_q^{-1}) \gamma_s^2 \gamma_q F_\Xi$$
$$+ 3(\lambda_s^3 - \lambda_s^{-3}) \gamma_s^3 F_\Omega \Big]. \tag{11.26}$$

In general, Eq. (11.25) must be equal to zero since strangeness is a conserved quantum number with respect to the strong interactions, and no strangeness is brought into the reaction. The possible exception is dynamic evolution with asymmetric emission of strange and antistrange hadrons. The grand-canonical condition,

$$\langle n_s \rangle - \langle n_{\bar{s}} \rangle = 0, \tag{11.27}$$

introduces an important constraint, i.e., it fixes λ_s in terms of λ_b (and charge λ_Q when the later is considered).

Equation (11.25) can be solved analytically when the contribution of multistrange particles is small:

$$\lambda_s|_0 = \lambda_q \sqrt{\frac{F_K + \gamma_q \lambda_q^{-3} F_Y}{F_K + \gamma_q \lambda_q^3 F_Y}}. \tag{11.28}$$

This relation between the strange chemical potential $\mu_s|_0 = T \ln \lambda_s|_0$ and the baryochemical potential $\mu_b = 3T \ln \lambda_q$ is shown for $\gamma_q = \gamma_s = 1$ in Fig. 11.4. To understand Fig. 11.4, we note that the term with $\lambda_q^{-3} = e^{-\mu_b/T}$ in Eq. (11.28) will tend to zero as μ_b gets larger and the term with λ_b will dominate in denominator. Thus, $\lambda_s \propto \lambda_b^{-2/3}$, i.e., $\mu_s \propto -\frac{2}{3}\mu_b$ for large μ_b. At small μ_b, in particular, for relatively small temperatures, the hyperon contribution is small and we see $\mu_s \propto \frac{1}{3}\mu_b$. Putting it differently, Eq. (11.28) knows that, in a baryon-rich HG phase, q$\bar{s}$ (K$^+$, K$_0$) kaons are the dominant carriers of $\bar{s}$ quarks, whereas qqs (Λ, Σ) hyperon states are the main carriers of s quarks at finite baryon density. We see that the competition between strangeness content in the four classes of strangeness carriers determines, at each temperature T, the location where one obtains a nontrivial $\mu_s = 0$ at finite μ_b, and the QGP property $\lambda_s = 1$ is accidentally present in the HG phase. There is no such nontrivial solution at sufficiently high temperature. For $T > 200$ MeV and $\gamma_q = \gamma_s = 1$, only negative strangeness chemical potential is seen in Fig. 11.4.

The line in the (μ_b–T) plane corresponding to $\mu_s = 0$ is the divide between positive and negative values of the strangeness chemical potential in a strangeness-balanced hadronic gas. The relation between μ_b and T corresponding to $\mu_s = 0$, i.e., $\lambda_s = 1$, arising from Eq. (11.26) when net strangeness vanishes, can be solved analytically allowing for the effect of multistrange baryons and antibaryons. First, we note that for $\lambda_s = 1$,

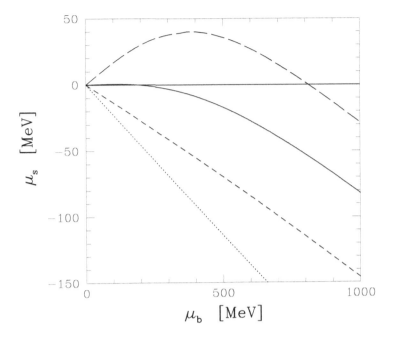

Fig. 11.4. The strange-quark chemical potential μ_s versus the baryon chemical potential μ_b in a strangeness-neutral grand-canonical chemically equilibrated HG. The long-dashed line corresponds to $T = 150$ MeV, the solid line to $T = 200$ MeV, and the short-dashed line to $T = 300$ MeV. The dotted line is the limiting curve for large T, computed here at $T = 1000$ MeV.

there is always an exact balance between Ω and $\overline{\Omega}$ and this term disappears. The coefficient of the hyperon F_Y contribution, when it is written in the form

$$\lambda_q^{-2} - \lambda_q^2 = \left(\lambda_q^{-1} - \lambda_q\right)\left(\lambda_q^{-1} + \lambda_q\right),$$

allows us to cancel out a common factor $\lambda_q^{-1} - \lambda_q$ present in all terms, along with $\gamma_q\gamma_s$. We obtain

$$\mu_b = 3T \ln(x + \sqrt{x^2 - 1}), \qquad 1 \le x = \frac{F_K - 2\gamma_s F_\Xi}{2\gamma_q F_Y}. \qquad (11.29)$$

This result is shown in Fig. 11.5. We have chosen to consider the nonequilibrium condition $\gamma_q = e^{m_\pi/(2T)}$ corresponding to the maximum entropy content in a hadronic gas, as could be emerging from hadronization of an entropy-rich QGP phase. The solid line is for $\gamma_s = \gamma_q$, while the dashed lines span the range $\gamma_s = 0.8$–2.8 in steps of 0.2, from right to left.

Below and to the left of this separation line in Fig. 11.5, we have positive strangeness chemical potential in a strangeness-balanced HG phase,

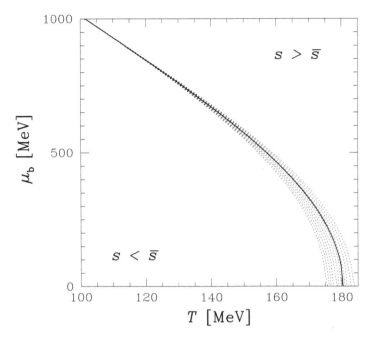

Fig. 11.5. The condition of vanishing strangeness at $\lambda_s = 1$ in a hadronic gas, evaluated for its maximum entropy content, i.e., with $\gamma_q = e^{m_\pi/(2T)}$. Solid line, $\gamma_s = \gamma_q$; dashed lines from left to right are for $\gamma_s = 0.8$–2.8 in steps of 0.2.

whereas above and to the right we have negative strangeness potential. The importance of this observation is that the relative yield $R_\Omega \equiv \Omega/\overline{\Omega}$, Eq. (11.21c), is strongly sensitive to the sign of μ_s. At present, we know that, at the SPS top energy, both in S–W and in Pb–Pb interactions, $R_\Omega \geq 1$ and thus $\lambda_s \geq 1$ and $\mu_s \geq 0$; the allowed range of T–μ_b is below and to the left in Fig. 11.5. Indeed, all analyses of the abundances of particles of which we are aware have yielded results in this domain of T and μ_b.

We further denote, in Fig. 11.5, the area below and to the left as $s < \bar{s}$, whereas the domain above and to the right is denoted as $s > \bar{s}$. What we indicate is that, for $\lambda_s = 1$, the resulting phase space of strange particles would add up to satisfy these conditions within these domains of T and μ_b. To recognize the importance of this condition consider that a QGP is evaporating hadrons. Below and to the left, with $s < \bar{s}$, the evaporation favors emission of antistrangeness and this allows the accumulation of an excess of strangeness in the evaporation remnant; this is the process called 'strangeness distillation' [134, 135].

The result of the distillation is the production of strangelets, drops of quark matter with an unusually high abundance of s quarks. Since strange quarks are negatively charged, such states would have unusually small charge relative to their mass; indeed in the limit of equal abundance of u, d, and s, a strangelet would be neutral. Many searches for long-lived (on the scale of strong interactions) strangelets have been performed, without success [44, 45, 232], suggesting that such states are not stable with respect to strong interactions. If they are produced, strangelets are dissociating into strange hadrons rather rapidly. One will note that, in the decay of hadronic strangelets, production of multistrange baryons, and in particular Ω, would be well above the normal expectations. Since the statistical yield of Ω is very small, even a small yield of strangelets could lead to visible distortions of the otherwise rarely produced Ω. An excess of Ω over the statistical-model expectations is seen at the top energy of the SPS; see section 19.3 and Fig. 8.11.

11.6 Exact conservation of flavor quantum numbers

We now consider, in more detail, what effect the exact conservation of quantum numbers, such as strangeness or baryon number, has on the size of the available particle phase space. As is intuitively clear, only when the yield numbers are small, can this lead to a noticeable effect. In the grand-canonical approach, flavor conservation, expressed by Eq. (11.27), is not exact. In other words, strangeness, even baryon number, is conserved on average but not exactly. We will focus our interest on the case of newly-produced flavors (strangeness and charm) since the number of pairs of quarks produced can be sufficiently small to warrant this. The exact conservation of the baryon number (or the light-quark flavors) is of particular interest in the study of the small collision systems.

When the number of strange-quark pairs is relatively small, Eq. (11.27) has to be replaced by the sharper 'canonical' conservation condition,

$$\langle n_{\rm s} - n_{\bar{\rm s}} \rangle = 0. \tag{11.30}$$

According to Eq. (11.30), the net strangeness vanishes exactly in each physical system we study. This introduces a correlation between the phase space of particles and of antiparticles and thus, in general, the chemical equilibrium yield of, e.g., the pairs of strange quarks evaluated under constraint Eq. (11.30), is smaller when compared with that expected when Eq. (11.27) is considered.

We are, in particular, interested in understanding under which conditions the canonical and grand-canonical yields are equal, and how the grand-canonical yields are altered by the physical constraint Eq. (11.30) [218]. For strangeness, this amounts to finding the yield for which we can study the colliding system as if it had infinite size. In the context of the production of charm, the yields (almost) always remain relatively small, and use of the canonical formulation is necessary in order to evaluate the expected chemical-equilibrium yields, even when the grand-canonical approach applies for all the other observables.

The grand partition function, Eq. (10.48), in the Boltzmann limit, can be written as a power series:

$$\mathcal{Z}_{cl} = e^{Z_f^{(1)}} = \sum_{n=0}^{\infty} \frac{1}{n!} \left(Z_f^{(1)} \right)^n. \tag{11.31}$$

To emphasize that any flavor (in particular s, c, and b) is under consideration here, we generalize slightly the notation s $\rightarrow$ f. The flavor and antiflavor terms within $Z_f^{(1)}$ are additive, and we consider at first only singly flavored particles in Eq. (11.19), adopting a simplified and self-explanatory notation:

$$Z_f^{(1)} = \gamma(\lambda_f \tilde{F}_f + \lambda_f^{-1} \tilde{F}_{\bar{f}}), \quad \tilde{F}_i = \frac{VT^3}{2\pi^2} F_i. \tag{11.32}$$

Combining Eq. (11.32) with Eq. (11.31), we obtain

$$\mathcal{Z}_{cl} = \sum_{n,k=0}^{\infty} \frac{\gamma^{n+k}}{n!k!} \lambda_f^{n-k} \tilde{F}_f^n \tilde{F}_{\bar{f}}^k. \tag{11.33}$$

When $n \neq k$, the sum in Eq. (11.33) contains contributions with unequal numbers of f and $\bar{f}$ terms. Only when $n = k$ do we have contributions with exactly equal number of f and $\bar{f}$ terms. We recognize that only $n = k$ terms contribute to the canonical partition function:

$$Z_{cl}^{f=0} = \sum_{n=0}^{\infty} \frac{\gamma^{2n}}{n!n!} (\tilde{F}_f \tilde{F}_{\bar{f}})^n = I_0 \left(2\gamma \sqrt{\tilde{F}_f \tilde{F}_{\bar{f}}} \right). \tag{11.34}$$

The modified Bessel function I_0 is well known, see Eqs. (8.23) and (8.27).

The argument of I_0 has a physical meaning, it is the yield of flavor pairs N_{pair}^{GC} in the grand-canonical ensemble, evaluated with grand-canonical conservation of flavor, Eq. (11.27). To see this, we evaluate

$$\langle N_f \rangle - \langle N_{\bar{f}} \rangle = \lambda_f \frac{\partial}{\partial \lambda_f} \ln \mathcal{Z}_{cl}^f = \gamma(\lambda_f \tilde{F}_f - \lambda_f^{-1} \tilde{F}_{\bar{f}}) = 0. \tag{11.35}$$

We obtain, see Eq. (11.28),

$$\lambda_f|_0 = \sqrt{\tilde{F}_{\bar{f}}/\tilde{F}_f}, \tag{11.36}$$

and thus

$$\ln \mathcal{Z}_{\text{cl}}^{\text{f}}\bigg|_{\lambda_{\text{f}}=\lambda_{\bar{\text{f}}}|0} = \langle N_{\text{f}}\rangle + \langle N_{\bar{\text{f}}}\rangle = 2\gamma\sqrt{\tilde{F}_{\text{f}}\tilde{F}_{\bar{\text{f}}}} \equiv 2N_{\text{pair}}^{\text{GC}}, \tag{11.37}$$

which is just the argument of the I_0 function in Eq. (11.34). In the grand-canonical-ensemble approach the (average) number of pairs $N_{\text{pair}}^{\text{GC}}$ is extensive in volume, since $\tilde{F}_i \propto V$.

In order to evaluate, using Eq. (11.34), the number of flavor pairs in the canonical-ensemble, we need to average the number n over all the contributions to the sum in Eq. (11.34). To obtain the extra factor n, we perform the differentiation with respect to γ^2 and obtain the canonical ensemble f-pair yield,

$$\langle N_{\text{f}}^{\text{CE}}\rangle \equiv \gamma^2 \frac{d}{d\gamma^2}\ln Z_{\text{cl}}^{\text{f}=0} = \gamma\sqrt{\tilde{F}_{\text{f}}\tilde{F}_{\bar{\text{f}}}}\,\frac{I_1\left(2\gamma\sqrt{\tilde{F}_{\text{f}}\tilde{F}_{\bar{\text{f}}}}\right)}{I_0\left(2\gamma\sqrt{\tilde{F}_{\text{f}}\tilde{F}_{\bar{\text{f}}}}\right)}$$

$$= N_{\text{pair}}^{\text{GC}}\,\frac{I_1(2N_{\text{pair}}^{\text{GC}})}{I_0(2N_{\text{pair}}^{\text{GC}})}. \tag{11.38}$$

where we have used Eq. (8.24). The first term is identical to the result we obtained in the grand-canonical formulation, Eq. (11.37). The second term is the effect of exact conservation of flavor.

The intuitive derivation of the canonical constraint we have presented follows the approach of [218]. This can be generalized to more complex systems using the projection method [229, 262]. This method can be applied to solve more complex situations, for example inclusion of multistrange hadrons, conservation of several 'Abelian' quantum numbers [61, 102] (such as strangeness, baryon number, and electrical charge), and the problem of particular relevance in this field, the exact conservation of color: all hadronic states, including QGP, must be exactly color 'neutral' [111, 112]. The solution of this 'nonabelian-charge' problem is most interesting but reaches well beyond the scope of this book.

For the case of 'Abelian' quantum numbers, e.g., flavor or baryon number, the projection method arises from the general relation between the grand-canonical and canonical partition functions implicit in Eq. (4.20):

$$\mathcal{Z}(\beta, \lambda, V)_{\text{cl}} = \sum_{n_{\text{f}}=-\infty}^{\infty} \lambda^{n_{\text{f}}} Z_{\text{f}}(\beta, V; n_{\text{f}}). \tag{11.39}$$

In the canonical partition function Z_{f}, some discrete (flavor, baryon) quantum number has the value $n_{\text{f}} \equiv \text{f}$. The inverse of this expansion is given in Eq. (4.21). On making the substitution $\lambda = e^{i\varphi}$ we obtain

$$Z_f(\beta, V; n_f) = \int_0^{2\pi} \frac{d\varphi}{2\pi} e^{-in_f\varphi} \mathcal{Z}(\beta, \lambda = e^{i\varphi}, V). \tag{11.40}$$

In the case of the Boltzmann limit, and including singly charged particles only, we obtain for the net flavor n_f from Eq. (11.33)

$$Z_f(\beta, V; n_f) = \sum_{n,k=0}^{\infty} \frac{\gamma^{n+k}}{n!k!} \int_0^{2\pi} \frac{d\varphi}{2\pi} e^{i(n-k-n_f)\varphi} \tilde{F}_f^n \tilde{F}_{\bar{f}}^k. \tag{11.41}$$

The integration over φ yields the $\delta(n - k - n_f)$ function. Replacing $n = k + n_f$, we obtain

$$Z_f(\beta, V; n_f) = \sum_{k=0}^{\infty} \frac{\gamma^{2k+n_f}}{k!(k+n_f)!} \tilde{F}_f^{k+n_f} \tilde{F}_{\bar{f}}^k. \tag{11.42}$$

The power-series definition of the modified Bessel function I_f is

$$I_{n_f}(z) = \sum_{k=0}^{\infty} \frac{(z/2)^{2k+n_f}}{k!(k+n_f)!}. \tag{11.43}$$

Thus we obtain

$$Z_f(\beta, V; n_f) = \left(\frac{\tilde{F}_f}{\tilde{F}_{\bar{f}}}\right)^{n_f/2} I_{n_f}\left(2\gamma\sqrt{\tilde{F}_f \tilde{F}_{\bar{f}}}\right). \tag{11.44}$$

The case of $n_f = 0$ which we considered earlier, Eq. (11.34), is reproduced. We note that, for integer n_f, we have $I_{n_f} = I_{-n_f}$, as is also evident in the integral representation Eq. (8.27). We used n_f as we would count the baryon number, thus, in flavor counting, n_f counts the flavored quark content, with quarks counted positively and antiquarks negatively. This remark is relevant when the factors $\tilde{F}_f$ and $\tilde{F}_{\bar{f}}$ contain baryochemical potential.

When the baryon number is treated in the grand-canonical approach, and strangeness in the canonical approach, there is potential for mathematical difficulties. These can usually be avoided by considering the meromorphic expansion of the partition function Eq. (11.39). Inserting the explicit form Eq. (11.32) we obtain

$$\mathcal{Z}_{cl} \simeq e^{\gamma(\lambda_f \tilde{F}_f + \lambda_f^{-1} \tilde{F}_{\bar{f}})} = \sum_{n_f=-\infty}^{\infty} \lambda_f^{n_f} \left(\frac{\tilde{F}_f}{\tilde{F}_{\bar{f}}}\right)^{n_f/2} I_{n_f}\left(2\gamma\sqrt{\tilde{F}_f \tilde{F}_{\bar{f}}}\right). \tag{11.45}$$

Multistrange particles can be introduced as additive terms in the exponent in Eq. (11.45). This allows us to evaluate their yields [148]. However, the canonical partition function is dominated by singly strange particles and we will assume, in the following, that considering only these suffices to obtain the effect of canonical conservation of flavor. In order to find yields of rarely produced particles such as, e.g., Ω(sss), we show the omega term explicitly:

$$Z_{\mathrm{f}}(\beta, V; n_{\mathrm{f}} = 0) = \int_0^{2\pi} \frac{d\varphi}{2\pi} e^{\tilde{F}_{\mathrm{f}} e^{i\varphi} + \tilde{F}_{\bar{\mathrm{f}}} e^{-i\varphi} + \lambda_\Omega e^{3i\varphi} \tilde{F}_\Omega + \cdots}. \tag{11.46}$$

The unstated terms in the exponent are the other small abundances of multiflavored particles. The fugacities not associated with strangeness, as well as the yield fugacity γ_{s}, are incorporated in Eq. (11.46) into the phase-space factors $\tilde{F}_i$ for simplicity of notation.

The number of Ω is obtained by differentiating $\ln Z_{\mathrm{f}}(\beta, V)$ with respect to λ_Ω, and subsequently neglecting the subdominant terms in the exponent,

$$\langle n_\Omega \rangle \simeq \frac{\tilde{F}_\Omega}{I_0} \int_0^{2\pi} \frac{d\varphi}{2\pi} e^{3i\varphi} e^{\tilde{F}_{\mathrm{f}} e^{i\varphi} + \tilde{F}_{\bar{\mathrm{f}}} e^{-i\varphi}}. \tag{11.47}$$

The result of the integration is easily read off the meromorphic expansion, Eq. (11.45), to be $Z_{\mathrm{f}}(\beta, V; n_{\mathrm{f}} = -3)$, Eq. (11.44). This result is easily understood, the three strange quarks in the particle observed are balanced by the background of singly strange particles (kaons and antihyperons),

$$\langle n_\Omega \rangle \simeq \tilde{F}_\Omega \left(\frac{\tilde{F}_{\mathrm{f}}}{\tilde{F}_{\bar{\mathrm{f}}}} \right)^{-3/2} \frac{I_3(2N_{\mathrm{pair}}^{\mathrm{GC}})}{I_0(2N_{\mathrm{pair}}^{\mathrm{GC}})}. \tag{11.48}$$

We recall that, according to Eq. (11.36), the middle term is just the fugacity factor λ_{s}^3. The first two factors in Eq. (11.48) constitute the grand-canonical yield, while the last term is the canonical Ω-suppression factor. A full treatment of the canonical suppression of multistrange particle abundances in small volumes has been used to obtain particle yields in elementary interactions [60].

Similarly, one finds that the suppression of Ξ abundance has the factor I_2/I_0, whereas, as discussed for the general example of the flavor-pair yield, the yield of single strange particles is suppressed by the factor I_1/I_0. The yield of all flavored hadrons in the canonical approach (superscript 'C') can be written as a function of the yield expected in the grand-canonical approach in the general form

$$\langle s^\kappa \rangle^{\mathrm{C}} = \tilde{F}_\kappa \left(\frac{\tilde{F}_{\mathrm{f}}}{\tilde{F}_{\bar{\mathrm{f}}}} \right)^{\kappa/2} \frac{I_{|\kappa|}(2N_{\mathrm{pair}}^{\mathrm{GC}})}{I_0(2N_{\mathrm{pair}}^{\mathrm{GC}})} = \langle s^\kappa \rangle^{\mathrm{GC}} \frac{I_{|\kappa|}(2N_{\mathrm{pair}}^{\mathrm{GC}})}{I_0(2N_{\mathrm{pair}}^{\mathrm{GC}})}, \tag{11.49}$$

with $\kappa = \pm 3, \pm 2$, and ± 1 for Ω, Ξ, and Y and K, respectively. On the left-hand side in Eq. (11.49) the power indicates the flavor content in the particle considered, with negative numbers counting antiquarks. We note, on inspecting the final form of Eq. (11.49), that the canonical suppression of particle and antiparticle abundances is the same, certainly so when we study systems with several pairs present. In very small systems, one may need to evaluate the quantum distributions including multistrange particles in order to obtain precise results. A particle/antiparticle asymmetry can occur if baryon/antibaryon asymmetry applies.

The simplicity of Eq. (11.49) originates from the assumption that the contributions of singly strange particles to conservation of strangeness are dominant. This assumption is consistent with the neglect of quantum statistics. In fact, on expanding the Bose distribution for kaons, one finds that the next-to-leading-order contribution, which behaves as strangeness $n_s = \pm 2$ hadrons, is dominating the influence of all multistrange hadrons. Our study is consistent with the Boltzmann statistics assumed here; more complex evaluation taking multistrange hadrons into account, but considering kaons as Boltzmann particles, is theoretically inconsistent.

11.7 Canonical suppression of strangeness and charm

The canonical flavor yield suppression factor,

$$
\eta \equiv \frac{I_1\left(2\gamma\sqrt{\tilde{F}_f\tilde{F}_{\bar{f}}}\right)}{I_0\left(2\gamma\sqrt{\tilde{F}_f\tilde{F}_{\bar{f}}}\right)} = \frac{I_1(2N_{\mathrm{pair}}^{\mathrm{GC}})}{I_0(2N_{\mathrm{pair}}^{\mathrm{GC}})} < 1, \tag{11.50}
$$

depends in a complex way on the volume of the system, or, expressing it alternatively, on the grand-canonical number of pairs, $N_{\mathrm{pair}}^{\mathrm{GC}}$. The suppression function $\eta(N) \equiv I_1(2N)/I_0(2N)$ is shown in Fig. 11.6 as a function of N. For $N > 1$, we see (dotted lines) that the approach to the grand-canonical limit is relatively slow; it follows the asymptotic form

$$
\eta \simeq 1 - \frac{1}{4N} - \frac{1}{128N^2} + \cdots, \tag{11.51}
$$

whereas for $N \ll 1$, we see a nearly linear rise:

$$
\eta = N - \frac{N^3}{2} + \cdots. \tag{11.52}
$$

Overall, when the the yield of particles is small, we have, using Eq. (11.52),

$$
N_{\mathrm{pair}}^{\mathrm{CE}} = (N_f^{\mathrm{GC}})^2. \tag{11.53}
$$

Hagedorn was puzzled by this quadratic behavior of the particle yield, being concerned about rarely occurring astrophysical processes of pair production. In his 1970/71 CERN lectures [141], he asked how the yield

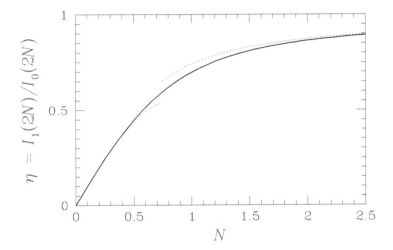

Fig. 11.6.   Solid line: the canonical yield-suppression factor as a function of the grand-canonical yield of particles N. Dotted lines: asymptotic expansion forms presented in the text.

of particles can be $Y \propto e^{-2m/T}$, when the threshold for production of a pair is relevant, and another time $Y \propto e^{-m/T}$, when the statistical yield is evaluated. This is the grand-canonical yield for $m > T$, as seen in section 10.4,

$$N^{\text{GC}} = \frac{g_{\text{f}}}{2\pi^2} T^3 V \sqrt{\frac{\pi m^3}{2T^3}} e^{-m/T}, \tag{11.54}$$

whereas when the yield of particles is small, e.g., when $m \gg T$, the canonical result applies:

$$N^{\text{CE}} = \frac{g_{\text{f}}^2}{8\pi^3} T^3 m^3 V^2 e^{-2m/T}. \tag{11.55}$$

We see that the Hagedorn puzzle has been resolved. The reaction volume is an important factor controlling which of the two results Eqs. (11.54) and (11.55) should be considered in a given physical situation.

We next consider whether there is any effect of QGP compared with HG in the study of canonical conservation of strangeness. The possible difference would arise from the different sizes of the phase space for strangeness in these two phases of matter. In the Boltzmann limit, the flavor and antiflavor phase space in the symmetric QGP is:

$$\tilde{F}_{\text{f}} = \tilde{F}_{\bar{\text{f}}} = g_{\text{f}} V \int \frac{d^3 p}{(2\pi)^3} e^{\frac{\sqrt{p^2 + m_{\text{f}}^2}}{T}} = \frac{3VTm_{\text{f}}^2}{\pi^2} K_2(m_{\text{f}}/T). \tag{11.56}$$

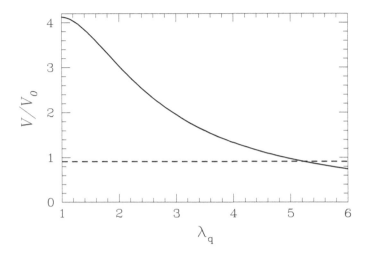

Fig. 11.7. Volume needed for one strange quark pair using grand-canonical counting as function of λ_q for $T = 160$ MeV, $\gamma_q = 1, \gamma_s = 1$, $V_h = (4\pi/3)\, 1\,\text{fm}^3$. Solid line: hadron gas phase space, dashed line: quark phase space with $m_s = 160$ MeV.

For the hadronic phase, it is derived from Eqs. (11.19) and (11.20). We have, counting strange-quark content as positively 'flavor charged' as before,

$$\tilde{F}_f = \lambda_q^{-1}\tilde{F}_K + \lambda_q^2\tilde{F}_Y, \tag{11.57}$$

$$\tilde{F}_{\bar{f}} = \lambda_q\tilde{F}_K + \lambda_q^{-2}\tilde{F}_Y. \tag{11.58}$$

All these quantities $\tilde{F}_i$ are proportional to the reaction volume.

The interesting result, seen in Fig. 11.6, is that the suppression of yield is at the level of 30% when one pair of particles would be expected to be present in grand-canonical chemical equilibrium; the suppression means that instead we find that the true phase-space yield is 0.7 pairs. Actually, in p–p interactions at 158 GeV/c projectile momentum, the analysis of experimental results yields 0.66 ± 0.07 strange pairs [277].

Pursuing this line of thought, but also to obtain a reference regarding the magnitudes involved for strangeness, we consider how big a volume we need in order to find (using grand-canonical-ensemble counting) one pair of strange particles. In the hadronic phase space, with λ_s chosen to conserve strangeness, we have

$$\frac{V}{V_h} = \frac{2\pi^2}{V_h T^3 \gamma_q \gamma_s \sqrt{(F_K + \lambda_q^3 F_Y)(F_K + \lambda_q^{-3} F_Y)}}. \tag{11.59}$$

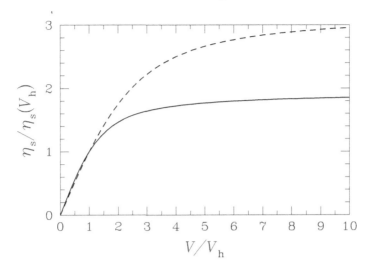

Fig. 11.8.   Canonical yield enhancement at large volumes compared with the case of unit hadron volume $V_h = \frac{4}{3}\pi$ fm^{-3}. Solid line, QGP phase; dashed line, HG.

For p–p interactions, we consider $T = 160$ MeV, and the elementary hadronic volume is chosen to be $V_h = \frac{4}{3}\pi$ fm^3. The applicable value of λ_q, if statistical methods are used, is close to unity. The result is shown as the solid line in Fig. 11.7, as a function of λ_q, for $\gamma_q = 1$ and $\gamma_s = 1$. The dashed line is the corresponding result for the QGP strange-quark phase space, which naturally does not depend on λ_q, and has been obtained by choosing $m_s = 160$ MeV. Just a little less than one hadronic volume suffices; one finds one pair in V_h for $m_s = 200$ MeV.

We show in Fig. 11.8 the canonical strangeness-suppression factor Eq. (11.50), both for a QGP (solid line) and for a HG (dashed line). We have converted the suppression η into an enhancement by normalizing at $\eta(V = V_h)$. For the QGP, we take $m_s = 160$ MeV, whereas, for a HG, we take $\mu_b = 210$ MeV. Both phases are considered at $T = 145$ MeV. Since the strangeness content in QGP is greater than that in HG, there is less 'catching up' to do and the overall yield is increased by factor 1.8, whereas for HG, we find an increase by a factor of three. Practically all of this enhancement occurs when the reaction volume increases to five, i.e., for rather small reaction systems.

We now look at the suppression of multistrange particle abundances by the factors $\eta_3(N) = I_3(2N)/I_0(2N)$, for Ω, and $\eta_2(N) = I_2(2N)/I_0(2N)$, for Ξ. For small values of N, we obtain

$$\eta_\kappa \equiv \frac{I_\kappa(2N)}{I_0(2N)} \rightarrow N^\kappa \frac{1}{\kappa!}\left(1 - \frac{\kappa}{\kappa+1}N^2\right). \qquad (11.60)$$

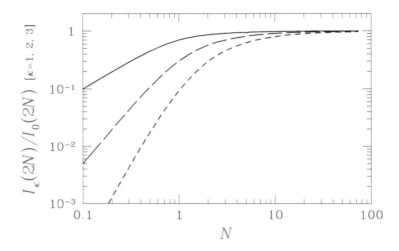

Fig. 11.9.   Canonical yield-suppression factors I_κ/I_0 as function of the grand-canonical particle yield N. Short-dashed line: the suppression of triply-flavored hadrons; long-dashed line: the suppression of doubly-flavored hadrons; and solid line, the suppression of singly-flavored hadrons.

This result is easily understood on physical grounds: for example, when the expected grand-canonical yield is three strangeness-containing pairs, it is quite rare that all three strange quarks go into an Ω. This is seen in Fig. 11.9 (short-dashed curve), and in fact this will occur about a tenth as often as we would expect from computing the yield of Ω, ignoring the canonical conservation of strangeness. The other lines in Fig. 11.9 correspond to the other suppression factors; the long-dashed line is $\eta_2(N) = I_2(2N)/I_0(2N)$ and the solid line is $\eta(N) = I_1(2N)/I_0(2N)$. They are shown to be dependent on the number of strange pairs expected in the grand-canonical equilibrium, denoted in Fig. 11.9 as N.

It has been proposed to exploit the canonical suppression which grows with strangeness content to explain the increase in production of strange hadrons seen in Fig. 1.6 on page 19, when the per-participant yield in A–A interactions is compared with that from p–Be interactions [228]. A direct comparison of the reduction factors η_κ is possible. Choosing as the reference point the yield $N \lesssim 1$, the claim is that one can come close to explaining the enhancement in production of three out of five strange hadrons seen in Fig. 1.6. The reader should notice that the enhancement effect is derived from the suppression of the base yield in the small reference system. We obtain this effect by rebasing the results shown in Fig. 11.9 to the strangeness yield observed in p–p reactions evaluated within canonical formulation; see Fig. 11.10. The three cases studied in Fig. 11.9 are seen, where the dotted lines are derived from

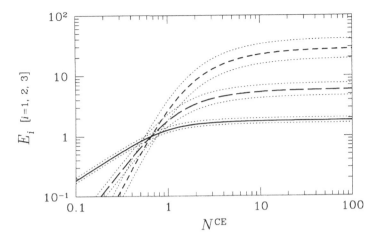

Fig. 11.10. Canonical yield-suppression factor of Fig. 11.9 expressed as enhancement factors E_i, $i = 1, 2, 3$ as functions of the canonical-pair-particle yield N^{CE}. Solid line: E_1, the enhancement of singly-flavored hadrons, relative to the yield 0.66 ± 0.07, expected in p–p reactions. Similarly, long-dashed line: E_2, enhancement of doubly-flavored hadrons; and short-dashed line: E_3, enhancement of triply-flavored hadrons. Dotted lines correspond to the errors arising from the error in the strangeness yield, to which the results are normalized.

the error in the reference yield of production of strangeness in p–p reactions.

We see, in quantitative terms, the strength of the canonical effect, especially for multistrange hadrons, and its rapid rise with the yield of strangeness [223]. The canonical-enhancement effect rises rapidly but smoothly and saturates at the grand-canonical yield in rather small systems. The grand-canonical chemical-equilibrium yield is reached for systems comprising ten strangeness pairs and for reaction systems about six times greater than the p–p system, considering that the yield of singly-strange particles is enhanced by a factor three, as is seen in Fig. 11.10. This result is inconsistent with the experimental results from the NA52 experiment [153], which reveal an abrupt threshold for enhancement of production of strangeness at $\simeq 50$ participants, just where the WA57 team recently reported a sudden onset of enhancement in yield of Ξ [108]. Given the sensitivity of the results shown in Fig. 11.10 to the strangeness reference yield, it is natural to conclude that the explanation of strange-hadron enhancement offered in [228] is based on a fine tuned p–Be strangeness yield, not cross-checked with the (at-present-unavailable) experimental yield.

We addressed, with such a great precision, the canonical chemical-equilibrium yields of strange particles expected to originate from small

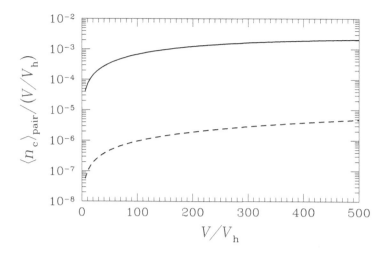

Fig. 11.11.   The canonical yield of pairs of open charm quarks $\langle n_c \rangle_{\mathrm{pair}}$ per unit volume as a function of volume, in units of $V_{\mathrm{h}} = \frac{4}{3}\pi$ fm^3. Solid line, QGP with $m_c = 1.3$ GeV; dashed line, HG at $\mu_b = 210$ MeV, both phases at $T = 145$ MeV.

systems, since hadron yields observed in p–p and p–$\bar{\mathrm{p}}$ interactions, and even in hadron jets produced in LEP e$^+$–e$^-$ reactions are remarkably close to the expectations for chemical equilibrium [62], allowing in the analysis for the canonical suppression and including the effects of quantum degeneracy. This thorough analysis results in a not completely satisfactory χ^2 per degree of freedom ($= 61/21$). Yet a reader of this thorough report will have the impression that a modern-day Maxwell's demon must be at work, generating canonical chemical equilibria for hadrons in all these elementary interactions, and abundances of strange quarks within a factor of two of absolute chemical equilibrium.

On the other hand, a demon that works for strangeness should also work for charm. The yield of charm in Pb–Pb interactions is estimated from the lepton background at 0.5 pairs per central collision [13]. We can use the small-N expansion, Eq. (11.60). The corresponding A–A canonical enhancement factor, compared with p–A, is $N_{\mathrm{AA}}/N_{\mathrm{pA}} \simeq 100A$. (Here N is now the yield of 'open' charm rather than strangeness.) The measured open-charm cross sections, however, scale with the number of participants, and there is no space for a large canonical enhancement/suppression of production of charm. To be more specific, we show, in Fig. 11.11, the specific yield of charm $\langle n_c \rangle_{\mathrm{pair}}$ per unit volume as a function of the volume. The canonical effect is the deviation from a constant value and it is significant, $\mathcal{O}(100)$. Even at $V = 500V_{\mathrm{h}}$ the infinite-volume grand-canonical limit is not yet attained, for the case of the larger phase space of QGP (solid line), the total yield of charm is just one charm pair. The absolute yield in both phases is strongly dependent on the temperature used, here $T = 145$ MeV. In QGP, we took $m_c = 1.3$ GeV. The phase space of a HG includes all known charmed mesons and baryons, with abundances of light quarks controlled by $\mu_b = 210$ MeV and $\mu_s = 0$.

Although, by choosing a slightly higher value of T, we can easily increase the equilibrium yield of charm in a HG to the QGP level [133], this does not eliminate the effect of canonical suppression of production of charm if chemical equilibrium is assumed for charm in the elementary interactions. We are simply so deep in the 'quadratic' domain of the yield, see Eq. (11.60), that playing with parameters changes nothing, since we are constrained in Pb–Pb interactions by experiment to have a yield of charm of less than one pair.

It is natural to argue that the very heavy charm quarks are not in chemical equilibrium, and that their production has to be studied in kinetic theory of collision processes of partons. However, this means that there is no twenty-first-century Maxwell's demon with control of charm, and, of course, also not of strangeness. The production and enhancement of charm and strangeness in heavy-ion collisions is in our opinion a kinetic phenomenon. To study it, we should explore a wide range of collision volume and energy. The objective is to determine boundaries of the high, possibly QGP-generated, yields.

12 Hagedorn gas

12.1 The experimental hadronic mass spectrum

One of the most striking features of hadronic interactions, which was discovered by Hagedorn [140], is the growth of the hadronic mass spectrum with the hadron mass. With the 4627 different hadronic states we have used in the study of properties of HG in section 11.1 [136], it is reasonable to evaluate the mass spectrum of hadronic states $\rho(m)$, defined as the number of states in the mass interval $(m, \, m + dm)$. We represent each particle by a Gaussian, and obtain $\rho(m)$ by summing the contributions of individual hadronic particles:

$$\rho(m) = \sum_{m^* = m_\pi, m_\rho, \dots} \frac{g_{m^*}}{\sqrt{2\pi}\sigma_{m^*}} \exp\left(-\frac{(m - m^*)^2}{2\sigma_{m^*}^2}\right). \tag{12.1}$$

Here, g_{m^*} is the degeneracy of the hadron of mass m^* including, in particular, spin and isospin degeneracy, and $\sigma = \Gamma/2$, $\Gamma = \mathcal{O}(200)$ MeV being the width of the resonance. The pion, with $m_\pi \simeq \sigma$ is a special case, and is set aside in such smoothing of the mass spectrum. Downward modification of its mass has a great impact on properties of HG and is thus not allowed.

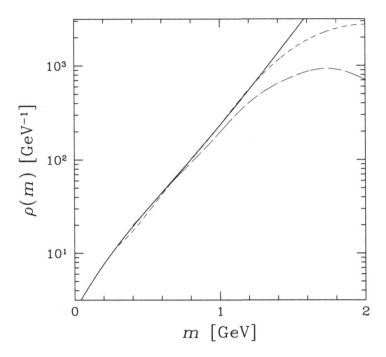

Fig. 12.1.   Dashed lines are the smoothed hadronic mass spectrum. The solid
line represents the fit Eq. (12.2) with $k = -3$, $m_0 = 0.66$ GeV, and $T_0 = 0.158$
GeV. Long-dashed line: 1411 states of 1967. Short-dashed line: 4627 states of
1996.

We compare the logarithm of the resulting smoothed mass spectrum
for the hadronic particles known in 1967 (long-dashed line) with that for
those known in 1996 (short-dashed line) in Fig. 12.1. We see that, in the
20 years following Hagedorn's last study of the phenomenon, the newly
classified hadron resonances have improved the exponential behavior. We
refer to a hadronic gas with an exponential mass spectrum as a Hagedorn
gas. The solid line in Fig. 12.1 represents a fit using the empirical shape

$$\rho(m) \approx c(m_0^2 + m^2)^{k/2} \exp(m/T_0) \tag{12.2}$$

with $k = -3$. This value is preferred in the statistical-bootstrap model,
section 12.2. However, many other values of k fit the mass spectrum well.
The inverse slope T_0 and the preexponential power k are correlated in a
fit of the mass-spectrum data and we present, in table 12.1, the results
for several choices of k. We shall show that the value of k determines
the behavior of the thermodynamic quantities of a gas of hadrons when
$T \to T_0$ and its value is of some relevance.

Table 12.1. Fitted parameters of Eq. (12.2) for given k

k	c	m_0	T_0
-2.5	0.83479	0.6346	0.16536
-3.0	0.69885	0.66068	0.15760
-3.5	0.58627	0.68006	0.15055
-4.0	0.49266	0.69512	0.14411
-5.0	0.34968	0.71738	0.13279
-6.0	0.24601	0.73668	0.12341
-7.0	0.17978	0.74585	0.11489

The mass spectra for fermions $\rho_F(m)$ and bosons $\rho_B(m)$ can differ, and, using these two functions, the generalization of Eq. (10.62) reads

$$\ln \mathcal{Z}_{HG} = \frac{\beta^{-3}V}{2\pi^2} \sum_{n=1}^{\infty} \rho_n(m)\frac{1}{n^4}(n\beta m)^2 K_2(n\beta m), \qquad (12.3)$$

where

$$\rho_n(m) \equiv \rho_B(m) - (-1)^n \rho_F(m), \qquad (12.4)$$

The Boltzmann approximation amounts to keeping in Eq. (12.3) the term with $n = 1$, in which case

$$\rho(m) \equiv \rho_1(m) = \rho_B(m) + \rho_F(m). \qquad (12.5)$$

To understand how the parameter k influences the behavior of the Hagedorn gas, we now introduce the asymptotic form Eq. (10.45) with the first term only, and consider the (classical, 'cl') Boltzmann limit,

$$\ln \mathcal{Z}_{HG}^{cl} = cV \left(\frac{T_0}{2\pi}\right)^{3/2} \int_{M_0}^{\infty} m^{k+3/2} e^{(m/T_0 - m/T)} \, dm + D(T, M_0), \quad (12.6)$$

where $M_0 > m_0$ is a mass above which the asymptotic form of K_2 holds, and where $D(T, M_0)$ is finite. Because of the exponential factor, the integral is divergent for $T > T_0$, and the partition function is singular at T_0 for a range of k.

The pressure and the energy density for $T \to T_0$ are

$$P(T) \to \begin{cases} \left(\frac{1}{T} - \frac{1}{T_0}\right)^{-(k+5/2)}, & \text{for } k > -\frac{5}{2}, \\[2mm] \ln\left(\frac{1}{T} - \frac{1}{T_0}\right), & \text{for } k = -\frac{5}{2}, \\[2mm] \text{constant}, & \text{for } k < -\frac{5}{2}; \end{cases} \qquad (12.7)$$

and

$$
\epsilon \rightarrow
\begin{cases}
\left(\dfrac{1}{T} - \dfrac{1}{T_0}\right)^{-(k+7/2)}, & \text{for} \quad k > -\tfrac{7}{2}, \\[2ex]
\ln\left(\dfrac{1}{T} - \dfrac{1}{T_0}\right), & \text{for} \quad k = -\tfrac{7}{2}, \\[2ex]
\text{constant}, & \text{for} \quad k < -\tfrac{7}{2}.
\end{cases}
\tag{12.8}
$$

The energy density goes to infinity for $k \geq -\tfrac{7}{2}$, when $T \to T_0$, and in this range falls the result of the statistical-bootstrap model with point hadrons, Eq. (12.35). Therefore T_0 appears as a limiting temperature for such a hadronic system [140].

Interestingly, the partition function and its derivatives may be singular at $T = T_0$ even when the volume of the system is finite, unlike the more conventional situation, with a true singularity expected only if the volume is infinite. However what is actually needed is an infinite number of participating particles, which in the conventional situation can occur only for $V \to \infty$. In relativistic statistical physics, particles are produced, and, for an exponential mass spectrum, an infinite number of particles arises already in a finite volume, for a sufficiently singular value of k and point-like hadrons, and $T \to T_0$. When hadrons of finite volume are considered, we find in section 12.3 that the energy density remains finite at $T = T_0$, independently of the value of the mass power k in the hadronic mass spectrum Eq. (12.2).

The reader will wonder whether the seemingly small difference between the exponential mass spectrum, and the so-far-known hadron mass spectrum, seen in Fig. 12.1 for $m > 1.5$ GeV, matters. We now compare the energy and pressure of HG evaluated using individual hadrons, thin lines in Fig. 12.2 (see also Fig. 11.1), with the results obtained using the analytical mass spectrum defined by Eq. (12.2), with parameters given in table 12.1. The vertical dotted line shows the limiting temperature for $k = -3$. Comparing the thick lines (exponential mass spectrum) with the results including known hadrons only (thin lines), we see in Fig. 12.2 significant differences both for ϵ/T^4 (a factor of four for $k = -3$) and P/T^4 (a factor of two for $k = -3$) in the physically relevant domain $T \simeq 150$ MeV. The various thick lines correspond to values of k listed in table 12.1 and can be assigned by noting at which value of temperature T_0 the singular behavior arises.

One would be tempted to conclude that, without full knowledge of the hadronic spectrum, we cannot use individual hadrons in the study of the properties of the HG, and hence evaluation of the total multiplicity of hadronic particles, as, e.g., is required in order to obtain Fig. 9.8 on page 169. There is, however, another effect, which counterbalances the

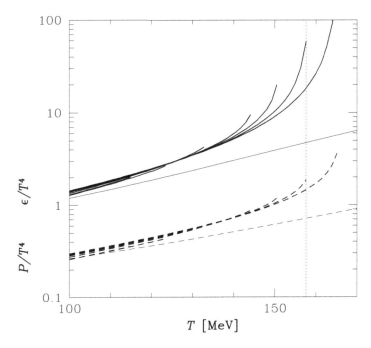

Fig. 12.2. The energy density ϵ/T^4 (solid lines) and pressure P/T^4 (dashed lines) (on a logarithmic scale) for a hadronic gas with a smoothed exponential mass spectrum, with values of $k = -2.5$ (the most divergent thick line) to $k = -7$ in steps of 0.5. The thin lines were obtained by using the currently known experimental mass spectrum. All fugacities γ and $\lambda = 1$.

effect of missing hadron resonances. When the finite size of a hadron is introduced, e.g., according to Eq. (11.1), significant decreases in magnitude of energy density, pressure, and number of particles at a given temperature ensue. For the value $\mathcal{B}^{1/4} = 190\,\text{MeV}$, corresponding to $4\mathcal{B} = 0.68$ GeV fm^{-3}, a value we introduce to reproduce lattice QCD results in section 16.2, we show in Fig. 12.3 that there is practical agreement between the exponential mass-spectrum properties with finite-volume correction (thick lines) and the point hadron gas evaluated using known hadrons (long-dashed thin line for ϵ/T^4 and dotted thin line for P/T^4). Considering that the population of very massive resonances is not going to rise to full chemical equilibrium in nuclear collisions, along with the uncertainties in the finite-volume correction, (e.g., choice of $\mathcal{B}$) the remaining 15%–20% difference between the resonance gas with finite-volume correction and the point gas of known hadrons is not physically relevant. On the other hand, this clearly is the level of precision (theoretical systematic error) of the current computation of abundances of hadrons using

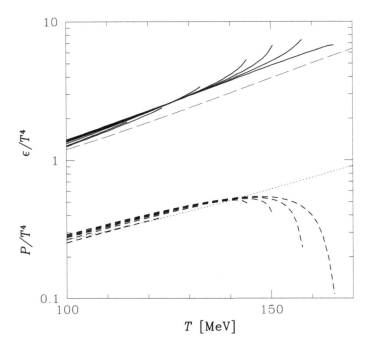

Fig. 12.3.   The same as Fig. 12.2 but thick lines now show the gas of hadrons with an exponential mass spectrum including the finite-volume correction with $\mathcal{B} = (190\,\mathrm{MeV})^4$.

the known hadron spectrum. Most of this remaining systematic error disappears when hadron-abundance ratios are evaluated.

   A cross check of the validity of the energy density and pressure obtained either by summing the physically known spectrum of point hadrons, or by employing the exponentially extrapolated spectrum of finitely sized hadrons is obtained by comparing them with lattice-gauge results. Results presented in section 15.5, in Fig. 15.3, show that, at the critical temperature, $\epsilon/T_c^4 \simeq 6.5$. On comparing this with results shown in Fig. 12.3, we see that this result is consistent with the exponentially extrapolated results for $-3.5 \leq k \leq -2.5$ corrected for finite hadron volume, with 150 MeV $\lesssim T_c \lesssim$ 165 MeV, the center of the range of lattice simulations. Using only the known point hadrons, a slightly larger value of $T_c \simeq 171$ MeV is found. Comparison of pressure, shown in Fig. 12.3, with the lattice result in Fig. 16.2 is more difficult but clearly the results are also in qualitative agreement.

   We have learned that the use in the field of heavy-ion collisions of a gas of point hadrons is justified because the contributions of probably still unknown hadronic resonances and the excluded-volume effect approximately cancel out. These remarks apply to all values of k we considered, though clearly the values $k = -3.5, -3$, and -2.5 are privileged by the comparison with lattice-gauge-theory results, and the value $k = -3$ is also central to the statistical-bootstrap model. We will now address this theoretical framework which leads to the exponential mass spectrum. We stress that our measure of singularity k refers in this book always to the point-particle theory; consideration of the finite hadronic volume removes the singular behavior of the hadronic energy density.

12.2 The hadronic bootstrap

To study interacting hadrons in a volume V, we first consider the N-particle level density $\sigma_N(E, V)$, a generalization of Eq. (4.37). σ_N generates the N-particle partition function,

$$Z_N(\beta, V) = \int \sigma_N(E, V) e^{-\beta E} dE. \tag{12.9}$$

For the non-interacting case, the number of states σ_N of N particles is obtained by carrying out the momentum integration Eq. (4.36) for each particle, keeping the total momentum $\vec{P} = 0$ and the energy E fixed. We divide by $N!$ for indistinguishable particles of degeneracy g and obtain

$$\sigma_N(E, V) = \frac{g^N V^N}{(2\pi)^{3N} N!} \prod_{i=1}^{N} \int \delta\left(\sum_{i=1}^{N} \varepsilon_i - E\right) \delta^3\left(\sum_{i=1}^{N} \vec{p}_i\right) d^3 p_i, \tag{12.10}$$

where the single-particle energy Eq. (4.31) is $\varepsilon_i = \sqrt{p_i^2 + m^2}$.

If an interaction between these particles is such that they form a bound state with mass m^* and nothing else happens, then the level density of this new system, including the effect of interaction, would be described as a mixture of ideal gases, one of mass m and the other of mass m^*. The logarithm of the partition function of such a system is additive, and the interaction in the gas of the mass m is accounted for by allowing for the presence of the second gas of mass m^*.

Beth and Uhlenbeck [64] formulated this argument more precisely for the case in which the interaction leads to the formation of a resonance in a scattering process, e.g.,

$$\pi + N \rightarrow \Delta \rightarrow \pi + N.$$

In such a case, the ℓth partial wave will be at large distances,

$$\psi_\ell(r, p) \sim \frac{1}{pr} \sin\left(pr - \frac{\ell\pi}{2} + \eta_\ell(p)\right), \tag{12.11}$$

where $\eta_\ell(p)$ is the phase shift due to scattering.

To simplify, we argue in a manner similar to the study of the level density above Eq. (4.37). We consider a large sphere of radius R. The wave function Eq. (12.11) should vanish at $r = R$:

$$pR - \frac{\ell\pi}{2} + \eta_\ell(p) = n_\ell\pi\,; \qquad n_\ell = 0,\, 1,\, 2,\, \ldots\,. \tag{12.12}$$

n_ℓ labels the allowed two-body spherical momentum states $\{p_0,\, p_1,\, \ldots\}$. The density of states of angular momentum ℓ at p is $\Delta n_\ell/\Delta p$ and

$$\frac{dn_\ell}{dp} = \frac{R}{\pi} + \frac{1}{\pi}\frac{d}{dp}\eta_\ell(p). \tag{12.13}$$

Without interaction, $\eta_\ell(p) \equiv 0$, and we recognize that the interaction changes the two-particle density of states by $(1/\pi)\,d\eta_\ell/dp$.

We recall that the presence of a resonance leads to a rapid phase shift by π over the width of the resonance. In what follows, we shall assume that hadronic resonances are narrow, thus

$$\frac{1}{\pi}\frac{d\eta_\ell(p')}{dp} \approx \sum_* \delta(p' - p^*). \tag{12.14}$$

Such a δ-function appearing in the density of states Eq. (12.13) is exactly equivalent to the introduction of additional particles with masses m^*, which can be obtained from the masses of scattered particles and the relative momentum of the resonance p^*.

Consider now the probability for an N-body final state in a collision,

$$P(E, N) = \int |\langle f|S|i\rangle|^2 \delta\!\left(E - \sum_{i=1}^{N} E_i\right) \delta^3\!\left(\sum_{i=1}^{N} \vec{p}_i\right) \prod_{i=1}^{n} d^3 p_i$$

$$\equiv \int |S|^2\, dR_N\!\left(E, m_1, m_2, \ldots, m_N\right), \tag{12.15}$$

where the second expression introduces a short-hand notation for the N-particle phase-space volume element dR_N. Note that, in the Fermi model [121], the S-matrix element $|\langle f|S|i\rangle|^2$ is taken to be constant. Now we use Eq. (12.13) with Eq. (12.14) assuming that there is just one resonance. Then

$$P(E, N) = \int |S'|^2\, dR_N\!\left(E, m_1, m_2, \ldots, m_N\right)$$

$$+ \int |S'|^2\, dR_{N-1}\!\left(E, m^*, m_3, \ldots, m_N\right). \tag{12.16}$$

The first term comes from R/π in Eq. (12.13) and the second term from $(1/\pi)\,d\eta_\ell(p')/dp'$ as given by Eq. (12.14) when there is a resonance in the

ℓth partial wave. We write S' instead of S to indicate that the part of the interaction responsible for the existence of a resonance between particles 1 and 2 is eliminated from S [63].

This manipulation was first done by Belenkij [63], but, until the work of Hagedorn [140], it was not pushed to its full consequences, involving many resonances, and the observation that the hadronic interactions are completely dominated by resonances. In that case, one can continue the process that led from Eq. (12.15) to Eq. (12.16) and the final hadronic state produced in any interaction is described by the sum over all possible N-particle phase spaces involving all possible hadronic states characterized by the mass spectrum $\rho(m)$, Eqs. (12.1) and (12.2).

Knowledge of all phase shifts in all channels, including, $2, 3, \ldots, n$-body phase shifts, is equivalent to the definition of the full S-matrix. If hadronic resonances characterize the phase shifts, then one can say that knowledge of the resonance spectrum determines the physics considered, or in reverse, hadronic interactions manifest themselves solely by the formation of resonances.

Can $\rho(m)$ be estimated in some way from the hypothesis that resonances dominate strong interactions? We follow the arguments of Hagedorn of 1965 [140], the statistical-bootstrap hypothesis. Consider the partition function given by Eq. (12.3) in the classical Boltzmann limit:

$$Z_{\mathrm{HG}}^{\mathrm{cl}}(V,T) = \exp\left[\frac{VT}{2\pi^2} \int_0^\infty \rho(m) m^2 K_2\left(\frac{m}{T}\right) dm\right]. \tag{12.17}$$

This equation expresses the partition function of the hadronic system of volume V at temperature T in terms of the hadrons whose hadronic mass spectrum is $\rho(m)$. Since we are looking for the asymptotic form of $\rho(m)$, we can replace the Bessel function $K_2(m/T)$ in Eq. (12.17) by its asymptotic form, Eq. (10.45), to obtain

$$Z_{\mathrm{HG}}^{\mathrm{cl}}(V,T) \simeq \exp\left[\int_0^\infty \left(\frac{mT}{2\pi}\right)^{3/2} \rho(m) e^{-m/T} dm\right]. \tag{12.18}$$

The Boltzmann factor $e^{-m/T}$ in the partition function shows that, with rising temperature, the contribution of resonance states of higher masses becomes more and more important.

On the other hand, the partition function of the same hadronic system can be written in terms of the density of all single-particle hadronic levels $\sigma_1(E, V)$:

$$Z_{\mathrm{HG}}^{\mathrm{cl}}(V,T) = \exp\left(\sum_i e^{-E_i/T}\right) = \exp\left(\int_0^\infty \sigma_1(E,V) e^{-E/T} dE\right). \tag{12.19}$$

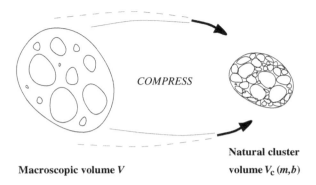

Fig. 12.4.   The statistical-bootstrap idea: a system compressed to the 'natural cluster volume' becomes itself a cluster (consisting of clusters (consisting of . . .)).

The partition function $Z^{\mathrm{cl}}_{\mathrm{HG}}(V,T)$ for the same hadronic system is expressed in two different ways, once in term of the mass spectrum of its *constituents* and once in term of the density of states of the system as a *whole*. The physical meanings of $\sigma(E,V)$ and of $\rho(m)$ must be clearly understood:

- $\sigma_1(E,V)dE$ is the number of states between E and $E+dE$ of an interacting system enclosed in an externally given volume V; and
- $\rho(m)\,dm$ is the number of different hadronic resonance states between m and $m+dm$ of an interacting system confined to its 'natural volume' V_{c}, i.e., to the volume resulting from the forces keeping interacting hadrons together as resonances.

Now, if we could compress a macroscopic hadron system to that small volume which would be the natural volume $V_{\mathrm{c}}(E)$ corresponding to the energy E, it would itself become another hadron, just one among the infinite number counted by the mass spectrum $\rho(m)$. This bootstrap idea is represented in Fig. 12.4. This hypothesis implies that Eq. (12.10) can now be written as an equation for the hadronic mass spectrum, which we cast into relativistically covariant form, akin to the form we discuss below, Eq. (12.63), including finite volume and baryon number

$$
\mathcal{H}\rho(p^2) = \mathcal{H}\delta_0(m^2 - m^2_{\mathrm{in}})
$$
$$
+ \sum_{N=2}^{\infty} \frac{1}{N!} \int \delta^4\!\left(p - \sum_{i=1}^{N} p_i\right) \prod_{i=2}^{N} \mathcal{H}\rho(p_i^2) d^4 p_i, \qquad (12.20)
$$

where $\mathcal{H} \propto V_{\mathrm{c}}$, and we have separated out the first 'input' term. As before, $\delta_0(p^2 - m^2) = \Theta(p_0)\delta(p^2 - m^2)$. Eq. (12.20) is the statistical-bootstrap equation for the hadronic mass spectrum. There are two input constants entering, namely $\mathcal{H}$ and the mass m_{in} in the single-particle term. All other hadrons are clusters in their respective volumes, generated by this single particle of mass m_{in}.

Interestingly, a semi-analytical solution of Eq. (12.20) is available. We consider the relativistic four-dimensional Laplace transform of Eq. (12.20),

$$\int e^{-\beta \cdot p} \, \mathcal{H}\rho(p^2) \, d^4p = \varphi(\beta) + \sum_{N=2}^{\infty} \frac{1}{N!}$$

$$\times \prod_{i=1}^{N} \int e^{-\beta \cdot p_i} \, \mathcal{H}\rho(p_i^2) \, d^4p_i, \qquad (12.21)$$

where there appears on the right-hand side, because of the δ-function in Eq. (12.20), the product of N identical independent integrals. Defining φ and G by

$$\varphi(\beta) = \int e^{-\beta \cdot p} \, \mathcal{H}\delta_0(p^2 - m_{\text{in}}^2) d^4p = \mathcal{H}2\pi m_{\text{in}}^2 \frac{K_1(\beta m_{\text{in}})}{\beta m_{\text{in}}}, \qquad (12.22)$$

and

$$G = \int e^{-\beta \cdot p} \, \mathcal{H}\rho(p^2) \, d^4p \qquad (12.23)$$

we see that Eq. (12.21) becomes,

$$G(\varphi) = \varphi + e^{G(\varphi)} - G(\varphi) - 1, \qquad (12.24)$$

or,

$$\varphi = 2G(\varphi) - e^{G(\varphi)} + 1. \qquad (12.25)$$

Given the Laplace transform of the hadronic mass spectrum $G(\varphi)$, Eq. (12.23), one can use an inverse Laplace transform to obtain $\rho(m)$ or, at least, to determine its asymptotic behavior, in which we are interested. How one can proceed to solve Eq. (12.25) is shown in Fig. 12.5. We draw on the left in Fig. 12.5(a) the curve $G(\varphi)$ and then invert it, here 'graphically' on the right in Fig. 12.5(b). We see that this solution branch satisfies

$$G(\varphi) \leq \ln 2 = G_0, \qquad (12.26)$$

and increases as a function of

$$\varphi \leq \varphi_0 = \ln(4/e), \qquad (12.27)$$

up to the point where it has a root singularity. We have for $\varphi \to \varphi_0$:

$$G(\varphi) \approx G_0 \pm \text{constant} \times \sqrt{\varphi_0 - \varphi} + \cdots. \qquad (12.28)$$

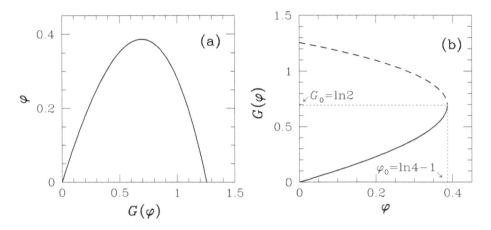

Fig. 12.5.   (a) $G(\varphi)$ according to Eq. (12.25); (b) the graphical solution of Eq. (12.24).

For $\varphi \geq \varphi_0$, $G(\varphi)$ becomes complex and there are non-physical branches of the solution.

It is the square-root singularity of $G(\varphi)$ which determines that the mass power k of the hadronic mass spectrum $cm^k \exp(m/T_0)$ is $k = -3$. We recall that φ is actually itself defined in terms of β by Eq. (12.22), and it is monotonically decreasing with β: there is a minimum value of β_0 corresponding to a maximum value $T_0 = 1/\beta_0$ such that

$$\varphi_0 = \ln(4/e) = \mathcal{H}2\pi m_{\text{in}}^2 \frac{K_1(\beta_0 m_{\text{in}})}{\beta_0 m_{\text{in}}}. \qquad (12.29)$$

This implies, because of Eq. (12.28), that the physical branch of $G(\varphi)$ behaves like

$$G(\beta \simeq \beta_0) = G_0 - \text{constant} \times \sqrt{\beta - \beta_0} \qquad (12.30)$$

near β_0.

However, $G(\varphi)$, Eq. (12.23), can have a singularity at β_0 only if

$$\rho(m^2) \to cm^k e^{\beta_0 m}, \qquad m \to \infty. \qquad (12.31)$$

The behavior of $G(\beta)$ can be made more explicit by introducing $1 = \int \delta_0(p^2 - m^2)dm^2$ in Eq. (12.23) followed by a change of the sequence of integrals,

$$G = \mathcal{H} \int \rho(m^2)2\pi m^2 \frac{K_1(\beta m)}{\beta m} dm^2. \qquad (12.32)$$

Combining Eq. (12.31) with Eq. (12.32) we find that, for $\beta \to \beta_0$,

$$G(\beta \simeq \beta_0) \approx G_0 + \text{constant} \times \int_{m_{in}}^{\infty} e^{-m(\beta - \beta_0)} m^{3/2+k} dm, \qquad (12.33)$$

which yields

$$G(\beta \to \beta_0) \approx G(\beta_0) + \text{constant}$$
$$\times \left(\frac{1}{\beta - \beta_0}\right)^{k+5/2} \Gamma(k + \tfrac{5}{2}, (\beta - \beta_0)m_{\text{in}}), \quad (12.34)$$

where Γ is the incomplete gamma function. To obtain a square-root singularity in Eq. (12.34), consistent with Eq. (12.30), one needs $k = -3$.

In summary, the statistical-bootstrap approach assumes that hadrons are clusters consisting of hadrons and that, for large mass, the compound hadrons have the same mass spectrum as that of the constituent hadrons, which leads to a hadronic mass spectrum of the asymptotic form

$$\rho(m^2) \propto m^{-3} e^{m/T_0}. \quad (12.35)$$

This spectrum, as we have seen in Fig. 12.1, describes the known part of the experimental hadronic mass spectrum. For point hadrons, this leads to a singularity of the partition function at T_0, which appears, in view of Eq. (12.8), as a limiting temperature, at which infinite energy density is reached, since $k = -3 > -\tfrac{7}{2}$.

12.3 Hadrons of finite size

In the first presentation of the SBM results and methods, we have considered point-like hadrons in an arbitrary volume V. For a dilute gas, this is a good approximation. However, when we formulated the bootstrap hypothesis, we dealt with a system that has the density of the 'inside' of a hadron. We now generalize this approach and introduce the volume of the constituent cluster in the spirit of the quark model of hadrons, and confinement; see section 13.1. In the following we will also allow for clusters of finite baryon number.

The natural volume $V(m)$ of a hadron cluster is to be proportional to the cluster mass,

$$V(m) = \frac{m}{4\mathcal{B}}, \quad (12.36)$$

where $\mathcal{B}$, which has the dimension of energy density, is the bag constant; see Eq. (13.9). This equation is valid in the restframe of the cluster. For a cluster with 4-momentum p^μ, Eq. (12.36) takes the form

$$V^\mu(m) = \frac{p^\mu}{4\mathcal{B}}, \quad (12.37)$$

which defines the proper 4-volume V^μ of the particle. In the cluster restframe, Eq. (12.37) reduces to Eq. (12.36) and therefore is its unique generalization. Each object of 4-momentum p_i can be given a volume

V_i^μ. All clusters have the same proper energy density $\epsilon_0 = 4B$. In a relativistically covariant formulation, we also consider the energy, and the inverse temperature in terms of 4-vectors [261]:

$$E \to p^\mu = (p^0, \vec{p}); \qquad p_\mu p^\mu = p \cdot p = m^2, \tag{12.38}$$

$$\frac{1}{T} \to \beta^\mu = (\beta^0, \vec{\beta}); \qquad \beta_\mu \beta^\mu = \beta \cdot \beta = \beta^2 = \frac{1}{T^2}. \tag{12.39}$$

Note that the four-dimensional vector product $p \cdot p = (p^0)^2 - (\vec{p})^2$ is recognized by the absence of the vector arrow. In this notation,

$$Z(V, T) = \int_0^\infty \sigma e^{-E/T} \, dE \to \int_0^\infty \sigma e^{-\beta \cdot p} d^4 p. \tag{12.40}$$

We now need to obtain the covariant form of the N-finite-sized-particle level density $\sigma_N(p, V, b)$ Eq. (12.10). These particles occupy 'available 4-volume'

$$\Delta^\mu = V^\mu - \sum_{i=1}^N V_i^\mu. \tag{12.41}$$

Δ is the volume in which the particles move as if they were point-like, while in reality they have finite proper volumes and move in V. The level density of extended particles in the volume V must be identical to that of the point-like particles in the available volume Δ. This means that, for a system with baryon number b and 4-momentum p,

$$\sigma_N(p, V, b) \equiv \sigma_{N\text{pt}}(p, \Delta, b), \tag{12.42}$$

where 'pt' refers to point-like particles. Equation (12.42) is, in spirit, a Van der Waals correction, which introduces a new repulsive interaction into the system of hadronic resonances.

The generalization to an invariant phase-space volume is

$$\frac{V \, d^3 p}{(2\pi)^3} \Rightarrow \frac{2 V_\mu p^\mu}{(2\pi)^3} \delta_0(p^2 - m^2) \, d^4 p. \tag{12.43}$$

To go back from the invariant form to the restframe we need

$$\delta_0(p^2 - m^2) \, d^4 p = \frac{d^3 p}{2 p_0}. \tag{12.44}$$

Then, in the restframe of the volume V, Eq. (8.17),

$$\boxed{\frac{2 V_\mu p^\mu}{(2\pi)^3} \delta_0(p^2 - m^2) \, d^4 p = 2 V p_0 \frac{d^3 p}{2 p_0} = \frac{V}{(2\pi)^3} \, d^3 p.} \tag{12.45}$$

Therefore, in the Boltzmann approximation, and assigning to each cluster the degeneracy $g \to \tau(m_i^2, b_i)dm_i^2$ with intrinsic baryon content b_i, Eq. (12.10) generalizes to [143]

$$\sigma_N(p, V, b) = \frac{1}{N!} \int \delta^4\left(p - \sum_{i=1}^{N} p_i\right) \delta_K\left(b - \sum_{i=1}^{N} b_i\right)$$

$$\times \prod_{i=1}^{N} \frac{2\Delta \cdot p_i}{(2\pi)^3} \tau(m_i^2, b_i) \delta_0(p_i^2 - m_i^2) \, d^4 p_i \, dm_i^2. \quad (12.46)$$

$\tau(m_i^2, b_i)$ is the mass spectrum of a cluster with baryon number b_i in the mass interval $[m_i^2, dm_i^2]$. It is the analog of $\rho(m_i^2)$ in Eq. (12.20). The discreet conservation of baryon number is assured by the Kronecker-δ_K function.

The micro-canonical Lorentz-invariant density of states of a system made of any number of clusters, each cluster having any baryon number b_i, with $-\infty < b_i < \infty$, reads

$$\sigma(p, V, b) = \sum_{N=1}^{\infty} \frac{1}{N!} \int \delta^4\left(p - \sum_{i=1}^{N} p_i\right) \sum_{\{b_i\}} \delta_K\left(b - \sum_{i=1}^{N} b_i\right)$$

$$\times \prod_{i=1}^{N} \frac{2\Delta \cdot p_i}{(2\pi)^3} \tau(m_i^2, b_i) \delta_0(p_i^2 - m_i^2) \, d^4 p_i \, dm_i^2. \quad (12.47)$$

In Eq. (12.47), the contributing states are subdivided into any number of subsets corresponding to any partition of the total 4-momentum p^μ and the total baryon number b.

The canonical partition function, for a fixed baryon number b, is the Laplace transform of the level density given by Eq. (12.47),

$$Z(T, V, b) = \int e^{-\beta \cdot p} d^4 p \sum_{N=1}^{\infty} \frac{1}{N!} \int \delta^4\left(p - \sum_{i=1}^{N} p_i\right) \sum_{\{b_i\}} \delta_K\left(b - \sum_{i=1}^{N} b_i\right)$$

$$\times \prod_{i=1}^{N} \frac{2\Delta \cdot p_i}{(2\pi)^3} \tau(m_i^2, b_i) \delta_0(p_i^2 - m_i^2) \, d^4 p_i \, dm_i^2, \quad (12.48)$$

from which we obtain the grand-canonical partition function, Eq. (4.20), defined by

$$\mathcal{Z}(T, V, \lambda) = \sum_{b}^{\infty} \lambda^b Z(T, V, b) = \sum_{b=-\infty}^{\infty} \lambda^b \int e^{-\beta \cdot p} \sigma(p, V, b) \, d^4 p, \quad (12.49)$$

where λ is the baryon-number fugacity corresponding to the baryonic chemical potential μ: $\lambda = \exp \mu / T$.

To implement Eq. (12.42), we postulate

$$\mathcal{Z}(T, V, \lambda) \rightarrow \mathcal{Z}(T, \langle V \rangle, \lambda) = \mathcal{Z}_{\mathrm{pt}}(T, \Delta, \lambda). \tag{12.50}$$

Equation (12.50) permits us to calculate everything for fictitious point particles in Δ and afterwards obtain the correct quantities by eliminating Δ in favor of a computed, average value $\langle V \rangle$. Use of $\langle V \rangle$ instead of V constitutes an approximation, and a lot of effort over the years, since this approach was first proposed [143], has gone into remedying this step in a consistent statistical-physics approach, and into generalizing the idea contained in Eq. (12.42). A state-of-the-art calculation is given, e.g., in [170]. However, the original and physically simple model presented here offers all the required understanding without the ballast of mathematical complexity, and yields sufficiently precise results.

$\mathcal{Z}_{\mathrm{pt}}(T, \Delta, \lambda)$ can be written in the form

$$\mathcal{Z}_{\mathrm{pt}}(T, \Delta, \lambda) = \sum_{N=1}^{\infty} \frac{1}{N!} \int \delta^4 \left(p - \sum_{i=1}^{N} p_i \right) e^{-\beta \cdot p} \, d^4 p \tag{12.51}$$

$$\times \sum_{b=-\infty}^{\infty} \lambda^b \sum_{\{b_i\}} \delta_{\mathrm{K}} \left(b - \sum_{i=1}^{N} b_i \right) \prod_{i=1}^{N} \frac{2\Delta \cdot p_i}{(2\pi)^3} \tau(p_i^2, b_i) \, d^4 p_i.$$

The momentum δ^4 function permits us to do the $d^4 p$ integration and the δ_{K} permits the summation over b. The integrand thereafter splits into N independent identical integrals, and the sum over N yields an exponential function. Taking its logarithm, we obtain

$$\ln \mathcal{Z}_{\mathrm{pt}}(T, \Delta, \lambda) \equiv \ln \mathcal{Z}(T, \langle V \rangle, \lambda) = \mathcal{Z}_1(T, \Delta, \lambda), \tag{12.52}$$

where

$$\mathcal{Z}_1(T, \Delta, \lambda) \equiv \int \frac{2\Delta \cdot p}{(2\pi)^3} \tau(p^2, \lambda) e^{-\beta \cdot p} \, d^4 p, \tag{12.53}$$

with

$$\tau(p^2, \lambda) = \sum_{b=-\infty}^{\infty} \lambda^b \tau(p^2, b). \tag{12.54}$$

All information about the interaction is contained in the 'grand-canonical' hadronic mass spectrum $\tau(m^2, \lambda)$.

We obtain now the relation between $\langle V \rangle$ and Δ in the restframe. We use Eq. (12.37) to find the expectation value of the volume:

$$\langle V^\mu \rangle = \Delta^\mu + \frac{p^\mu}{4B} \rightarrow \Delta + \frac{\langle E \rangle}{4B} \bigg|_{\mathrm{restframe}}. \tag{12.55}$$

The energy density $\epsilon(\beta, \lambda)$ can be obtained from Eq. (12.52), for the energy:

$$\langle E \rangle = -\frac{\partial}{\partial \beta} \ln \mathcal{Z}(\beta, \langle V \rangle, \lambda) = -\frac{\partial}{\partial \beta} \ln \mathcal{Z}_{\text{pt}}(\beta, \Delta, \lambda). \tag{12.56}$$

Since $\ln \mathcal{Z}_{\text{pt}}$ is linear in Δ, the last term is equal to $\Delta \epsilon_{\text{pt}}(\beta, \lambda)$; hence,

$$\epsilon(\beta, \lambda) = \frac{\Delta \epsilon_{\text{pt}}(\beta, \lambda)}{\langle V \rangle}. \tag{12.57}$$

Inserting Eq. (12.55) into Eq. (12.57) and solving for $\langle E \rangle$, we find

$$\boxed{\epsilon(\beta, \lambda) = \frac{\epsilon_{\text{pt}}(\beta, \lambda)}{1 + \epsilon_{\text{pt}}(\beta, \lambda)/(4\mathcal{B})},} \tag{12.58}$$

which we have used in Eq. (11.1).

We can use Eq. (12.58) in Eq. (12.55) to obtain a more explicit relationship between the volume V and the available volume Δ:

$$\langle V \rangle = \Delta \left(1 + \frac{\epsilon_{\text{pt}}(\beta, \lambda)}{4\mathcal{B}} \right), \tag{12.59a}$$

$$\Delta = \langle V \rangle \left(1 - \frac{\epsilon(\beta, \lambda)}{4\mathcal{B}} \right). \tag{12.59b}$$

This procedure can be followed for the baryon density, pressure and, in principle, other statistical quantities:

$$\boxed{\nu(\beta, \lambda) \equiv \frac{\langle b \rangle}{\langle V \rangle} = \frac{\nu_{\text{pt}}(\beta, \lambda)}{1 + \epsilon_{\text{pt}}(\beta, \lambda)/(4\mathcal{B})},} \tag{12.60}$$

$$\boxed{P(\beta, \lambda) = \frac{P_{\text{pt}}(\beta, \lambda)}{1 + \epsilon_{\text{pt}}(\beta, \lambda)/(4\mathcal{B})}.} \tag{12.61}$$

12.4 Bootstrap with hadrons of finite size and baryon number

As explained in section 12.2 a system of total mass m, when it is compressed to its natural volume $V_c(m)$, becomes one of the particles counted in the hadronic mass spectrum (see Fig. 12.4). By the same token, a nuclear cluster with baryon number b compressed to its natural volume $V_c(m, b)$ becomes a cluster appearing in the mass spectrum $\tau(m^2, b)$. The bootstrap hypothesis can now be expressed by writing

$$\sigma(p, \Delta, b)|_{\langle v \rangle \to v_c(m, b)} \iff \tau(p^2, b), \tag{12.62}$$

where $\Longleftrightarrow$ means 'corresponds to' in a way to be specified.

With the condition Eq. (12.62), the statistical-bootstrap-model equation for τ arises from Eq. (12.47):

$$
\mathcal{H}\tau(p^2, b) = \mathcal{H}g_b\delta_0(p^2 - m_b^2) + \sum_{N=2}^{\infty} \frac{1}{N!} \int \delta^4\left(p - \sum_{i=1}^{N} p_i\right)
$$
$$
\times \sum_{\{b_i\}} \delta_K\left(b - \sum_{i=1}^{N} b_i\right) \prod_{i=1}^{N} \mathcal{H}\tau(p_i^2, b_i)\, d^4 p_i, \tag{12.63}
$$

where

$$
\mathcal{H} \equiv \frac{2m_0^2}{(2\pi)^3 4\mathcal{B}}. \tag{12.64}
$$

This equation is obtained, by first separating the 'input particle' (corresponding to $N = 1$) in Eq. (12.47), and then making the following replacement, where Eq. (12.37) is used:

$$
\sigma(p, V_c, b) \Rightarrow \frac{2V_c(m, b) \cdot p}{(2\pi)^3}\tau(p^2, b) \Rightarrow \frac{2m_0^2}{(2\pi)^3 4\mathcal{B}}\tau(p^2, b), \tag{12.65a}
$$

$$
\frac{2\Delta \cdot p_i}{(2\pi)^3}\tau(p_i^2, b_i) \Rightarrow \frac{2m_0^2}{(2\pi)^3 4\mathcal{B}}\tau(p_i^2, b_i). \tag{12.65b}
$$

The factors m^2 and m_i^2 have been absorbed into the definition of $\tau(p_i^2, b_i)$. Either $\mathcal{H}$ or m_0 may be taken as the new free parameter of the model.

The first term in Eq. (12.63), the 'input-particle' term, comes from the cluster structure: if clusters consist of clusters, which consist of clusters, and so on, this should end at some 'elementary' particles, here a hadron of baryon number b and of mass m_b. Typically, the input consists of the pion for the $b = 0$ term and the nucleon for $b = \pm 1$. The similarity of Eq. (12.63) to Eq. (12.20) allows us to repeat all the steps we made in solving Eq. (12.20), to obtain from Eq. (12.63) the asymptotic form of the hadronic mass spectrum. We introduce two functions $\varphi(\beta, \lambda)$ and Φ:

$$
\varphi(\beta, \lambda) \equiv \int e^{-\beta \cdot p} \sum_{b=-\infty}^{\infty} \lambda^b \mathcal{H}g_b\delta_0(p^2 - m_b^2)\, d^4 p,
$$
$$
= 2\pi\mathcal{H} \sum_{b=-\infty}^{\infty} \lambda^b g_b m_b^2 \frac{K_1(m_b\beta)}{m_b\beta}, \tag{12.66}
$$

and

$$
\Phi(\beta, \lambda) \equiv \int e^{-\beta \cdot p} \sum_{b=-\infty}^{\infty} \lambda^b \mathcal{H}\tau(p^2, b)\, d^4 p. \tag{12.67}
$$

Once the set of 'input particles' is introduced, $\varphi(\beta, \lambda)$ is a known function, while $\Phi(\beta, \lambda)$ is unknown. By applying the double Laplace transform (integration over p and summation over b) used in the definition of $\varphi(\beta, \lambda)$ and $\Phi(\beta, \lambda)$ to the Eq. (12.63), we obtain

$$\Phi(\beta, \lambda) = \varphi(\beta, \lambda) + e^{\Phi(\beta,\lambda)} - \Phi(\beta, \lambda) - 1. \tag{12.68}$$

This implicit equation for Φ can be solved again without regard to the actual dependence on β and λ. Writing

$$G(\varphi) \equiv \Phi(\beta, \lambda), \tag{12.69}$$

we obtain

$$\varphi = 2G(\varphi) - \exp[G(\varphi)] + 1, \tag{12.70}$$

which is Eq. (12.25). The graphical solution found in section 12.2 shows that $G(\varphi, \lambda)$ has a square-root singularity at

$$\varphi(\beta, \lambda) \to \varphi_0 = \ln(4/e), \tag{12.71}$$

which defines a critical curve $\beta_{cr}(\lambda)$ in the (β, λ) plane. In the vicinity of this curve,

$$G(\varphi) \approx G_0 + \text{constant} \times \sqrt{\varphi_0 - \varphi}, \tag{12.72}$$

and, therefore,

$$\Phi(\lambda, \beta \simeq \beta_{cr}) = \Phi_0 - C(\lambda)\sqrt{\beta - \beta_{cr}}. \tag{12.73}$$

As we have shown in section 12.2, this square-root singularity fixes the power k of m in the hadronic mass spectrum at $k = -3$ and we obtain

$$\tau(m^2, \lambda) \propto m^{-3} e^{\beta_{cr}(\lambda)m}. \tag{12.74}$$

For $\lambda = 1$, $\beta_{cr} = \beta_0$ and we recover the usual form of the hadronic mass spectrum. However, the generalization obtained gives a solution for any value of λ.

Given the mass spectrum in Eq. (12.52), we can calculate all the usual thermodynamic quantities. For this, we need to write down $\ln \mathcal{Z}_{pt}$. The formal similarity of Eq. (12.53) and Eq. (12.67) yields a relation between $\ln \mathcal{Z}_{pt}$ and Φ that is best expressed in the restframe of Δ and β:

$$\boxed{\ln \mathcal{Z}_{pt}(T, \Delta, \lambda) = -\frac{2\Delta}{(2\pi)^3 \mathcal{H}}\frac{\partial}{\partial\beta}\Phi(\beta, \lambda).} \tag{12.75}$$

The point-like quantities are derived from $\ln \mathcal{Z}_{pt}$ as given by Eq. (12.75):

$$\epsilon_{pt}(\beta, \lambda) = \frac{2}{(2\pi)^3 \mathcal{H}} \frac{\partial^2}{\partial \beta^2} \Phi(\beta, \lambda), \tag{12.76}$$

$$\nu_{pt}(\beta, \lambda) \equiv \frac{\lambda}{\Delta} \frac{\partial}{\partial \lambda} \ln \mathcal{Z}_{pt}(\beta, \Delta, \lambda) = -\frac{2\lambda}{(2\pi)^3 \mathcal{H}} \frac{\partial}{\partial \lambda} \frac{\partial}{\partial \beta} \Phi(\beta, \lambda), \tag{12.77}$$

$$P_{pt}(\beta, \lambda) \equiv \frac{T}{\Delta} \ln \mathcal{Z}_{pt}(\beta, \Delta, \lambda) = -\frac{2T}{(2\pi)^3 \mathcal{H}} \frac{\partial}{\partial \beta} \Phi(\beta, \lambda). \tag{12.78}$$

All the above point-particle quantities involve derivatives of $\Phi(\beta, \lambda)$; they become singular at $\varphi = \varphi_0$. Explicitly, using Eq. (12.72), which contains a square-root singularity, we have:

$$\frac{\partial}{\partial \beta} \Phi(\beta, \lambda) = \frac{dG}{d\varphi} \frac{\partial \varphi}{\partial \beta} \rightarrow \text{constant} \times \frac{\partial \varphi}{\partial \beta} \frac{1}{\sqrt{\varphi_0 - \varphi}}. \tag{12.79}$$

Therefore, $\varphi \rightarrow \varphi_0$ implies point-particle infinities for all of the above quantities, with the second derivative required in ϵ_{pt} being the most singular. On comparing the degrees of divergence of the numerator and denominator in Eqs. (12.58), (12.60), and (12.61), we see that the energy density and the baryon density are finite, while the pressure vanishes, on the critical curve. The overcompensation of the pressure is seen already in Fig. 12.3, which was evaluated with a model hadron mass spectrum.

This behavior of the pressure reflects the fact that we have counted only the pressure generated by the clusters and, as we shall see in the following subsection, all clusters coalesce on the critical curve, and hence the pressure of a single large cluster vanishes. This of course is an artifact, since at that point we should have included the internal cluster pressure, since the single cluster we find is in the QGP-type state. We will not introduce in this book the required generalization which can be found in [213].

12.5 The phase boundary in the SBM model

We have seen that the singular point of the solution to the bootstrap equation is located at the value $\varphi_0 = \ln(4/e)$ and that the critical curve in the (β, λ) plane is defined by

$$\varphi(\beta, \lambda) = \varphi_0 = \ln(4/e). \tag{12.80}$$

Its position depends, of course, on the actual content of $\varphi(\beta, \lambda)$, i.e., on the fundamental set of 'elementary particles' $\{m_b, g_b\}$ and the value of the constant $\mathcal{H}$, Eq. (12.64). In the case of three elementary pions $(\pi^+ \pi^0 \pi^-)$

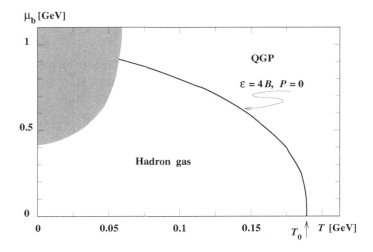

Fig. 12.6. The SBM critical curve in the (μ_b, T) plane. In the shaded region, the theory is not valid because we have neglected Bose–Einstein and Fermi–Dirac statistics.

and four elementary nucleons (spin $\otimes$ isospin) and four antinucleons, we obtain, from Eq. (12.66), the relation

$$\varphi_0(\beta_{cr}, \lambda_{cr}) = 2\pi\mathcal{H}T_{cr}\left[3m_\pi K_1\left(\frac{m_\pi}{T_{cr}}\right)\right.$$
$$\left. + 4\left(\lambda_{cr} + \frac{1}{\lambda_{cr}}\right)m_N K_1\left(\frac{m_N}{T_{cr}}\right)\right]. \qquad (12.81)$$

The condition of Eq. (12.81), written in $T_{cr} = 1/\beta_{cr}$ and $\mu_{cr} = T_{cr}\ln\lambda_{cr}$, yields the critical curve shown in Fig. 12.6, drawn for $\mathcal{H} = 0.724$ GeV^{-2}. For $\mu = 0$, the curve ends at $T = T_0$, which becomes the maximum phase transition temperature instead of a limiting temperature.

Our system consists, for small T and μ, of nucleons and nuclei. For increasing T, creation of pions sets in and finally also creation of baryon–antibaryon pairs, as well as (not included here) creation of strange hadrons. If the latter is taken into account, the input set of 'elementary particles' has to be enlarged. This changes slightly the position of the critical curve and the equations of state of hadron matter, since T_0 is of the order of the pion mass, while the other particles have larger masses and make little contribution to $\varphi(\beta, \lambda)$. More precisely, each new conserved quantum number (strangeness, charm, ...) gives rise to another corresponding fugacity λ; hence the singularity is defined by $\varphi(\beta, \lambda_1, \lambda_2, \lambda_3, ..., \lambda_n) = \varphi_0$ as a hypersurface in an ($n + 1$)-dimensional space. Since, however, in physical situations, generally only the baryon number is different from zero, we have to consider only the intersection of this hypersurface with the (T, μ_b) plane. That procedure yields the curve which was said to be little different from the one shown in Fig. 12.6.

What does the SBM hadron gas do when it approaches the critical curve? As the point-particle quantities ϵ_{pt}, ν_{pt}, and p_{pt} diverge, one sees (by comparing degrees of divergence when $\varphi \to \varphi_0$) that

$$\epsilon(\beta_{cr}, \lambda_{cr}) \to 4\mathcal{B}, \tag{12.82a}$$

$$\nu(\beta_{cr}, \lambda_{cr}) \to \nu_{cr}(\beta_{cr}, \lambda_{cr}) \neq 0, \tag{12.82b}$$

$$P(\beta_{cr}, \lambda_{cr}) \to 0, \tag{12.82c}$$

$$\Delta(\beta_{cr}, \lambda_{cr}) \to 0, \quad \text{if } \langle V \rangle \neq 0, \tag{12.82d}$$

$$\langle V(\beta_{cr}, \lambda_{cr}) \rangle \to \infty, \quad \text{if } \Delta \neq 0, \tag{12.82e}$$

where β_{cr} and λ_{cr} are the values along the critical curve.

As noted already, see Eq. (12.36), the energy density of our clusters was constant and always equal to $4\mathcal{B}$. Equation (12.82a) suggests that, on the critical curve, the whole hadron system has condensed into one giant cluster, witnessed by the vanishing of the pressure; one can explicitly see that, for any given external volume $\langle V \rangle$, the number $\langle N \rangle$ of particles (clusters) contained in it goes to zero on the critical curve: indeed, introducing the fugacity ξ relative to the number of clusters, Eq. (12.52) can be written:

$$\mathcal{Z}_{pt}(\beta, \Delta, \lambda) = \mathcal{Z}_{pt}^{(\xi)}(\beta, \Delta, \lambda, \xi)|_{\xi=1} \equiv \sum_{N=0}^{\infty} \frac{1}{N!}(\xi \ln \mathcal{Z}_{pt})^N. \tag{12.83}$$

Hence, with Eq. (12.75),

$$\langle N \rangle = \xi \frac{\partial}{\partial \xi} \ln \mathcal{Z}_{pt}^{(\xi)}\bigg|_{\xi=1} = \ln \mathcal{Z}_{pt} = -\frac{2\Delta}{(2\pi)^3 \mathcal{H}} \frac{\partial}{\partial \beta} \Phi(\beta, \lambda), \tag{12.84}$$

and, with Eq. (12.59a),

$$\frac{\langle N \rangle}{\langle V \rangle} = \frac{\mathcal{B}}{\pi^3 \mathcal{H}} \frac{\partial \Phi(\beta, \lambda)/\partial \beta}{1 + \epsilon_{pt}(\beta, \lambda)} \underset{\text{critical curve}}{\Longrightarrow} 0, \tag{12.85}$$

because ϵ_{pt} contains a second derivative of $\Phi(\beta, \lambda)$. It follows that, from Eqs. (12.61), (12.78), and (12.85),

$$P\langle V \rangle = \langle N \rangle T, \tag{12.86}$$

that is, our hadron gas obeys, formally, the ideal-gas equation of state for the average number of clusters $\langle N \rangle$; $\langle N \rangle$ is not a constant, but a function of β and λ.

In the bootstrap model of hadronic gas, our finding is that the critical curve limits the HG phase; approaching it, all hadrons dissolve into a giant

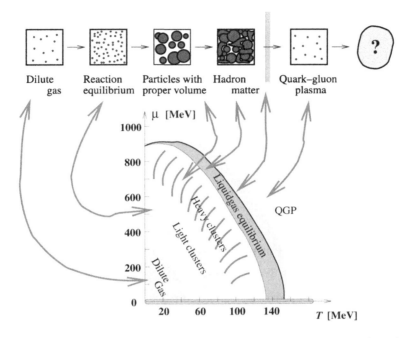

Fig. 12.7.   The physical interpretation of the different regions of the $(T,\ \mu)$ plane according to the statistical-bootstrap model of hadronic matter.

cluster. The gradual change in the structure of hot hadronic matter is illustrated in Fig. 12.7; at low T, μ we have a dilute pion gas of essentially point-like pions. With an increase in T and/or μ, progressively denser hadron matter is formed, and hadron proper volume becomes relevant. Near the phase boundary hadrons coalesce into large clusters comprising drops of QGP.

In SBM the singular curve is reached with finite energy density $4\mathcal{B}$. In the hadron phase, $\epsilon(\beta, \lambda) < 4\mathcal{B}$ and $\beta \le \beta_{\mathrm{cr}}$. For $\epsilon > 4\mathcal{B}$, we enter into a region that cannot be described by the thermodynamics of the SBM. Indeed in this region, $\beta \ge \beta_{\mathrm{cr}}$ and the partition function $\mathcal{Z}_{\mathrm{pt}}(\beta, \Delta, \lambda)$ and all densities become complex. This region cannot be described without making assumptions about the inner structure and dynamics of the 'elementary particles' $\{m_{\mathrm{b}},\ g_{\mathrm{b}}\}$ – here pions and nucleons – entering into the input function $\varphi(\beta, \lambda)$. In other words, to continue, we need to consider the hot hadron interior made of quarks and gluons. Assuming that we have a phase transition between a HG and a QGP, the evolution of the system in the P–V diagram is qualitatively illustrated in Fig. 3.2 on page 49. In order to make this picture quantitative we need to explore, within the realm of quantum chromodynamics, the hadron structure and the behavior of a gas of quarks and gluons with color interactions.

V

QCD, hadronic structure and high temperature

13 Hadronic structure and quantum chromodynamics

13.1 Confined quarks in a cavity

A hadronic particle, according to section 3.1, is a quark-filled bubble, a 'swiss-cheese' hole, in the structured vacuum. The highly excited drop of QGP is indeed much akin to the picture of an individual, colorless hadron, except that it is the thermal pressure that acts against the vacuum pressure, not the quantum pressure. As a first step in a more detailed discussion of the QGP phase, we briefly discuss how this approach allows us to understand properties of individual hadrons.

In the quark-bag model of hadronic structure, colorless qqq baryons or $\bar{q}q$ mesons are embedded in the structured vacuum sea. In a calculational framework proposed by Bogoliubov [78], independent quarks confined by a static Lorentz-scalar potential with infinite walls were considered. This is ensuring permanent 'confinement' of the constituents within a given volume. The interest in this approach grew only after it was understood that the confining potential is not to be derived from quark–quark interactions, but that it arises from the repulsion of colored quarks by the structured QCD vacuum state.

The structure of hadrons emerges on considering a static spherical state in which residual quark–quark interactions are introduced. This MIT-bag model is able to capture most features of the hadron spectrum [92, 93, 99, 151]. Our limited objective is to extract from a study of the hadronic spectrum information about the latent heat of the vacuum $\mathcal{B}$, and the mass of the strange quark m_{s}. To accomplish this we will not need to introduce in this book improvements addressing the restoration of translational invariance, and the absence of chiral symmetry; see section 3.3. For further details on chiral symmetry and the bag model, we refer the interested reader to [94, 256, 258].

Assuming that quarks are moving independently inside a region of space, the mass M_h of the bound-state system of quarks (bag) comprises the kinetic energy of confined quarks, as well as the volume energy of the disturbance in the vacuum,

$$M_h(R_h) = \frac{\sum_i \varepsilon_i}{R_h} + \frac{4\pi}{3} R_h^3 \mathcal{B} + \delta E_V + \Delta M_{mag}, \tag{13.1}$$

where ε_i is the (dimensionless) eigenvalue energy coefficient of the ith quark in a static cavity of radius R_h. Since we are considering the lowest possible hadronic states, this term becomes proportional to the number $n = 2$ or 3 of valence quarks and antiquarks in mesons or baryons, respectively:

$$\frac{\sum_i \varepsilon_i}{R_h} \to n \frac{\varepsilon_0}{R_h}. \tag{13.2}$$

We have further introduced, in Eq. (13.1), the finite-volume correction to the vacuum energy

$$\delta E_V = \frac{z_0}{R_h} \tag{13.3}$$

in $M_h(R_h)$. Since this is a term independent of the number of quarks in the bag, not requiring a dimensioned constant, a judicious choice of the number z_0 allows for many other effects.

A residual interaction must be introduced in order to describe the energy splitting between the baryon octet with $j = \frac{1}{2}^+$ and the decuplet with $j = \frac{3}{2}^+$, see Fig. 2.1, and similarly between the nonets of pseudo-scalar $j = 0^-$ and the vector mesons $j = 1^-$; see Fig. 2.2 on page 27. A Coulomb-like interaction could not accomplish this, since it can not distinguish among the different angular-state multiplets. Akin to the magnetic field which splits the spin states, an interaction of magnetic (hyperfine) type is needed,

$$\Delta M_{mag} = \sum_{i>j} \left\langle \frac{\alpha_s}{r_{ij}} t_i^a t_j^a \vec{\sigma}_i \cdot \vec{\sigma}_j \right\rangle = \frac{1}{R_h} \sum_{i>j} c_{ij} h_{ij}. \tag{13.4}$$

This is the usual form with the spin-Pauli matrices $\vec{\sigma}_i$, Eq. (13.27), and specific to the color interaction, $SU(3)$ generators t^a, Eq. (13.58). The important feature of this color-magnetic hyperfine interaction is that it does reflect correctly the signs, and even the magnitude, of the splittings between various hadronic multiplets.

The coefficients c_{ij}, in Eq. (13.4), are found by evaluating

$$c_{ij} = \langle h \,|(t^a \vec{\sigma})_i \cdot (t^a \vec{\sigma})_j|\, h \rangle. \tag{13.5}$$

This is done by methods developed in the study of hyperfine QED interactions, and we defer this discussion. h_{ij}, in Eq. (13.4), are the transition-matrix elements of magnetic moments, which incorporate the coupling α_s, but where we have taken out the dominant dimensional factor $1/R_h$.

The mass of the hadronic state is dependent on the size parameter R_h. In the absence of any other dimensioned constants (such as quark masses), but with a constant a_h containing quark and interaction contributions specific to each hadron state, we have

$$M_h(R_h) = \frac{4\pi}{3} R_h^3 \mathcal{B} + \frac{a_h}{R_h}. \tag{13.6}$$

There is a clearly defined minimum in Eq. (13.1) as a function of R_h, at which the forces associated with the vacuum and quarks balance. The physical state has a mass associated with this minimum:

$$\frac{\partial M_h}{\partial R_h} = 0 . \tag{13.7}$$

This condition,

$$\frac{a_h}{4\pi R_h^4} - \mathcal{B} = 0, \tag{13.8}$$

is equivalent to the pressure equilibrium point between the internal Fermi pressure and the exterior vacuum pressure (negative pressure seen from the interior).

Reinserting the result of Eq. (13.7) into Eq. (13.6), we find that the volume and mass of a quark-bound state are related:

$$\boxed{M_h = 4\mathcal{B}V_h, \quad R_h M_h = \tfrac{4}{3} a_h, \quad M_h = \tfrac{4}{3} a_h^{3/4} (4\pi\mathcal{B})^{1/4}.} \tag{13.9}$$

Aside from the bag constant, $\mathcal{B}$, the hadron-state-specific value of a_h determines the value of each hadron mass. To determine a_h in the study of hadronic spectra based on quark-cavity states, section 13.2, the five parameters ε_q and ε_s (the energies of light and strange quarks), and h_{qq}, h_{qs}, and h_{ss} (the (transition) magnetic moments seen in Eq. (13.1)) are set to the values expected from the structure of the unperturbed bag model, and the values of the elementary parameters α_s and m_s are fitted, along with z_0; thus one looks at four parameters. However, much more precise information on the magnitude of the bag constant, $\mathcal{B}$, is obtained in an approach in which one takes Eq. (13.1) as the starting point. In Eq. (13.1), aside from $\mathcal{B}$, six more parameters enter: z_0, ε_q, and ε_s; and h_{qq}, h_{qs}, and h_{ss}. One of the results of such an approach is the verification of how well hadronic structure is described by quark-cavity states.

Owing to the coincidence that the strange-quark mass and hadron radius have the same scale $m_s R_h \simeq 1$, there is considerable sensitivity to the exact value of the strange quark m_s, which determines the energy of the strange quark:

$$\varepsilon_s^{\text{fit}} = \sqrt{m_s^2 + \frac{x_s^2}{R_h^2}}. \tag{13.10}$$

Considering the running of the mass of the strange quark predicted by QCD, see Fig. 17.4 on page 328, the effective mass of the strange quark should vary with the hadronic size, e.g., according to

$$m_s(R_h) \equiv m_s^0 \ln(\pi R_h \Lambda_h). \tag{13.11}$$

The variation of the mass of the strange quark with the quark momentum introduces the eighth parameter Λ_h.

These considerations are quite successful at describing the hadronic spectrum [20]. Aside from the effect of the quark mass in the strange hadrons, the differences in mass between the various hadronic multiplets arise from the differences in the quantum numbers which influence the value of c_{ij}, Eq. (13.5). Thus, the value of a_h, Eq. (13.6), depends both on the quantum numbers of the multiplet and on the quark-flavor content.

A unique fit with a significant confidence level arises, yielding

$$\varepsilon_q^{\text{fit}} = 1.97 \pm 0.02. \tag{13.12}$$

The value of $\varepsilon_q^{\text{fit}}$ is close to the massless-quark value $x_0 = 2.04$ derived from solution of the Dirac equation in a cavity, see section 13.2. This result is thus providing an empirical foundation for the bag model of hadrons.

The values $m_s^0 = 234 \pm 14\,\text{MeV}$ and $\Lambda_h = 240 \pm 20\,\text{MeV}$ are found [20]. The effective mass of the strange quark $m_s(R_h)$ varies between 170 MeV in kaons, which are relatively small ($R_K = 0.5\,\text{fm}$), and about 320 MeV in strange baryons. Using a fixed mass for the strange quark, $m_s = 280\,\text{MeV}$ [99], was obtained. The QCD-motivated variability with hadron size of m_s in each hadron leads to a much better fit to the hadronic spectrum, and yields for the value of the vacuum energy $\mathcal{B}$

$$\boxed{\mathcal{B} = (171\,\text{MeV})^4, \quad 4\mathcal{B} = 0.45\,\text{GeV fm}^{-3}.} \tag{13.13}$$

The bag constant is larger than the original MIT result, $\mathcal{B}^{\text{MIT}} = (145\,\text{MeV})^4$. The main reason for the difference from the MIT fit arises from the allowed variation in mass of the strange quark with the size of the hadron. There is a remaining systematic dependence of the results for $\mathcal{B}$ on the here assumed behavior, Eq. (13.11). However, Eq. (13.13) is in much better agreement with the value of $\mathcal{B}$ required to describe the lattice-pressure results, for which we need a still larger value of $\mathcal{B}$, see Eq. (16.12).

Since the matrix elements h_{qq}, h_{qs}, and h_{ss} comprise α_s, a value for α_s is not determined within this procedure. We note that in the cavity bag model, $\alpha_s = 0.55$ is found.*

13.2 Confined quark quantum states

Since the short-range interactions among quarks can be ignored in the first instance, the 'independent' quark wavefunction ψ_q obeys the Dirac equation,

$$i\gamma^\mu \partial_\mu \psi_q - M\psi_q + (M - m)\Theta_V \psi_q = 0, \tag{13.14}$$

where $\Theta_V = 1$ inside the quark bag and $\Theta_V = 0$ outside. Inside the volume, the dynamics of quarks is governed by the (small) mass m, while outside it is subject to the mass M. Since, despite a great effort, we have not discovered free quarks, M must be very large, and the idea of color confinement indeed requires $M \to \infty$, in order to have quarks confined forever.

However, that limit is not trivial as it turned out. In a series of publications in 1974–75, the MIT-bag model was developed [92, 93, 151] in a way that creates the framework for quark dynamics with confinement, $M \to \infty$. Novel physical concepts were introduced since the confinement condition broke conservation of energy–momentum at the confinement boundary. Namely, in the limit $M \to \infty$, there is no quark quantum wave outside of the hadron volume, and, in order to have a stable physical state, the internal quantum pressure of the confined quarks must be balanced at the confinement boundary by some new external pressure pointing inward.

In a Lorentz-covariant formulation, along with pressure, there is also energy density, which had to be lower outside of the bag than inside, in order to have the inward-pointing pressure. The improvement of the energy–momentum tensor inside the volume region occupied by quarks includes the covariant bag term:

$$T^{\mu\nu} = T^{\mu\nu}_{\text{fields}} + g^{\mu\nu}\mathcal{B}. \tag{13.15}$$

On comparing this with, e.g., Eq. (6.6), we indeed see that the bag term increases the energy density, and decreases the pressure within the volume occupied by quantum particles.

On physical grounds, the size of the system is determined by balancing the internal pressure against the external pressure, and thus $\mathcal{B}$ has to enter into the dynamics of quark fields. We follow, in the next few lines,

---

* In some literature, a four-fold greater value of $\alpha_s = 2.2$ is quoted; the difference arises from the factor $\frac{1}{2}$ which relates λ^a to the generator t^a in Eq. (13.56).

the summary of the situation presented in [256]. We will need the surface Δ_S-function, which arises from the volume function in a familiar way,

$$\Delta_S = -n_\mu \partial^\mu \Theta_V, \tag{13.16}$$

where n^μ is the outward space-like normal to the surface of the bag. The static spherical cavity, in spherical coordinates, reads as usual:

$$\delta(R_h - r) = \frac{d\Theta(R_h - r)}{dr}. \tag{13.17}$$

The action which fully accounts for all physical aspects is

$$S = \int d^4x \left[\left(\sum_q \tfrac{1}{2}\overline{\psi_q} i\gamma^\mu \overleftrightarrow{\partial}_\mu \psi_q - B \right) \Theta_V - \tfrac{1}{2}\left(\sum_q \overline{\psi_q}\psi_q \right) \Delta_S \right], \tag{13.18}$$

where $\overline{\psi_q} = \psi_q^\dagger \gamma_0$, and γ^μ are Dirac matrices, Eq. (13.26).

To obtain the dynamic equations, we perform variation of the action seeking its stationary point:

$$\psi_q \to \psi_q + \delta\psi_q, \quad \overline{\psi_q} \to \overline{\psi_q} + \delta\overline{\psi_q}. \tag{13.19}$$

Furthermore, the geometry-defining volume Θ_V and surface Δ_S functions change under variation,

$$\Theta_V \to \Theta_V + \epsilon\Delta_S, \quad \Delta_S \to \Delta_S - \epsilon n_\mu \partial^\mu \Delta_S. \tag{13.20}$$

For a spherical cavity with $n^\mu = (0, \hat{r})$, the following three equations of motion give the stationary point of the action:
the Dirac equation,

$$i\gamma^\mu \partial_\mu \psi_q(x) = 0, \qquad x \in V; \tag{13.21}$$

the linear boundary condition,

$$i\gamma^\mu n_\mu \psi_q(x) = \psi_q(x), \qquad x \in S; \tag{13.22}$$

the quadratic boundary condition,

$$\tfrac{1}{2}n^\mu \partial_\mu \left(\sum_q \overline{\psi_q(x)}\psi_q(x) \right) - B = 0, \qquad x \in S. \tag{13.23}$$

A solution satisfying the boundary condition, Eq. (13.22), satisfies the requirement that the normal flow of quark current through the surface of the bag vanishes. To see this, we write this condition, along with the adjoint form:

$$i\gamma^\mu n_\mu \psi_q \,|_S = \psi\,|_S, \qquad -i\overline{\psi_q}\gamma^\mu n_\mu\,|_S = \overline{\psi}\,|_S . \tag{13.24}$$

To obtain the adjoint form, we used $\gamma^{\mu\dagger} = \gamma^0\gamma^\mu\gamma^0$. We construct the outward current at the surface:

$$n_\mu j^\mu \,|_S = n_\mu(\overline{\psi}\gamma^\mu\psi)\,|_S = \pm\overline{\psi}\psi\,|_S = 0. \tag{13.25}$$

Since the right-hand side of Eq. (13.25) is both positive and negative of the same value, as can be seen by using one of the two forms of Eq. (13.24), it must be zero.

Since there is no flow of probability through the surface, the linear boundary condition guarantees confinement of quarks. Moreover, this boundary condition allows us to determine the eigenenergies of quarks in a cavity of a given size R_h, as we shall discuss. On the other hand, the quadratic boundary condition Eq. (13.23) allows us to find the size of the system, establishing a balance of forces at the surface; see Eq. (13.8).

We now proceed to obtain examples of solutions of the bag-model dynamic Eq. (13.21), in the static-cavity approximation. γ^μ are the usual covariant Dirac matrices and we use the Bjørken–Drell conventions [74]:

$$\gamma^0 \equiv \beta \equiv \begin{bmatrix} I_2 & 0 \\ 0 & -I_2 \end{bmatrix}, \quad \gamma^i \equiv \beta\alpha^i \equiv \begin{bmatrix} 0 & \sigma^i \\ -\sigma^i & 0 \end{bmatrix}. \tag{13.26}$$

Here, I_2 is a unit 2×2 matrix, and $\vec{\sigma} = (\sigma^1, \sigma^2, \sigma^3)$ are Pauli's spin matrices:

$$\sigma^1 = \begin{bmatrix} 0 & 1 \\ 1 & 0 \end{bmatrix}, \quad \sigma^2 = \begin{bmatrix} 0 & -i \\ i & 0 \end{bmatrix}, \quad \sigma^3 = \begin{bmatrix} 1 & 0 \\ 0 & -1 \end{bmatrix}. \tag{13.27}$$

We are interested in the lowest-energy solution of the Dirac equation, Eq. (13.14), for a stationary spherical cavity. We consider the wave function

$$\psi_q(\vec{r}, t) = q_n(\vec{r})e^{-i\omega_n t}\tau_q. \tag{13.28}$$

τ_q is the flavor (e.g., isospin or $SU_f(3)$) part of the independent particle wave function, and ω_n is the nth-state eigenenergy. The stationary quark 4-spinor wave function $q(\vec{r})$ satisfies the equation (suppressing the quantum number(s) subscript 'n')

$$(\vec{\alpha} \cdot \vec{p} + \beta m - \omega)q = 0,$$

$$\begin{pmatrix} (m - \omega)I_2 & \vec{\sigma} \cdot \vec{p} \\ \vec{\sigma} \cdot \vec{p} & (m + \omega)I_2 \end{pmatrix} \begin{pmatrix} q^u \\ q^d \end{pmatrix} = 0, \tag{13.29}$$

where q^u are the upper and q^d the lower quark 2-spinor components.

When there is no potential inside the bag, each of the four components 'k' of the spinor q has to satisfy the energy–momentum condition obtained by 'squaring' the Dirac equation, i.e., the Klein–Gordon equation:

$$[\omega^2 - m_q^2 - (i\,\vec{\nabla})^2]q^k = 0. \tag{13.30}$$

We recall the spherical decompositions

$$\vec{\nabla} = \hat{r}\frac{\partial}{\partial r} - i\frac{\hat{r}}{r} \times \vec{L}, \qquad \vec{L} = \vec{r} \times i\vec{\nabla}, \tag{13.31}$$

$$\vec{\nabla}^2 = \frac{d^2}{dr^2} + \frac{2}{r}\frac{d}{dr} - \frac{\vec{L}^2}{r^2}, \tag{13.32}$$

and recognize that the components of the Dirac cavity solution have the form

$$q^k = N j_l(x\,r/R_h) \sum_{\mu=-l}^{l} c_{jj_z}^{l\mu}(k)Y_{l\mu}(\Omega), \tag{13.33}$$

where x is obtained from an eigenvalue condition, and $Y_{l\mu}(\Omega)$ are the usual spherical functions of fixed angular momentum l, μ. The Clebsch–Gordan coefficients $c_{jj_z}^{l\mu}$ are fixed by construction of a spinor spherical wave of good total angular momentum j and its z-axis projection, $j \leq j_z \leq -j$. Equation (13.30) now implies that

$$\omega \equiv \frac{\varepsilon}{R_h} = \sqrt{\frac{x^2}{R_h^2} + m_q^2}. \tag{13.34}$$

The no-node, lowest-energy quark-cavity solution is

$$q_0^u = N_0 j_0(x\,r/R_h)\chi_s, \quad j_0(z) = \frac{\sin z}{z}. \tag{13.35}$$

We use the 2-spinor χ_s, $s = \pm\frac{1}{2}$ for spin-up and -down quarks:

$$\chi_{1/2} = \begin{pmatrix} 1 \\ 0 \end{pmatrix}, \qquad \chi_{-1/2} = \begin{pmatrix} 0 \\ 1 \end{pmatrix}. \tag{13.36}$$

To obtain the corresponding lower components of the Dirac spinor, we use Eq. (13.29):

$$q^d = \frac{1}{\omega + m}\vec{\sigma} \cdot i\vec{\nabla}q^u. \tag{13.37}$$

The spherical decomposition Eq. (13.31), along with $\vec{L}q_0^u = 0$, and the spherical-Bessel-function property,

$$j_1 = -\frac{d}{dz}j_0 = \frac{\sin z}{z^2} - \frac{\cos z}{z}, \tag{13.38}$$

yields the lowest angular $(j = \frac{1}{2})$ quark wavefunction Eq. (13.28):

$$\psi_q(\vec{r}, t) = N \begin{pmatrix} j_0(xr/R_h)\chi_s \\ \dfrac{x}{R_h(\omega + m_q)} j_1(xr/R_h)i\sigma_r\chi_s \end{pmatrix} e^{-i\omega t}\tau_q. \qquad (13.39)$$

The radial spin matrix has been introduced:

$$\sigma_r = \hat{r} \cdot \vec{\sigma}, \qquad \sigma_r^2 = I. \qquad (13.40)$$

Equation (13.39) can easily be cast into the form Eq. (13.33), with the lower components q^d having $l = 1$.

The linear boundary condition, Eq. (13.22), at the surface of the bag, that is at a given radius R_h of the bag, takes the form

$$-i(\vec{\gamma} \cdot \vec{n})\psi \,|_s = \psi \,|_s, \qquad (13.41)$$

which eigenvalue condition will now be used to fix the value of x. The surface-normal vector for a spherical bag is $\vec{n} = \hat{r}$. The boundary condition reads

$$-i\sigma_r q^d \chi_s \,|_{r=R_h} = q^u \chi_s \,|_{r=R_h}, \quad i\sigma_r q^u \chi_s \,|_{r=R_h} = q^d \chi_s \,|_{r=R_h}. \qquad (13.42)$$

Using Eq. (13.39), we obtain

$$j_0(x) = \frac{x}{\sqrt{x^2 + (m_q R_h)^2} + m_q R_h} j_1(x), \qquad (13.43)$$

which takes the explicit form [93]

$$1 - x\cot x = \sqrt{x^2 + (m_q R_h)^2} + m_q R_h. \qquad (13.44)$$

When $m_q R_h \to 0$, the lowest eigenvalue is $x_0 = 2.04$. The (kinetic) energy of a massless quark in the confining radius R_h is

$$\omega_{bag}(m_q = 0) = \frac{2.04}{R_h}. \qquad (13.45)$$

The first radial excitation found, on solving Eq. (13.44) for the second-lowest eigenvalue, is relatively high, with $x_1 = 5.40$, more than doubling the kinetic energy.

For massive quarks we have

$$\omega_{bag} = \frac{\sqrt{x_0^2 + (m_q R_h)^2}}{R_h}, \qquad 2.04 < x_0 < \pi. \qquad (13.46)$$

As indicated in Eq. (13.46), with increasing $m_q R_h$, x_0 increases, never reaching the singularity of $\cot x$ at $x_0(m_q R_h \to \infty) = \pi$.

The quark-wave-function normalization N, in Eq. (13.39), is easily obtained:

$$\int d^3r \, \overline{\psi_q} \gamma_0 \psi_q = 4\pi N^2 R_h^3 \int_0^1 dz \, z^2 \left(j_0^2(xz) + \frac{x}{1 - x \cot x} j_1^2(xz) \right).$$

We have used the eigenvalue condition Eq. (13.44) and substituted $z = r/R_h$. The quadratic boundary condition Eq. (13.23) is

$$B = \frac{1}{2} \frac{d}{dr} \left(\sum_q \overline{\psi_q} \psi_q \right) \Bigg|_{R_h}. \tag{13.47}$$

We leave it to the reader as an exercise to show that Eq. (13.47) is indeed equivalent to the intuitive requirement that the total energy contained in the bag volume be at a minimum with respect to the radius of the bag, Eq. (13.7).

13.3 Nonabelian gauge invariance

The color hyperfine magnetic interaction, Eq. (13.4), is the residual force between quarks in the perturbative vacuum. It defines the hadron spectrum. To understand this force properly, we need to understand quantum chromodynamics and, in particular, the quark–quark interaction better. Akin to spin, the color charge of quarks is an internal quantum number, but it resembles in its properties more the electrical charge, so much so that one also speaks of 'color charge': like electrical charge, color charge is thought to be the source of a force field. Since we have so far not been able to build an apparatus to distinguish among the three fundamental colors, akin to the way electro-magnetic fields differentiate the spin and charge states, all colors must appear on an exactly equal footing.

Therefore, the theory of color forces, i.e., quantum chromodynamics (QCD), is based on the principle of gauge invariance extended to include invariance under arbitrary rotations (redefinition of color-principal 'axis') in the three-component color space. The resulting theory of strong interactions based on color forces has been called quantum chromodynamics in order to underline its formal similarity to quantum electrodynamics (QED). The form of the QCD Lagrangian is the generalization of QED required when one is considering invariance under local gauge color transformations.

Before introducing QCD, let us recall how, within QED, the principle of gauge invariance operates. It allows for changes in the electro-magnetic potential, leaving the electro-magnetic fields $F^{\mu\nu}$, i.e., $\vec{E}$ and $\vec{B}$, unchanged. The electro-magnetic potential $A^\mu = (A_0, \vec{A})$ is thus not defined uniquely. We further recall that a measurement, in general, does not allow one to observe the phase factor of a quantum wave.

We now show how a change in the choice of the quantum phase is related to the change in gauge of the potential A^μ. The effect of a local change in the phase of the wave function,

$$\psi \to \psi' = e^{-i\alpha}\psi, \tag{13.48}$$

$$\partial_\mu \psi \to \partial_\mu(e^{-i\alpha}\psi) = e^{-i\alpha}[\partial_\mu\psi - (i\partial_\mu\alpha)\psi], \tag{13.49}$$

can be compensated for by the simultaneous gauge transformation of the electro-magnetic potential A_μ:

$$A_\mu(x) \to A'_\mu(x) = A_\mu(x) + \frac{1}{e}(\partial_\mu\alpha). \tag{13.50}$$

This occurs if the quantum fields and potential are 'minimally coupled':

$$[(\partial_\mu + ieA_\mu)\psi]' = (\partial_\mu + ieA'_\mu)\psi' = e^{-i\alpha}(\partial_\mu + ieA_\mu)\psi. \tag{13.51}$$

We see that the generalized covariant derivative,

$$\partial_\nu \to D_\nu = \partial_\nu + ieA_\nu, \tag{13.52}$$

remains gauge invariant, up to an overall phase factor $e^{-i\alpha}$.

We will now generalize the principle of gauge invariance to the case of QCD. The additional difficulty is the non-commutative, i.e., *nonabelian*, aspect of the transformation, which is associated with the fact that there is not just one but several charges, i.e., colors. which a particle can carry: the usual 4-spinor wave function of a spin-$\frac{1}{2}$ particle, ψ, becomes in our case a component of a 12-spinor in color space:

$$\Psi = \begin{pmatrix} \psi_r \\ \psi_g \\ \psi_b \end{pmatrix}. \tag{13.53}$$

As long as the RGB (red, green, blue) 'color' is not observable, the color-gauge transformation generalizing Eq. (13.48) can, apart from introducing a phase, also rotate (mix) the color components of the wave function. The arbitrariness of the quantum wave is now expressed by the transformation

$$\Psi \to \Psi' = V\Psi, \qquad V^\dagger = V^{-1}, \qquad \det(V) = 1. \tag{13.54}$$

Since the complex rotations of a three-dimensional spinor-vector are described by unitary 3×3 matrices $V = (v_{ik})$ of unit determinant, the symmetry group of the gauge transformations is $SU_c(3)$, where the subscript reminds us of color, and will be omitted where confusion with flavor symmetry $SU_f(3)$ is unlikely.

The flavor $SU_f(3)$ symmetry arising from the near degeneracy in energy of the three 'light' u, d, and s quarks is quite different in its nature. The color symmetry permits us to rename *locally* the color of quarks. The flavor symmetry is a global symmetry: once a definition of flavor has been chosen at CERN it applies at the BNL, as long as both laboratories belong to a region of the Universe occupying the same vacuum state. However, even this situation has an exception, namely when, after chiral symmetry has locally been restored in a high-energy heavy-ion collision, the chiral-symmetry-breaking vacuum is reformed, and the dynamic processes that are occurring could lead to a physical vacuum state in which the definition of flavor is different from that already established in the remainder of the world. We will not further discuss in this book these disoriented chiral states, which are a current topic of research.

Any unitary matrix V with $\det(V) = 1$, Eq. (13.54), can be written as the imaginary exponential of a hermitian traceless matrix L:

$$V = \exp(iL), \qquad L^\dagger = L, \qquad \mathrm{tr}(L) = 0. \tag{13.55}$$

All traceless hermitian 3×3 matrices can be expressed as linear combinations of the eight generators t_a of the Lie group $SU(3)$, using eight real variables θ_a:

$$L = \frac{1}{2} \sum_{a=1}^{8} \theta_a \lambda_a, \qquad t_a \equiv \frac{1}{2} \lambda_a. \tag{13.56}$$

The 'fundamental' (Gell-Mann) 3×3-matrix representation of the $SU(3)$ algebra is well known. There are $n_c^2 - 1 = 8$, with $n_c = 3$ for $SU_c(3)$, $\lambda_a, a = 1, \ldots, 8$, matrices:

$$\lambda_1 = \begin{pmatrix} 0 & 1 & 0 \\ 1 & 0 & 0 \\ 0 & 0 & 0 \end{pmatrix}, \quad \lambda_2 = \begin{pmatrix} 0 & -i & 0 \\ i & 0 & 0 \\ 0 & 0 & 0 \end{pmatrix}, \quad \lambda_3 = \begin{pmatrix} 1 & 0 & 0 \\ 0 & -1 & 0 \\ 0 & 0 & 0 \end{pmatrix},$$

$$\lambda_4 = \begin{pmatrix} 0 & 0 & 1 \\ 0 & 0 & 0 \\ 1 & 0 & 0 \end{pmatrix}, \quad \lambda_5 = \begin{pmatrix} 0 & 0 & -i \\ 0 & 0 & 0 \\ i & 0 & 0 \end{pmatrix}, \quad \lambda_6 = \begin{pmatrix} 0 & 0 & 0 \\ 0 & 0 & 1 \\ 0 & 1 & 0 \end{pmatrix},$$

$$\lambda_7 = \begin{pmatrix} 0 & 0 & 0 \\ 0 & 0 & -i \\ 0 & i & 0 \end{pmatrix}, \quad \lambda_8 = \frac{1}{\sqrt{3}} \begin{pmatrix} 1 & 0 & 0 \\ 0 & 1 & 0 \\ 0 & 0 & -2 \end{pmatrix}. \tag{13.57}$$

The λ_a have been constructed in close analogy to the Pauli matrices σ_i, Eq. (13.27), and the t_a are analogous to the spin matrices $s_i = \sigma_i/2$. The first three λ_a correspond (up to the added third trivial row and column) to the Pauli matrices. While λ_8 is the second traceless diagonal 3×3

matrix we can construct, the pairs λ_4, λ_5 and λ_6, λ_7 are generalizations of the $SU(2)$ σ_1 and σ_2 matrices, which are similar to λ_1 and λ_2.

The following commutation and anticommutation relations can be used to define the algebra of the $SU(3)$ group:

$$[t_a, t_b] = i \sum_c f_{abc} t_c, \tag{13.58}$$

$$\{t_a, t_b\} = (1/3)\delta_{ab} I_3 + \sum_c d_{abc} t_c, \tag{13.59}$$

where I_3 is the 3×3 unit matrix. f_{abc} and d_{abc} are the antisymmetric and symmetric 'structure constants', respectively, of the Lie group $SU(3)$, which can be determined in a straightforward fashion from Eqs. (13.58) and (13.59). One of their frequently used properties is

$$\sum_{k,l=1}^{n_c^2-1} f_{ikl} f_{jkl} = n_c \delta_{ij}. \tag{13.60}$$

Another important relation, which is related to the definition of the color charge, is found by considering the trace of Eq. (13.59):

$$\mathrm{tr}(t_a t_b) = \tfrac{1}{2}\delta_{ab}. \tag{13.61}$$

A second frequently needed representation of the $SU(3)$ algebra in terms of the 8×8 matrices is called 'adjoint' representation. This representation plays with regard to the eight-component glue field a similar role to that which the fundamental representation plays with regard to the quark field. Pushing the analogy to the spin, we are looking for the equivalent of spin-1 representation. The matrix representation of generators in the adjoint representation is

$$(T_a)_{bc} = -i f_{abc}, \ a, b, c = 1, \ldots, 8. \tag{13.62}$$

T_a satisfy the same algebra as the generators t_a of the fundamental representation, Eqs. (13.58) and (13.59). The trace of the product of two T_a, an analogous result to Eq. (13.61), follows from Eq. (13.60) in view of Eq. (13.62):

$$\mathrm{tr}(T_a T_b) = n_c \delta_{ab}. \tag{13.63}$$

We now make the color-rotation matrix V space-dependent, $V \rightarrow V(x)$, allowing that the eight real parameters ϑ_a, in Eq. (13.56), depend on x,

$$V(x) = e^{-i\vartheta_a(x) t_a}, \tag{13.64}$$

with summation over a repeated group index such as a being implicitly understood henceforth. In analogy to Eqs. (13.48) and (13.49), the local nonabelian gauge transformation of matter fields is

$$\Psi \rightarrow \Psi' = V\Psi, \tag{13.65}$$

$$\partial_\mu \Psi \rightarrow \partial_\mu \Psi' = V[\partial_\mu \Psi + V^\dagger(\partial_\mu V)\Psi], \tag{13.66}$$

where we have used Eq. (13.54). Since $V^\dagger V = 1$, we have

$$(\partial_\mu V^\dagger)V = -V^\dagger(\partial_\mu V). \tag{13.67}$$

Instead of the term $(i\partial_\mu \alpha)$, in Eq. (13.49), we have now a matrix term $V^\dagger(\partial_\mu V)$; the entire expression is multiplied by a matrix V, Eq. (13.64), rather than a phase factor $e^{i\alpha}$.

13.4 Gluons

We introduce now the dynamic color-potential field A_μ. To proceed in analogy to QED, we have to couple the color potential to a product of quark–antiquark spinors. Thus, A_μ must be a 3×3 matrix, given the color structure of the spinor Ψ. The three-component quark wave function forms a triplet (fundamental) representation of the color group $SU(3)$, while the wave function of an antiquark forms an antitriplet. From the product of a triplet and an antitriplet, intuitively, we can understand that one can form an $SU(3)$ singlet; what remains is an octet of states. In analogy to spin, for which the product of two spin-$\frac{1}{2}$ particles can be a singlet $(S = 0)$ or a triplet $(S = 1)$ state, we write

$$3_c \times \bar{3}_c = 8_c \oplus 1_c. \tag{13.68}$$

Since the color-gauge-field quantum must have the same quantum numbers as a quark–antiquark pair, if it is to be able to be the product of their annihilation, it must contain at least eight non-matrix fields; the ninth singlet field corresponds to the colorless case. In analogy to the real field A_μ in QED, we now choose a hermitian matrix A_μ to represent the massless gluons, in the form of a linear combination of eight Gell-Mann matrices with eight real Yang–Mills fields $A_\mu^a(x)$:

$$A_\mu(x) = \tfrac{1}{2}A_\mu^a(x)\lambda_a = A_\mu^a(x)t_a. \tag{13.69}$$

We proceed in analogy to Eq. (13.50): if the potential A_μ changes under a local color-gauge transformation according to

$$A_\mu \rightarrow A' = VA_\mu V^\dagger + \frac{1}{g}V(\partial_\mu V^\dagger), \tag{13.70}$$

the minimally coupled derivative, $\partial_\mu + igA_\mu$, remains invariant in form under the gauge transformation, up to the nonabelian phase factor

$$(\partial_\mu + igA'_\mu)\Psi' = V(\partial_\mu + igA_\mu)\Psi. \tag{13.71}$$

To show this, we have to use Eq. (13.66) and remember Eq. (13.67).

It is often helpful to check the form of equations, and in particular relative signs, by remembering that the transition from QED to QCD equations can be effected with the introduction of a gauge-covariant derivative, both for quark and glue fields, defined by

$$D_\nu = \partial_\nu + igt_a A^a_\nu = \partial_\nu + igA_\nu. \tag{13.72}$$

In view of Eqs. (13.66) and (13.71), we have shown that this covariant derivative transforms under nonabelian gauge transformations as

$$D_\nu \to D'_\nu = VD_\nu V^\dagger. \tag{13.73}$$

Similarly, the eight-component field-strength tensor,

$$F_{\mu\nu} = t_a F^a_{\mu\nu}, \tag{13.74}$$

defined as,

$$F_{\mu\nu} = \partial_\mu A_\nu - \partial_\nu A_\mu + ig\,[A_\mu, A_\nu]\,, \tag{13.75}$$

or equivalently,

$$F^a_{\mu\nu} = \partial_\mu A^a_\nu - \partial_\nu A^a_\mu - gf_{abc}A^b_\mu A^c_\nu, \tag{13.76}$$

remains up to a phase form-invariant under a local color-gauge transformation, i.e.,

$$F_{\mu\nu} \to F'_{\mu\nu} = VF_{\mu\nu}V^\dagger. \tag{13.77}$$

This transformation property is most easily proved once we realize that we can define the field-strength tensor using the commutator of the covariant derivative, Eq. (13.72),

$$[D_\mu, D_\nu] \equiv igF_{\mu\nu}, \tag{13.78}$$

which verifies Eq. (13.77) in view of Eq. (13.73), remembering properties of V, Eq. (13.54).

13.5 The Lagrangian of quarks and gluons

The complete gauge-invariant Lagrangian of quantum chromodynamics is then

$$\mathcal{L}_{\text{QCD}} = \sum_f \overline{\Psi}_f \gamma^\mu (i\partial_\mu - g A_\mu) \Psi_f - m_f \overline{\Psi}_f \Psi_f - \tfrac{1}{2} \operatorname{tr}(F_{\mu\nu} F^{\mu\nu}). \quad (13.79)$$

Here, the summation over the different quark flavors f has been made explicit. The form similarity to the Lagrangian of QED,

$$\mathcal{L}_{\text{QED}} = \overline{\psi} \gamma^\mu (i\partial_\mu - e A_\mu) \psi - m\overline{\psi}\psi - \tfrac{1}{4} F^{\mu\nu} F_{\mu\nu}, \quad (13.80)$$

is evident.

In view of Eq. (13.61), we have an exact correspondence between the gluon and photon terms:

$$\frac{1}{4} F^{\mu\nu} F_{\mu\nu} = \frac{1}{2} (B^2 - E^2) \Big|_{\text{QED}},$$

$$\frac{1}{2} \operatorname{tr}(F^{\mu\nu} F_{\mu\nu}) = \frac{1}{4} \sum_a F^{\mu\nu}_a F^a_{\mu\nu} = \frac{1}{2} \sum_a (B_a B^a - E_a E^a) \Big|_{\text{QCD}}. \quad (13.81)$$

The key difference is in the nonlinear term entering into the definition Eq. (13.76) of $F^a_{\mu\nu}$, which is quadratic in the color potentials A^a_μ.

The analog of the Maxwell equations may be derived for the color field from the QCD Lagrangian, Eq. (13.79):

$$[D_\nu, F^{\mu\nu}] = g j^\mu. \quad (13.82)$$

The color indices of the matter fields Ψ define the matrix current j^μ obtained from $\mathcal{L}_{\text{QCD}}$,

$$j^a_\mu = \overline{\Psi} \gamma^\mu t_a \Psi. \quad (13.83)$$

We can write Eq. (13.82) in the more conventional form

$$\partial_\nu F^{\mu\nu}_a = g j^\mu_a + g J^\mu_a, \quad (13.84)$$

where we encounter the gluon current J^μ_a:

$$J^\mu_a = f_{abc} A^b_\nu F^{\mu\nu}_c, \quad J^\mu = -i[A_\nu, F^{\mu\nu}]. \quad (13.85)$$

The nonlinear term, Eq. (13.85), placed on the right-hand side of Eq. (13.84), shows that the color field acts also as its own source. In other words, the quanta of the color field, *gluons*, carry color charge themselves. This is the source of the substantial physical difference between QCD and QED. From Eq. (13.84), one can further see that j^a_μ does not obey a continuity equation, which means that the color charge of quarks alone is not conserved. This is not surprising since quarks can emit or absorb gluons, which carry color. Only if we add the color charge of the gluon field, represented by the second term on the right-hand side of Eq. (13.84), is a conserved color current obtained.

In view of the similarity in form of $\mathcal{L}_{QCD}$ to $\mathcal{L}_{QED}$, many of the well-known other formal properties carry over. We refrain from systematically developing this here, even though we will call upon these similarities as needed in further developments.

14 Perturbative QCD

14.1 Feynman rules

The nonabelian gauge theory of quarks and gluons, proposed in section 13.5 and called QCD, has widely been accepted as the fundamental theory of strong interactions, with both quarks and gluons being the carriers of the strong-interaction charge [123]. The evidence for the validity of QCD as a dynamic theory governing hadronic reactions is overwhelming, and this is not the place where this matter should be argued. Rather, we will show how QGP-related practical results can be derived from the complex theoretical framework. There are many books dealing with more applications of QCD and the interested reader should consult these for further developments [110, 194, 280].

Akin to QED, QCD is a 'good' renormalizable theory. QCD is known to be also an *asymptotically free* theory, viz., the running coupling constant α_s, see Eq. (14.12), is a diminishing function as the energy scale increases. Therefore, the high-energy, or, equivalently, the short-distance behavior is amenable to a perturbative expansion. On the other hand, perturbative QCD has 'fatal' defects at large distances, which are signaled by the growth and the ultimate divergence of α_s as the scale of energy diminishes (infrared 'slavery'). Consequently, at any reasonable physical distance of relevance to the 'macroscopic' QGP, we have to deal with an intrinsically strongly coupled, nonperturbative physical system. A perturbative treatment ignores this, and, in principle, must be unreliable in problems in which the confinement scale becomes relevant. The question of when exactly this occurs will be one of the important issues we will aim to resolve, using as criterion $\alpha_s \leq 1$.

The perturbative approach, which applies to short-distance phenomena, has been tested extensively in high-energy processes. When the 'short distance' grows and approaches 0.5 fm, the perturbative expansion of QCD may still apply insofar as its results are restricted to the physics occurring in the deconfined, viz., QGP, phase. The rules of perturbative QCD follow the well-known Feynman rules of QED, allowing for the glue–glue interactions. However, unlike QED, in which the expansion parameter is $\alpha/\pi = \mathcal{O}(10^{-3})$, we must deal in case of QCD with 'strong' coupling, which is nearly 30–100 times greater than that in QED. Therefore, even when a perturbative description is suitable, some effort to reach the required precision is often necessary, involving the inclusion of higher-order Feynman diagrams.

The quadratic terms 'ψ^2' and 'A^2' in the Lagrangian $\mathcal{L}_{\text{QCD}}$ define free quark and gluon fields which are described by propagators of the same form as those for electrons and photons in QED. The terms of third and fourth order in $\mathcal{L}_{\text{QCD}}$ give rise to interaction vertices among the free propagators of quarks and gluons. There is a quark–gluon vertex, and three-gluon and four-gluon vertices. Propagators and vertices can be combined to generate Feynman diagrams in all possible ways.

One technical difference arises between QED and QCD, which is associated with the difficulties of gauge theories with regard to gauge fixing. To ensure gauge invariance of QCD, it is convenient to introduce into the perturbative expansion fictitious (virtual) particles called Fadeev–Popov (FP) ghosts, which never appear in physical states, but are to be included in all virtual processes. FP fields carry color, and satisfy Fermi statistics even though they propagate like spin-zero particles – hence the name 'ghost'. The complete Lagrangian in a Lorentz-covariant gauge reads

$$\mathcal{L}_{\text{QCD}} = -\frac{1}{4}(F_a^{\mu\nu}F_{\mu\nu}^a)^2 + \xi(\partial_\mu A_\mu^a)^2 + \bar{\phi}(\partial^\mu \delta_{ab} + g f^{asb} A_\mu^s)\partial^\mu \phi^b$$
$$+ \sum_f \overline{\Psi}_f\left[\gamma^\mu(\partial_\mu + gA_\mu^s) - m_f\right]\Psi_f, \tag{14.1}$$

where the summation over color indices is implied and the second term in Eq. (14.1) is the gauge-fixing term and the third term formed with the scalar fields ϕ introduces the Fadeev–Popov ghosts. We note that it is possible to work in a non-Lorentz-covariant gauge and to obtain results without introducing ghosts, instead using longitudinal and transverse gluons as identifiable degrees of freedom, with longitudinal gluons not present in any asymptotic physical states. For further discussion, we refer to relevant textbooks [110, 194, 280].

We skip the technical details regarding the development of the perturbative QCD, and only collect for the convenience of the reader and further reference the building blocks of the perturbative expansion required for evaluation of Feynman diagrams, i.e., propagators and vertices of QCD, presented in a self-explanatory notation. Latin indices refer to color degrees of freedom, Greek indices to space, q to quarks, g to gluons, and FP to the ghost field.

We have
the quark propagator,

$$(S_{\alpha\beta}(p))^{ab} = \left(\frac{\delta_{ab}}{\gamma p - m + i\varepsilon}\right)_{\alpha\beta}, \tag{14.2}$$

the gluon propagator,

$$i(D_{\mu\nu}(k))^{ab} = \frac{\delta_{ab}g_{\mu\nu}}{k^2 + i\varepsilon}, \tag{14.3}$$

the quark–antiquark–gluon vertex,

$$(\Gamma^{\mu,c}_{qqg})^{ab}_{\alpha\beta} = g(t^c)^{ab}(\gamma^\mu)_{\alpha\beta}, \tag{14.4}$$

the three-gluon vertex,

$$(\Gamma^{\mu\nu\sigma}_{g3})_{abc} = gf_{abc}[g^{\mu\nu}(k-p)^\sigma + \text{cyclic permutations}], \tag{14.5}$$

the four-gluon vertex,

$$(\Gamma^{\mu\sigma\nu\tau}_{g4})_{abcd} = g^2 f_{iab}f_{icd}[(g^{\mu\sigma}g^{\nu\tau} - g^{\mu\tau}g^{\nu\sigma}) + \text{cyclic permutations}], \tag{14.6}$$

the ghost propagator,

$$(G_{\mathrm{FP}}(k))^{ab} = \frac{\delta_{ab}}{k^2 + i\varepsilon}, \tag{14.7}$$

and the gluon–ghost–ghost vertex,

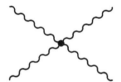

$$(\Gamma^\mu_{\mathrm{gFP}})_{abc} = -gf_{abc}p^\mu. \tag{14.8}$$

For any given process, for which a Feynman diagram is drawn using the lines and vertices as illustrated above, the above list allows one to compose the mathematical expression for the amplitude of the process. Very few additional rules need to be remembered, such as integration over the 'spare' momentum variables in the diagram, an overall coefficient (-1) for each fermion and ghost loop, and the absence of ghost propagators that do not begin and end in a vertex.

While the forms of the above-stated propagators and vertices change in a finite-temperature environment, which is mainly being addressed in this book, the structure of the perturbative expansion generated by these quantities remains the same. The construction of a matrix element and cross section requires wave-function-normalization factors, and flux factors that are all quite standard in this context, and available in numerous introductory textbooks. In these aspects, there is no difference between QCD and any other theoretical framework, such as QED. However, we recall that, in order to obtain a cross section, we average over initial states, and sum over final states, which now will include, in particular, the color degree of freedom.

14.2 The running coupling constant

The free-gluon propagator, which, like in QED, is proportional to $1/q^2$, implies that the 'free' color force falls off like $1/r$. The gluon propagator is, even in the perturbative vacuum, modified by scattering from virtual quark–gluon fluctuations. This 'dressing' of the propagator leads to the running, q^2-dependent coupling constant, and the dressed physical gluon propagator is expected to significantly differ from the free one. In order to see how this comes about, we consider the loop diagrams corresponding to the virtual and momentary creation of a pair of colored particles in the vacuum. For the contribution of fermions to the polarization loop, this process is very similar to the case of QED.

There are two more elementary processes that contribute to the covariant form of perturbative QCD, namely the gluon loop and the ghost loop, The vacuum-polarization loop $\Pi(q^2)$ comprises these three terms:

It is customary to sum the chain of higher-order diagrams, containing series of all different loops,

$$D(q^2) = D_0 + D_0 \Pi D_0 + D_0 \Pi D_0 \Pi D_0 + \cdots$$

$$D(q^2) = \frac{D_0(q^2)}{1 - \Pi(q^2)D_0(q^2)},$$ (14.9)

such that the gluon is dressed by the consecutive interactions with the vacuum polarization. The effect of the quark, glue, and ghost loops combines in the coefficient b_0, Eq. (14.14), of the vacuum-polarization function. Introducing a high momentum (ultraviolet) cutoff to secure the convergence, for massless quarks and gluons, $\Pi(q^2)$ takes the form

$$\Pi(q^2) = \frac{g^2 b_0}{8\pi}(-q^2)\ln\left(\frac{-q^2}{\mu^2}\right),$$ (14.10)

where μ is a reference momentum absorbing the cutoff, and defined by the renormalization condition: $\Pi(q^2 = -\mu^2) = 0$.

The value of μ^2 introduces, in general, a dependence on renormalization. In this order, this dependence can be absorbed into the choice of the value of the coupling constant at a given transfer of momentum; see Eq. (14.13) below. The overall sign of Π, Eq. (14.10), is related to the sign of b_0, and is opposite to that of the polarization function of fermions alone (QED) as long as the flavor number $n_f < (11/2)n_c$. The sign of the polarization function is quite important, as we shall see.

With the help of Eq. (14.10), we obtain

$$g^2 i D(q^2) = \frac{1}{q^2}\frac{g^2}{1 + \dfrac{g^2 b_0}{8\pi}\ln\left(\dfrac{-q^2}{\mu^2}\right)}.$$ (14.11)

The last factor acts as a momentum-dependent modification of the strong coupling constant g^2. It is therefore convenient to introduce the 'running' coupling constant $\alpha_s(q^2)$,

$$\boxed{\alpha_s(q^2) = \frac{2}{b_0 \ln(-q^2/\Lambda^2)},}$$ (14.12)

with

$$\Lambda^2 = \mu^2 \exp\left(-\frac{8\pi}{b_0 g^2}\right), \quad \alpha_s = \frac{g^2}{4\pi}.$$ (14.13)

The above expression applies for positive b_0, and, in this case, we see that, for an increasing q^2, the physical coupling constant $\alpha_s(q^2)$ decreases; QCD is asymptotically free.

In the case of QED, the sign of b_0 is opposite, and the effective coupling constant is finite at large distances, i.e., at small q^2, and increases with q^2. The reference scale μ^2 in Eq. (14.11) can be chosen to be at zero momentum (infinite distance) and this corresponds to the usual definition of the electron charge. This choice is not possible in case of QCD, since the interaction strength becomes infinite for $\mu^2 \to 0$.

Λ is the dimensional parameter which emerges in perturbative QCD. The original dimensionless coupling constant g^2 has been absorbed in the scale Λ governing the change of the running coupling constant α_s, in the process of transition from Eq. (14.11) to Eq. (14.12). This so-called 'transmutation' of the dimensionless scale of strength into a dimensioned strength parameter of the interaction also absorbs the scale dependence introduced by the choice of μ^2.

In the limit that all quark masses vanish, $m_i \to 0$, Λ is the only dimensional parameter of the theory of strong interactions. It is believed that the world of hadrons (except the pion) is not decisively dependent on the scale of the quark mass. Thus, it seems that Λ alone controls the mass and the size of the massive hadrons (nucleons and heavy non-strange mesons). To understand this, we would need to express the vacuum structure in terms of Λ, a problem which has not been resolved.

The *measurable* dimensioned parameter Λ^2 determines the strength of the interaction at a given momentum scale. This approach applies quite accurately for energy–momentum scales above the mass of the b quark, as we shall see later in Fig. 14.1, where the value of $\Lambda \simeq 90 \pm 15$ MeV applies. At small q^2, i.e., at 'large' distances, the coupling constant Eq. (14.12) diverges within the perturbative approach. The magnitude of the strong charge must be defined at some finite momentum scale, which has in recent years, been chosen to be the mass of the Z^0 boson, $\mu \equiv M_{Z^0} \simeq 91.19$ GeV.

Since we have more than one quark, the important coefficient b_0 is composed of a term proportional to the number of 'active' flavors n_f, i.e., those with $|q^2| > 4m_f^2$. The number of colors $n_c = 3$ enters the glue loop: in the gluon loop diagram, each external leg requires the triple-glue vertex, Eq. (14.5), which invokes relation Eq. (13.63) or equivalently Eq. (13.60) for two external gluon legs. b_0 for $SU_c(3)$ assumes the form

$$b_0 = \frac{1}{2\pi}\left(\frac{11}{3}n_c - \frac{2}{3}n_f\right). \tag{14.14}$$

The spin s of particles contributing to the vacuum polarization is found to be the key ingredient controlling the sign of b_0 [149],

$$b_0^s = \frac{(-)^{2s}}{2\pi}\left((2s)^2 - \frac{1}{3}\right), \tag{14.15}$$

which leads to Eq. (14.14), introducing $s = 1$ for gluons and $s = \frac{1}{2}$ for quarks. Equation (14.15) shows why for $s = 0, \frac{1}{2}$ the same (negative) sign appears, whereas for gluons with $s = 1$ there is a change of sign. Photons do not interact with photons and hence this issue did not arise in QED.

14.3 The renormalization group

The question of what happens as we carry out the same procedures in higher orders in perturbation theory now arises. A considerable amount of effort went into designing a scheme for computing the observable effects in QED, and this experience has been generalized to the more complex case of QCD. We will restrict ourselves to a few elements of the *renormalization-group* approach relevant to our presentation, sidestepping many interesting and intricate questions, which are addressed in, e.g., [110, 194, 280].

The renormalization-group approach allows us to understand the variation of the physical observables in terms of the momentum dependence of the coupling constant α_s. A functional dependence is found demanding that the result of a physical measurement (say a cross section σ) be invariant with respect to the process of renormalization, and, in particular, the observable (cross section) can not depend explicitly on the choice of the '(re)normalization' point μ^2,

$$\mu \frac{d}{d\mu} \sigma(p_i; \alpha_s, m; \mu) = 0. \tag{14.16}$$

Accounting for both a direct and an indirect dependence on μ in Eq. (14.16),

$$\left(\mu \frac{\partial}{\partial \mu} + \mu \frac{\partial \alpha_s}{\partial \mu} \frac{\partial}{\partial \alpha_s} + \mu \frac{\partial m}{\partial \mu} \frac{\partial}{\partial m} + \cdots \right) \sigma = 0. \tag{14.17}$$

It is convenient to define

$$\mu \frac{\partial \alpha}{\partial \mu} \equiv \beta(\alpha_s) \tag{14.18}$$

and

$$-\frac{\mu}{m} \frac{\partial m}{\partial \mu} \equiv \gamma(\alpha_s), \tag{14.19}$$

and thus:

$$\left(\mu \frac{\partial}{\partial \mu} + \beta(\alpha_s) \frac{\partial}{\partial \alpha_s} - \gamma(\alpha)m \frac{\partial}{\partial m} + \cdots \right) \sigma = 0. \tag{14.20}$$

Equation (14.20) allows us to understand the behavior of the observable σ, and it establishes the behavior of σ under simultaneous changes of reference scale μ, coupling constant, and mass.

Equations (14.18) and (14.19) establish how the parameters of QCD vary once they are known at some given scale. Therefore, precise knowledge of the renormalization functions β and γ is, to a large degree, equivalent to finding a practical 'solution' of QCD. For this reason, these quantities have attracted a lot of attention. We will look at the perturbative results only, in terms of a power expansion in α_s [255]:

$$\beta^{\text{pert}} = -\alpha_s^2 \left[\, b_0 + b_1\alpha_s + b_2\alpha_s^2 + \cdots \right], \qquad (14.21)$$

$$\gamma_{\text{m}}^{\text{pert}} = \alpha_s \left[\, w_0 + w_1\alpha_s + w_2\alpha_s^2 + \cdots \right]. \qquad (14.22)$$

For the $SU(3)$ gauge theory with n_f fermions only, the first two terms (two 'loop' orders) are renormalization-scheme-independent. When a dependence on the renormalization scheme arises, this means that compensating terms, which remove scheme dependence, are obtained on evaluating in the same scheme the physical process considered. For what follows in this book, this so-called two-loop level of perturbative expansion for β^{pert} and $\gamma_{\text{m}}^{\text{pert}}$ is sufficient.

We have

$$b_0 = \frac{1}{2\pi} \left(11 - \frac{2}{3}n_f \right), \qquad b_1 = \frac{1}{4\pi^2} \left(51 - \frac{19}{3}n_f \right), \qquad (14.23)$$

$$w_0 = \frac{2}{\pi}, \qquad w_1 = \frac{1}{12\pi^2} \left(101 - \frac{10}{3}n_f \right). \qquad (14.24)$$

The number n_f of 'active' fermions depends on the scale μ. Assuming that the two lightest quarks are effectively massless,

$$n_f(\mu) = 2 + \sum_{i=s,c,b,t} \sqrt{1 - \frac{4m_i^2}{\mu^2}} \left(1 + \frac{2m_i^2}{\mu} \right) \Theta(\mu - 2m_i), \qquad (14.25)$$

with values of m_i evaluated, in principle, for the energy scale being considered. There is a very minimal impact of the values of quark-mass thresholds in Eq. (14.25), on the running behaviors of the coupling constant and quark masses.

14.4 Running parameters of QCD

For the purpose of QGP studies, we are interested in understanding how the strength of the QCD interaction and the quark mass change with the energy scale. The simplest way to obtain this result is to integrate the first-order differential equation, Eqs. (14.18) and (14.19), given initial values of $\alpha_s(M)$ and $m_i(M)$, using the perturbative definition of the functions β and γ, Eqs. (14.21) and (14.22), in terms of the perturbative expansion Eqs. (14.23) and (14.24).

For the determination of the coupling constant, it has become common to refer to the value of $\alpha_s(M_Z = 91.19\,\text{GeV})$. We use, in Fig. 14.1, the value [136] $\alpha_s(M_Z) = 0.1182 \pm 0.002$ (thick solid lines). The thin solid

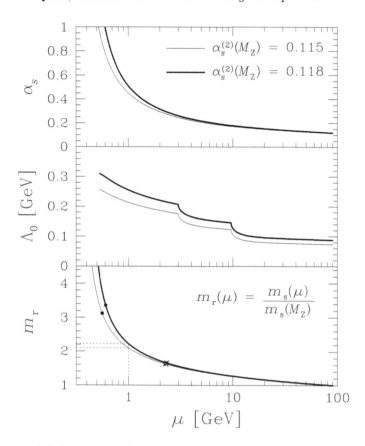

Fig. 14.1. $\alpha_{\rm s}(\mu)$ (top section), the equivalent parameter Λ_0 (middle section), and $m_{\rm r}(\mu) = m(\mu)/m(M_{\rm Z})$ (bottom section) as functions of the energy scale μ. The initial values are $\alpha_{\rm s}(M_{\rm Z}) = 0.118$ (thick solid lines) and $\alpha_{\rm s}(M_{\rm Z}) = 0.115$ (thin solid lines). In the bottom section, the dots indicate the strangeness-pair-production thresholds for $m_{\rm s}(M_{\rm Z}) = 90\,{\rm MeV}$, while crosses indicate charm-pair-production thresholds for $m_{\rm c}(M_{\rm Z}) = 700\,{\rm MeV}$.

lines are for $\alpha_{\rm s}(M_{\rm Z}) = 0.115$ arising from analysis of decays of heavy quarkonium (b$\bar{\rm b}$), and addressing an energy scale closer to our direct interest in this book.

    As can be seen in the top portion of Fig. 14.1, the variation of $\alpha_{\rm s}^{(2)} < 1$ (the upper index indicates the level of perturbative expansion used; see Eq. (14.21)) with the energy scale is substantial. We note the rapid change in $\alpha_{\rm s}(\mu)$ at, and below, $\mu = 1$ GeV. This is not an unexpected result. However, the fact that a solution with $\alpha_{\rm s} \leq 1$ governs the energy scale 1 GeV is important, since the formation of the strange quark flavor occurs in a hot QGP phase formed in experiments at 160–200A GeV (SPS–

CERN) at around 1 GeV. We can thus use the methods of perturbative QCD to study this process.

Among the parameters in Eq. (14.25), only the mass of the bottom quark plays a (hardly) noticeable role; the results shown were obtained for $m_s = 0.16\,\mathrm{GeV}$, $m_c = 1.5\,\mathrm{GeV}$, and $m_b = 4.8\,\mathrm{GeV}$ [225]. When m_b is changed by 10%, the error on a low energy scale is barely visible. Other quark masses have less significance since the error has less opportunity to 'accumulate' in the solution of the differential equation as the energy scale decreases in the integration of Eq. (14.18).

As expected and seen in the top portion of Fig. 14.1, in the soft QGP domain $0.8\,\mathrm{GeV} < \mu < 3\,\mathrm{GeV}$, it is impossible to use a constant value of α_s. More surprisingly, the frequently used approximate inverse-logarithm form Eq. (14.12) for α_s is nearly equally inappropriate. To see this, we define a quantity $\Lambda_0(\mu)$,

$$\alpha_s(\mu) \equiv \frac{2b_0^{-1}(n_f)}{\ln \mu^2/\Lambda_0^2(\mu)}, \quad \Lambda_0(\mu) = \mu \exp\left(-\frac{1}{b_0\alpha_s(\mu)}\right), \qquad (14.26)$$

where $\alpha_s(\mu)$ is obtained in two-loop or higher-order perturbation expansion of the β-function using Eq. (14.18). The form of Eq. (14.26) is chosen to be identical to the one-loop form, compare with Eq. (14.13). Using the result for α_s shown in the top portion of Fig. 14.1, we obtain $\Lambda_0(\mu)$ seen in the middle section of the figure.

If the one-loop form Eq. (14.12) with a constant Λ_0 were a good approximation to α_s, we should see a sequence of step functions, dropping at each heavy-quark threshold. In fact, above the bottom threshold, for $\mu > 2m_b$, this is nearly the case. However, below the charm threshold, for $\mu < 2.5$ GeV, where practically all QGP action is occurring, we see a rather rapid change in $\Lambda_0(\mu)$, which drops from a value near $\Lambda_0(1\,\mathrm{GeV}) \simeq 300$ MeV toward $\Lambda_0(3\,\mathrm{GeV}) \simeq 200$ MeV.

It is common to refer to the number of active quarks by using an upper index on $\Lambda_0(\mu)$, thus $\Lambda_0^{(3)}$ refers to the range $1\,\mathrm{GeV} < \mu < 2m_c$, and $\Lambda_0^{(5)}$ refers to $\mu > 2m_b$, and below the top threshold. We also see, in Fig. 14.1, that $\Lambda_0^{(5)} \simeq 90 \pm 15$ MeV. This value of $\Lambda_0^{(5)}$ derived from a comparison with the one-loop solution should not be mixed up with $\Lambda^{(5)} = 205 \pm 25$ MeV, which is the value required to describe α_s in the analytical two-loop solution, Eq. (14.28) [136].

To understand how quark masses depend on the energy scale, given α_s, we integrate Eq. (14.19). Substituting $m(\mu) = m_r(\mu)m(M_Z)$, we recognize that $m_r(\mu)$ is a multiplicative factor applicable to all quark masses. $m_r(\mu)$ is shown in the bottom portion of Fig. 14.1. All quark masses 'run' according to this result. A quark mass given at the scale $\mu = M_Z$ increases by factor 2.2 at scale $\mu = 1$ GeV, as the dotted lines drawn to guide the eye show. Near to $\mu \simeq 1\,\text{GeV}$, the quark-mass factor $m_{\text{r}}(\mu)$ is driven by the rapid change of α_{s}. For each of the different functional dependences $\alpha_{\text{s}}(\mu)$, a different function m_{r} is found, and two results are presented, corresponding to the two cases considered in the top section of Fig. 14.1.

Since α_{s} refers to the scale of $\mu_0 = M_{\text{Z}}$, it is a convenient reference point also for quark masses. The value $m_{\text{s}}(M_{\text{Z}}) = 90 \pm 18\,\text{MeV}$ corresponds to strange-quark mass $m_{\text{s}}(1\,\text{GeV}) \simeq 195 \pm 40\,\text{MeV}$, i.e., $m_{\text{s}}(2\,\text{GeV}) \simeq 150 \pm 30\,\text{MeV}$, at the upper limit of the established range seen in table 1.1 on page 7. Similarly, we consider $m_{\text{c}}(M_{\text{Z}}) = 700 \pm 50\,\text{MeV}$, for which value we find the low-energy mass $m_{\text{c}}(1\,\text{GeV}) \simeq 1550 \pm 110\,\text{MeV}$, i.e., $m_{\text{c}}(2\,\text{GeV}) \simeq 1200 \pm 85\,\text{MeV}$, which is also at the upper limit of the accepted range, table 1.1.

For quark-pair production, the intuitive energy scale to consider is a range near to twice the (running) quark mass. Since, below $\sqrt{s} = 1$ GeV, the mass of the strange quark increases rapidly, the pair-production threshold is considerably greater than $2m_{\text{s}}(1\,\text{GeV}) \simeq 400$ MeV. The dots in the bottom portion of Fig. 14.1 show where the strangeness threshold is found, and this is at $2m_{\text{r}}(2m_{\text{s}})m_{\text{s}} = 611\,\text{MeV}$ for $\alpha_{\text{s}}(M_{\text{Z}}) = 0.118$. The strangeness threshold is where $\alpha_{\text{s}} \simeq 1$ and we can expect, considering that the phase space for pair production opens up at about $3m$, that strangeness is produced predominantly in an energy domain accessible to perturbative treatment.

For charm, the threshold shift due to running mass occurs in the opposite direction: since the mass of charmed quarks for $\mu = 1\,\text{GeV}$ is above 1 GeV, the production-threshold mass is smaller than $2m_{\text{c}}(1\,\text{GeV}) \simeq 3.1$ GeV; the production threshold is found at $\sim 2m_{\text{c}}^{\text{th}} \simeq 2.3$ GeV, and the corresponding values of m_{r} are indicated by crosses in the bottom portion of Fig. 14.1. In other words, we expect that, near the threshold, there is a slight enhancement in production of charm related to a reduction of the threshold, while the coupling strength is at $\alpha_{\text{s}}(2m_{\text{c}}) \simeq 0.3$.

The inclusion of higher-order terms in the perturbative expansion Eq. (14.21) does not influence the behavior of α_{s}. This is shown in Fig. 14.2, in which a study of α_{s} is shown. To obtain the solid line, the full current 'scheme-dependent' knowledge about the perturbative β-function is employed. The four-loop β-function obtained in the modified minimum-subtraction scheme ($\overline{\text{MS}}$) was used [136]. On the other hand, Eq. (14.21) demonstrates that there is a considerable sensitivity to the initial value $\alpha_{\text{s}}(M_{\text{Z}})$. If $\alpha_{\text{s}}(M_{\text{Z}})$ were to increase, the evaluation of the coupling strength in the 'low'-energy domain $\mu \lesssim 1$ GeV of interest here would become impossible, or at best unreliable, see the dotted lines in Fig. 14.2 above the solid line. In fact, we do not present many re-

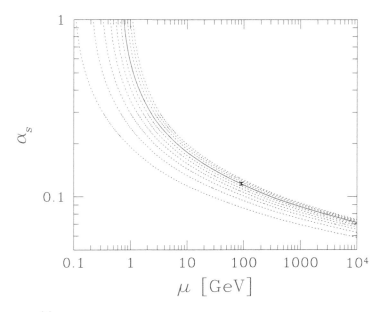

Fig. 14.2. $\alpha_s^{(4)}(\mu)$ as a function of the energy scale μ for a variety of initial conditions. Solid line, $\alpha_s(M_Z) = 0.1182$ (see the experimental point, which includes the error bar at $\mu = M_Z$); dotted lines, sensitivity to variation of the initial condition.

sults in this domain since the renormalization-group-evolution equation, Eq. (14.18), becomes numerically unstable when the four-loop perturbative β-function is used.

Interestingly, a 20% reduction in $\alpha_s(M_Z)$ leads to a 'good' $\alpha_s(0.1 \, \text{GeV})$. The distance scale $1/\mu$ at which QCD becomes unstable is not just 1 fm, but, as this study shows, an intricate functional of the strength of the fundamental interaction, which has reliably been established only in recent years. An essential prerequirement for the perturbative theory of the production of strangeness in QGP, which we will develop in section 17.3, is the relatively small value $\alpha_s(M_Z) \simeq 0.118$.

For studying thermal processes in QGP at temperature T, the proposed interaction scale is, see Eq. (16.11),

$$\mu \equiv 2\pi T \simeq 1 \, \text{GeV} \, T/T_c,$$

for $T_c \simeq 160 \, \text{MeV}$. We can expect considerable sensitivity in this low range of μ to the exactness of the functional form of $\alpha_s(\mu)$, and it is necessary to use the precise function $\alpha_s(\mu)$. In Fig. 14.3, the solid line bounded by error lines corresponds to the exactly computed two-loop α_s with physical quark thresholds, Eq. (14.25), and with $\alpha_s(M_Z) = 0.1181 \pm$

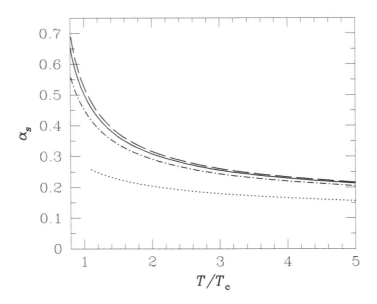

Fig. 14.3. $\alpha_{\mathrm{s}}(2\pi T)$ for $T_{\mathrm{c}} = 0.160\,\mathrm{GeV}$. Dashed line, $\alpha_{\mathrm{s}}(M_{\mathrm{Z}}) = 0.119$; solid line $= 0.1181$; chain line $= 0.1156$. Dotted line, approximate two-loop solution, given in Eq. (14.28), with the choice $\Lambda = 150\,\mathrm{MeV}$.

0.002, evaluated for the thermal scale, and expressed in terms of T/T_{c}. The range of experimental uncertainty in $\alpha_{\mathrm{s}}(T)$, due to uncertainty in $\alpha_{\mathrm{s}}(M_{\mathrm{Z}})$, is delimited by dashed and chain lines bordering the solid line in Fig. 14.3. A good approximation is obtained fitting $\alpha_{\mathrm{s}}(T)$ with a logarithmic form,

$$\alpha_{\mathrm{s}}(T) \simeq \frac{\alpha_{\mathrm{s}}(T_{\mathrm{c}})}{1 + C\ln(T/T_{\mathrm{c}})}, \quad C = 0.760 \pm 0.002, \text{ for } T < 5T_{\mathrm{c}}. \quad (14.27)$$

The value $\alpha_{\mathrm{s}}(T_{\mathrm{c}}) = 0.50^{-0.05}_{+0.03}$ applies in the two-loop description with $\mu = 2\pi T$ and $T_{\mathrm{c}} = 0.16\,\mathrm{GeV}$ (see Fig. 14.3).

A popular approximation of α_{s}, which incorporates the next term beyond the one-loop logarithmic term Eq. (14.26), is

$$\bar{\alpha}_{\mathrm{s}}^{(2)}(\mu) \simeq \frac{2}{b_0\bar{L}}\left(1 - \frac{2b_1}{b_0^2}\frac{\ln\bar{L}}{\bar{L}}\right), \quad \bar{L} \equiv \ln(\mu^2/\Lambda^2). \quad (14.28)$$

$\bar{\alpha}_{\mathrm{s}}^{(2)}$ agrees, using the standard value $\Lambda^{(5)} = 205 \pm 25\,\mathrm{MeV}$, with the exact solution shown at the top of Fig. 14.1, but only for $\mu > 2m_{\mathrm{b}}$. On the other hand, when one is studying thermal properties of a QGP at a low energy scale the use of Eq. (14.28) below $\mu = 2m_{\mathrm{b}}$ introduces considerable error, as can be seen in Fig. 14.3. Equation (14.28) is represented by the dotted line, and misses the exact result by a factor of two, for $T_c < T < 1.75T_c$, the effective range of observables emerging from SPS and RHIC experiments. The experimental error in determination of α_s is today considerably smaller. This large difference between exact and approximate result arises, in part, because the value of $\Lambda^{(5)}$ used to obtain the thermal behavior was adjusted to be $\Lambda^{(4)} = 0.95T_c \simeq 0.15$ GeV. This value would be correct if T_c were indeed around 210 MeV, as has been thought for some time.

The high sensitivity of physical observables to α_s, makes it imperative that we do not rely on this approximation. Yet a fixed value $\alpha_s = 0.25$ (instead of $\alpha_s = 0.5$) derived from this approximation is still often used in studies of the phase properties of QGP, loss of energy by jets of partons, thermalization of charmed quarks, thermal production of strange quarks, etc. Such a treatment of thermal QCD interaction underestimates by as much as a factor of four the interaction with the QGP phase, and thus the speed of these processes. In most cases, this mundane factor matters, and we see that an accurate evaluation of α_s at the appropriate physical scale is required in order to establish the correct magnitude of these results.

15 Lattice quantum chromodynamics

15.1 The numerical approach

The perturbative approach to QCD lacks the capability to describe the long-distance behavior, which is essential for understanding the QGP–HG transformation. We need a more rigorous approach in order to characterize the physical mechanisms at the origin of color confinement, and the transition to the deconfined state of hadronic matter. A suitable nonperturbative approach is the numerical study of QCD on a lattice (L-QCD).

L-QCD is a vast field that is evolving very actively. We will limit our presentation to a pedestrian guide to the language used in this field, along with a report on a few key results of greatest importance to us. We will not be embarking on a thorough introduction to the theoretical and numerical methods. For a survey of the historical developments until the early eighties we refer to the monograph by Creutz [97], and for a summary of recent theoretical advances, and many numerical results addressing hot QCD, we refer the reader to the recent survey by Karsch [159].

The particular usefulness of the lattice-gauge-theory formulation is that it allows one to numerically carry out Feynman path integrals which represent expectation values of quantum-field-theory operators. Specifically, the expectation value of an operator $\mathcal{O}$, including both glue and quark fields, is

$$\langle \mathcal{O} \rangle = \frac{\int \left[d\mathrm{A}_\mu \, d\bar{\psi} \, d\psi \right] \mathcal{O} e^{-i \int \mathcal{L}(\mathrm{A}, \bar{\psi}, \psi) \, d^4 x}}{\int \left[d\mathrm{A}_\mu \, d\bar{\psi} \, d\psi \right] e^{-i \int \mathcal{L}(A, \bar{\psi}, \psi) \, d^4 x}} . \tag{15.1}$$

These integrals are over all values of the gluon and quark fields at all points in space and time. However, most parts of the domain of the integral are unimportant – in 'weak coupling' (i.e., when perturbative expansion makes good sense), only paths close to the classical paths (classical-field solutions) are important. In order to do a path integral efficiently, it is essential to sample more densely the domains that give large contributions, see Eq. (15.2) below, and this obviously then poses a practical challenge. The integration measure of the path integral is indicated by $\left[d\mathrm{A}_\mu \, d\bar{\psi} \, d\psi \right]$. The goal of computations, in lattice QCD, is to evaluate Eq. (15.1) numerically by evaluating the integrand at selected lattice points representing its domain.

The functional integral in Eq. (15.1), expressed on the lattice, means that we are integrating the fields at each lattice site and lattice link, and the domain of the integral has accordingly a very high dimensionality. The method of choice for doing such integrals numerically is the Monte-Carlo (random-choice) method. However, a considerable complication in applying this method arises since $e^{-i \int \mathcal{L} \, d^4 x}$ is not in general a positive real number: aside from the i factor in the exponent, it is a functional of quark fields, which have to be represented by anticommuting numbers: $\psi_x \psi_y = -\psi_y \psi_x$. This problem can be solved since the dependence on ψ and $\bar{\psi}$ of Eq. (15.1) has the form of a polynomial times a Gaussian. Therefore, the quark portion of the path integral can be done analytically. This integral yields a 'Fermi determinant' *FD*, which changes for each configuration of the gauge (gluon) fields considered. We will address the form of *FD* in section 15.4.

In order to allow a Monte-Carlo integration procedure for the gauge fields, the explicit i in the exponent in Eq. (15.1) is combined with dt, and the integral is considered in 'imaginary' time, or, as it is usually said, Euclidian space. It is generally believed that some, if not all, physical results can be analytically continued from the real- to the imaginary-time axis. Even so, the Fermi determinant remains real only for zero chemical potential, and we can use as a probability for sampling the Monte-Carlo integral

$$\rho(A) = FD_I e^{-\int \mathcal{L}_a \, d^4 x_I} . \tag{15.2}$$

$\mathcal{L}_a$ is the gluon part of the Lagrangian in imaginary time and subscript I indicates that the quantities have been suitably modified by the transformation $t \rightarrow it$.

The physical system is restricted to a finitely sized box, which introduces an infrared (long-distance) cutoff at the size of the box, L. The continuous space and time is represented by a lattice, which introduces an ultraviolet cutoff (i.e., distance) at the lattice spacing[†] ℓ. In going from the continuum to the lattice, derivatives are replaced by finite differences. This replacement must be done in a gauge-invariant way, and hence one often refers to lattice gauge theory (LGT) or, in our context, lattice quantum chromodynamics (L-QCD).

In what follows, we will describe how to deal with dynamic gluons and quarks. Since the presence of quarks requires that the Fermi determinant *FD* be evaluated, this imposes a need for much more computational effort than in the case of the 'pure-gauge' lattice which comprises gluons only. In an intermediate step we can study quark operators in a non-fluctuating gauge-field background; this is the 'quenched'-quark approximation, which excludes the contributions of particle–antiparticle pairs. The full calculation then has 'dynamic' quarks.

15.2 Gluon fields on the lattice

Replacement of continuous space–time x_μ by a lattice $x_\mu = \ell n_\mu$ must be accomplished in a gauge-invariant manner, and, as with any other regulator, in order to be able to interpret the results, the regulator, i.e., lattice spacing, must be removed ($\ell \to 0$) after a finite result has been obtained. In other words, contact with the real physical world exists only, in the continuum limit, when the lattice spacing is taken to zero; this limit must be reached in natural fashion in any formulation. Moreover, we must always be aware that, on the lattice, we sacrifice Lorentz invariance, and have to be vigilant about the fates of all internal symmetries, which we desire to preserve. A suitable approach was devised by Wilson [273].

The action, an integral over the Lagrangian, is replaced by a sum over sites:

$$\beta S = \int dx\, \mathcal{L} \to \ell^4 \sum_n \overset{\cdot}{\mathcal{L}}_n. \tag{15.3}$$

β reminds us that all calculations are carried out in a four-dimensional Euclidian world, and β corresponds to the time dimension, or, as we shall see for equilibrium thermodynamics, the usual relation $\beta = 1/T$ applies. The generating functional used to obtain many of the results implied by

---

[†] It is common to call the lattice spacing a. To avoid conflicts of notation with the color indices of QCD, we chose the symbol ℓ, which is not used as often, though it should not be confounded with the angular-momentum eigenvalue employed earlier.

Eq. (15.1) is now an ordinary integral over all lattice fields and sites ϕ_n^i:

$$Z = \int \left(\prod_i \prod_n d\phi_n^i \right) e^{-\beta S}. \tag{15.4}$$

In the specific case of interest to us, quantum chromodynamics, this integral does not comprise the gauge fields A_μ^a; these are represented by fundamental variables $U_\mu(n)$, which live on the links connecting point x_n and $x_n + \ell\mu$ of a $d = 4$ dimensional space [273],

$$U_\mu(n) \equiv e^{ig\ell t^a A_\mu^a(n)}, \tag{15.5}$$

which form arises from Eqs. (13.64) and (13.69). We have $U_\mu(x_n + \ell\mu)^\dagger \equiv U_\mu(n + \mu)^\dagger = U_\mu(x_n) \equiv U_\mu(n)$. t_a are generators of the $SU_c(3)$ gauge group, $U_\mu(n)$ are elements of this group. The quark fields $\Psi(n)$ remain 'attached' to the lattice sites x_n; see below. Under the gauge transformation, the site variables (quark fields) transform as in Eq. (13.65) and link variables, which, as we will see, represent a generalization of field strengths, transform under gauge transformations in generalization of Eq. (13.77),

$$U_\mu(x) \rightarrow V(x)U_\mu(x)V^\dagger(x + \hat{\mu}). \tag{15.6}$$

An action for gauge fields involves a gauge-invariant product of U_μ's around some closed contour, a 'plaquette'. Since, for almost any closed contour, the leading term in the expansion is proportional to $F_{\mu\nu}^2$ in the continuum limit, there is considerable arbitrariness in the definition of gluon lattice action. The simplest contour has a perimeter of four links. In $SU(N)$

$$\beta S^W \equiv \frac{2N}{g^2} \sum_n \sum_{\mu > \nu} \text{Re tr} \{1 - U_\mu(n)U_\nu(n + \hat{\mu})U_\mu^\dagger(n + \hat{\nu})U_\nu^\dagger(n)\}. \tag{15.7}$$

βS^W is called the 'Wilson action'. The volume element in Eq. (15.4) is simply an integral over the group elements:

$$\left[\prod_i \prod_n d\phi_n^i \right] \rightarrow \left[\prod_n dU_n \right]. \tag{15.8}$$

Summation over all group elements amounts to a projection of the argument in the integral onto its color-singlet component.

15.3 Quarks on the lattice

The Euclidian fermion action in the continuum (in four dimensions) is

$$S = \int d^4x \, [\bar{\psi}(x)\gamma^\mu \partial_\mu \psi(x) + m\bar{\psi}(x)\psi(x)]. \tag{15.9}$$

A 'naive' lattice formulation is obtained by replacing the derivatives by symmetric differences:

$$S_{\text{L}}^{\text{naive}} = \frac{1}{2\ell} \sum_{n,\mu} \bar{\psi}_n \gamma^\mu (\psi_{n+\mu} - \psi_{n-\mu}) + m \sum_n \bar{\psi}_n \psi_n. \tag{15.10}$$

The elementary solution of the associated dynamic equations, i.e., the propagator, is

$$\begin{aligned} G(p) &= \frac{\ell}{i\gamma^\mu \sin(p_\mu \ell) + m\ell} \\ &= \frac{-i\gamma^\mu \ell \sin(p_\mu \ell) + m\ell^2}{\sum_\mu \sin^2(p_\mu \ell) + m^2 \ell^2} \rightarrow \frac{1}{i\gamma^\mu p_\mu + m}. \end{aligned} \tag{15.11}$$

We identify the physical spectrum through the poles in the propagator at $p_0 = iE$:

$$\sinh^2(E\ell) = \sum_j \sin^2(p_j \ell) + m^2 \ell^2. \tag{15.12}$$

The lowest-energy solutions, as expected yielded for $p = (0,0,0)$ the usual $E \simeq \pm m$, but there are many other degenerate solutions yielding this value of E, at $\ell p = (\pi, 0, 0)$, $(0, \pi, 0,)$, ..., (π, π, π). This is a model for 16 light fermions, not one. More generally, when fermions are discretized in this way on a d-dimensional lattice, they double and produce 2^d species.

Initially, two ways to deal with this problem were developed. The 'Wilson fermions' [273], and the 'Kogut–Suskind (staggered) fermions' [168]; more recently, also a five-dimensional formulation with 'domain-wall fermions' [155] has been considered. Wilson fermions are implemented by adding a second-derivative-like term,

$$S^{\text{W}} = -\frac{r}{2\ell} \sum_{n,\mu} \bar{\psi}_n (\psi_{n+\mu} - 2\psi_n + \psi_{n-\mu}), \tag{15.13}$$

to S^{naive}, Eq. (15.10). The parameter r must lie between 0 and 1; $r = 1$ is almost always used and '$r = 1$' is implied when one speaks of using 'Wilson fermions'. The propagator is

$$G(p) = \frac{-i\gamma_\mu \sin(p_\mu \ell) + m\ell - r \sum_\mu [\cos(p_\mu \ell) - 1]}{\sum_\mu \sin^2(p_\mu \ell) + \left\{ m\ell - r \sum_\mu [\cos(p_\mu \ell) - 1] \right\}^2}. \tag{15.14}$$

It has one pair of 'low-energy' poles at $p_\mu \simeq (\pm im, 0, 0, 0)$. The other poles are at $p \simeq r/\ell$. In the continuum limit, these states become infinitely massive.

This makes all but one of the fermion species heavy (with heavy masses close to the cutoff $1/\ell$), and we have, in principle, the required discretization method. However, for the n_f flavor QCD addition of S^W, Eq. (15.13) breaks the $SU(n_f)_L \times SU(n_f)_R$ chiral symmetry; see section 3.3. The size of the symmetry breaking is proportional to the lattice spacing and only close to the continuum limit $\ell \to 0$ does the explicit chiral-symmetry breaking become small. At any finite lattice spacing a proper representation of the chiral symmetry of the massless theory becomes a subtle fine-tuning process.

Despite the computational problems related to implementation of chiral symmetry, there is also some advantage with this formulation. Wilson fermions are closest to the continuum formulation – there is a four-component spinor on every lattice site for every color and/or flavor of quark. Therefore, the usual rules apply to the formulation of currents, and states. Explicitly, the Wilson-fermion action for an interacting theory is

$$\ell S^W = \sum_n \bar{\Psi}_n \Psi_n - \kappa \sum_{n\mu} \left(\bar{\Psi}_n (r - \gamma_\mu) U_\mu(n) \Psi_{n+\mu} + \bar{\Psi}_n (r + \gamma_\mu) U_\mu^\dagger \Psi_{n-\mu} \right).$$

(15.15)

We have rescaled the fields $\psi = \sqrt{2\kappa}\Psi$, and have introduced the 'hopping parameter' $\kappa^{-1} = 2(m\ell + 4r)$.

In studies of properties of QGP, another description of quarks on a lattice has been used more extensively. In the staggered-fermion method, a one-component staggered-fermion field rather than the four-component Dirac spinors is used. The name staggered is used since Dirac spinors and quark flavors are constructed by combining appropriate single-component fields from different lattice sites. Staggered fermions also break the chiral symmetry, but there remains a $U(1) \times U(1)$ symmetry, which comprises much of the physics of chiral symmetry. Moreover, explicit chiral symmetry is present for $m_q \to 0$, even for finite lattice spacing, as long as all flavor masses are degenerate. On the other hand, flavor symmetry and translational symmetry are mixed together, which poses problems, since in the real world, the flavor symmetry is broken.

Since exact chiral symmetry and broken flavor symmetry are important physical phenomena influencing the physics of high-temperature QCD, see Fig. 3.4 on page 54, a third approach to place quarks on a lattice is currently being developed. The domain-wall formulation of lattice fermions is expected to support accurate chiral symmetry, even at finite lattice spacing. In this new fermion formulation, it seems that it will be possible to more easily simulate two-flavor, finite-temperature QCD near the chiral phase transition. For further theoretical details, we refer to [265], and the first exploratory hot-QCD calculations are reported in [90].

Another area of rapid development is the search for a most appropriate 'improved' action for each of the important applications of L-QCD. The original Wilson action for the gauge fields is not unique, since the principle of gauge invariance leads to the building block, the plaquette, but not to the actual form of the action made from plaquettes. Consequently, the form of Wilson-fermion action can also be 'improved'. Improvements can perform better in one area than in another, since they address the problems encountered in extracting physics from extensive numerical calculations.

15.4 From action to results

Once we have quark fields on the lattice, as noted at the end of section 15.1, we must deal with their anticommuting nature. We carry out the integral over the Fermi fields in the path integral Eq. (15.4). For n_f degenerate flavors of staggered fermions

$$Z = \int [dU][d\psi][d\bar{\psi}] \exp\left(-\beta S(U) - \sum_{i=1}^{n_f} \bar{\psi} M(U) \psi\right) \qquad (15.16)$$

$$= \int [dU] \left(\det M(U)\right)^{n_f/2} \exp(-\beta S(U)). \qquad (15.17)$$

In order to make explicit the positive-definite nature of the Fermi determinant FD appearing as the preexponential factor in Eq. (15.17), we will be writing it as

$$FD = \det(M^\dagger M)^{n_f/4}.$$

Recalling that a determinant is a product of eigenvalues, we can express its logarithm as a sum of logarithms of eigenvalues, i.e., a trace, and we write

$$Z = \int [dU] \exp\left(-\beta S(U) - \frac{n_f}{4} \operatorname{tr} \ln(M^\dagger M)\right). \qquad (15.18)$$

The major computational problem dynamic fermion simulations face is inverting the fermion matrix M for any change in any of the gluon-link fields U. M has eigenvalues with a very large range – from 2π to $m_q \ell$ – and, in the physically interesting limit of small m_q, the matrix becomes ill-conditioned. Just a few years ago, it had been possible only to study quenched fermions, i.e., to proceed ignoring the second term in the exponent in Eq. (15.18). Today, it is possible to compute at relatively heavy values of the quark mass and to extrapolate to $m_q = 0$.

We will not enter further into practical discussion of how to do the high-dimensional Monte-Carlo integrals; neither shall we discuss the many ingenious algorithms that are in use. This is clearly a field of its own merit, and would fill more than this volume. However, even if somebody did provide all numerical answers, we still would need to take a few further steps. Perhaps the most important physical consideration is the approach to the continuum limit, $\ell \to 0$. Any calculation by necessity will be done with a finite lattice spacing ℓ. The lattice spacing ℓ is an ultraviolet cutoff, and it is the variable which now provides the scale to the QCD coupling constant g and quark masses m_i.

The continuum limit, in which we are interested, requires the limit $\ell \to 0$, holding physical quantities fixed, not the input ('bare') action parameters. In section 14.3, we have shown that the input parameter g, the bare coupling of quantum chromodynamics, is replaced by the running coupling constant α_s measured at some given scale (typically M_Z). When only one dimensioned parameter is present, in the absence of quarks or when all quark masses are set to zero, the situation is simple. For example, in order to evaluate hadron masses on the lattice, one computes the dimensionless combination $\ell m(\ell)$. One can determine the physical meaning of the lattice spacing by fixing one hadron mass from experiment. Then other dimensional quantities can be predicted.

Consider, as an example, the ratio of two hadron masses:

$$\frac{\ell m_1(\ell)}{\ell m_2(\ell)} = \frac{m_1(0)}{m_2(0)} + \mathcal{O}(\ell) + \cdots. \tag{15.19}$$

The leading term does not depend on the value of the ultraviolet-cutoff. One of the goals of a lattice calculation aiming to obtain the physical properties is to separate an ultraviolet-cutoff-distance scale-dependent remainder from the physical observable. One says that the calculation 'scales' if the ℓ-dependent terms in Eq. (15.19) are zero or small enough that one can extrapolate to $\ell = 0$. All the ℓ-dependent terms are 'scale violations'. To be able to make extrapolations, the results for several values of lattice spacing ℓ are required.

The precision with which we can extract the physics will obviously depend on how small the lattice cutoff is. However, the lattice must cover a sufficiently large region of space–time for the physical question we are addressing. We cannot study the proton of size $R = 1$ fm without having a few lattice distances within its radius, and a lattice of a few fermis. Repeating this basic cell domain infinitely using periodic boundary conditions helps, but cannot much reduce these requirements.

A summary of the lattice-gauge-theory conditions and procedures which we have introduced is presented in table 15.1.

For the study of hadron masses, as long as fundamental symmetries are respected, the physical size of the lattice should be $\ell \simeq 0.1$–0.2 fm, and the repeating cell ought to have the size $\simeq 5$ fm. The computer power of today

Table 15.1. A summary of the procedure for L-QCD simulations

Do the path integral:
      quark integrals analytically
      gluon + Fermi-determinant integrals numerically
Restrictions:
      Imaginary time
      Chemical potential = 0
         quark–antiquark symmetry makes the determinant real
Approximations:
      Restrict volume $L \ll \infty$ (infrared cutoff)
      Introduce lattice $\ell \geq 0$ (ultraviolet cutoff)
         Need: $\ell \ll l_{\text{typical}} \ll L$
         limit: scale $\ell \to 0$
      Need low quark masses,
         Difficulty: $t_{\text{computer}} \sim (m_{\text{q}})^{-2}$
         limit: scale $m_{\text{q}} \to 0$

allows $32^3 \times 16$, or $24^3 \times 48$, lattices. In view of the physical difference between time and space, the time, i.e., inverse-temperature, dimension of the lattice can be chosen to be smaller than the spatial extent of the lattice. On such lattices, the hadron spectra that emerge nowadays are quite realistic; see section 15.5 below.

However, such lattices may not be large enough to describe precisely many of the interesting properties of the QGP. We need to describe two different quark mass scales (u and d, and s) while maintaining chiral symmetry in the light-quark sector, and treating an odd number of quark flavors (both staggered and Wilson quarks favor the presence of an even number of flavors). This task has not been resolved completely, and the properties of QGP we will discuss retain a systematic uncertainty. Moreover, a many-body system such as QGP should have many collective modes of excitation. Given the size of the physical lattices studied, collective oscillations with wave lengths greater than a few fermis are not incorporated. Although this does not influence in a critical way the properties of the equations of state, such long-range oscillations are often carriers of flows, e.g., of heat. Therefore, the study of transport properties of the QGP phase on the lattice is not yet possible.

The euclidian lattice is indeed ideal for simulating high-temperature QCD since, in this case, there is a direct correspondence between the imaginary time and temperature – the path-integral weight is, in fact, the partition function with $\ell N_t = 1/T = \beta$. The statistical-physics properties are the operators which are fairly simple, e.g., the energy density involves only one point in space–time. One has to remember that other lattice studies of the hadron mass spectrum involve two point functions, whereas weak-interaction matrix-element computations typically involve three point functions since we have to create a hadron from the vacuum at $t = 0$, act on the state with an operator at $t = t_1$, and then annihilate the hadron at $t = t_2$.

Among the quantities which are studied in high-temperature QCD are the Polyakov loop [273], the chiral condensate $\bar{\psi}\psi$, Eq. (3.22), the energy and pressure, screening lengths of color-singlet sources, the potential between static test quarks, and the response of the quark density to an infinitesimal chemical potential. Of these, the Polyakov loop and $\bar{\psi}\psi$ are the most intensively studied. $\bar{\psi}\psi$ is the order parameter for chiral symmetry breaking. It is nonzero under ordinary conditions $T \to 0$, and we expect it to vanish when chiral symmetry is restored for $T > T_c$. Loosely speaking, the Polyakov loop has the value $e^{-F/T}$, where F is the free energy of a static test quark. In pure $SU_c(3)$ gauge QCD, the Polyakov loop is zero at low temperatures, indicating confinement of the test quark, and nonzero at high temperature.

It is understood today that dynamic quarks make a big difference in high-temperature QCD, and the 'quenched' approximation has been found to be in general unsuitable. Looking at the energy of free quarks and gluons (the Stefan–Boltzmann law, see section 4.6) even with $n_f = 2$ flavors of light quarks, we find that the $16 = 8_c \times 2_s$ gluon degrees of freedom are dominated by $21 = 2_f \times 3_c \times 2_s \times \frac{7}{4}$ equivalent quark degrees of freedom. The thermal properties of quarks dominate those of gluons. Quenched quarks are known to exclude the particle–antiparticle-pair fluctuations in the vacuum. Thus, if quenched quarks are used, some important physical processes are forbidden. For example, consider a quark–antiquark pair connected by a string of color flux. With quenched quarks, when the distance grows, we encounter an ever-growing linear potential – if pair fluctuations are excluded, this string never breaks. In the presence of dynamic quarks, when the string is long enough, there is enough energy to create a quark–antiquark pair, which breaks the string, forming two mesons.

It is easy to find, in the numerical Monte-Carlo integration, the transition (crossover) to some novel high-temperature behavior in a lattice simulation, though it is very difficult to ascertain the nature of the transition. To vary the temperature with a fixed number of lattice spacings in the time direction, the lattice spacing ℓ is changed by varying the coupling g. This works because g is the coupling constant defined on the scale of the lattice spacing. In an asymptotically free theory, the coupling decreases for shorter length scales. Therefore, decreasing

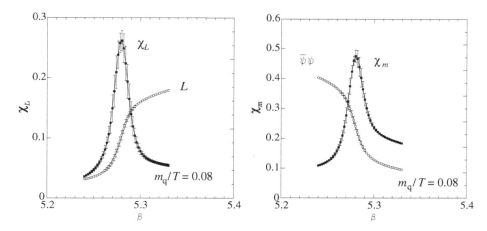

Fig. 15.1.   Deconfinement and restoration of chiral symmetry in two-flavor L-QCD. Open circles: on the left-hand side the Polyakov loop L, which is the order parameter for deconfinement in the pure gauge limit ($m_q \to \infty$); and on the right-hand side quark chiral condensate $\bar{\psi}\psi$, which is the order parameter for chiral-symmetry breaking in the chiral limit ($m_q \to 0$), shown as a function of the coupling $\beta = 6/g^2$. Also shown are the corresponding susceptibilities $\chi_L \propto (\langle L^2 \rangle - \langle L \rangle^2)$ (left) and $\chi_m = \partial \langle \bar{\psi}\psi \rangle / \partial m$ (right) which peak at the same value of the coupling [159].

g, or increasing $6/g^2$ ($6 = 2n_c$), makes ℓ smaller and the temperature, $T = (N_t \ell(g))^{-1}$, higher. As the temperature is increased through the crossover, $\bar{\psi}\psi$ drops and the Polyakov loop increases, see Fig. 15.1. The Polyakov loop and $\bar{\psi}\psi$ change at the same temperature, indicating that 'deconfinement', and restoration of chiral symmetry are happening at the same temperature.

Little is known with certainty about the nature of the crossover between the confined (frozen) phase and the new phase suggested by Fig. 15.1. In particular, we cannot yet be sure what kind of phase transition or transformation is encountered, see Fig. 3.4 on page 54 and the related discussion. It is fairly well established, from lattice simulations, that there is a first-order phase transition in the pure gauge limit, and for three-massless quarks. As a quark mass is lowered from infinity this transition disappears, and there may be a continuous crossover from the low-temperature regime to the high-temperature regime. But even if there is no phase transition, the crossover is fairly sharp. This can be seen by considering the inverse screening lengths for $q\bar{q}$ sources with the quantum numbers of the pion π and its opposite-parity partner σ. At high temperatures, they become very close, with the remaining small difference being due to the explicit breaking of symmetry originating from the quark mass. This and other quantities show that the high-temperature regime does indeed have the expected characteristics of the QGP.

Only recently have calculations progressed far enough to lead to firm results about properties of the QGP. To find the temperature of the crossover in physical units, the lattice spacing must be determined by computing some physical quantity, such as the ρ-meson or the nucleon masses. The π mass is not a good choice of scale, since it can be made arbitrarily small by making the quark mass small, as we have seen in section 3.3. Though the value of the crossover temperature one finds is still in doubt for real-life quarks, at present the opinion of experts we shall discuss in the next section is biased toward a value of 160 MeV [160], near the Hagedorn temperature; see chapter 12.

15.5 A survey of selected lattice results

There are many lattice results related to QCD properties we study, addressing diverse questions such as hadronic masses, matrix elements, and physical properties of hot QCD. Given the rapid development of the field which promises to render any presentation quite rapidly obsolete, we focus our attention on 'stable' results that are most relevant in the context of this book and, in particular, the study of equations of state of QGP. We will not further discuss in this section the intricate extrapolations (continuum limit, massless-quark limit) which form part of the process of evaluation of the bare numerical results, and which we described above in section 15.4.

The running of the gauge coupling constant has now been tested for the case of two massless Wilson fermions [76]. The lattice results compare very well with the renormalization-group result, as can be seen in Fig. 15.2. These results are, at present, still mainly of academic interest and are different in detail from Fig. 14.1, since, in the range $\alpha_s < 0.4$, we actually have to include s, c, and b quarks, in order to compare with experiment. On the other hand, the fact that the running is seen as expected in the theoretical evaluation of two-light-flavor QCD reassures us regarding the validity of the findings we presented in Fig. 14.1.

The study of hadronic masses allows one to draw conclusions about the input quark masses. The CP-PACS collaboration has recently completed an extensive evaluation using its dedicated (peak) 614-GFLOPS (giga-floating-point-operations) computer [33]. The lattice action and simulation parameters were chosen with a view to carrying out a precise extrapolation to the continuum limit, as well as scaling in the chiral $m_q \to 0$ limit for dynamic up- and down-quark masses. However, the strange quarks were considered in a quenched approximation, which entails an 'uncontrolled' error.

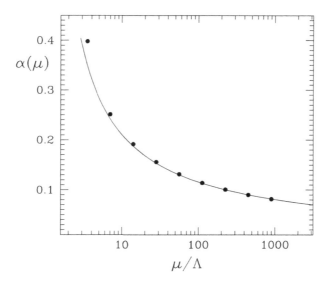

Fig. 15.2. Dots, lattice-gauge theory with two light fermions for $\alpha_s(\mu)$, compared with the perturbative three-loop result (line); parameters are chosen such that, for $\Lambda = 1$ GeV, $\alpha_s(M_Z) = 0.118$; results of the ALPHA collaboration [76].

Evaluation of the masses of kaons and/or ϕ allows one to determine the mass of the strange quark. The results found using dynamic u and d quarks are $m_s^{\overline{MS}}(2\,\text{GeV}) = 88^{+4}_{-6}$ MeV (K input) and $m_s^{\overline{MS}}(2\,\text{GeV}) = 90^{+5}_{-11}$ MeV (ϕ input), which are about 25% lower than the values found with quenched u and d quarks. The low value for the mass of the strange quark is well within the accepted range; see table 1.1 on page 7.. The consistency of these two results is quite remarkable. Moreover, using the mass of the K meson to fix the strange-quark mass, the difference from experiment for the mass of the K^* meson is $0.7^{+1.1}_{-1.7}\%$, and that for the ϕ meson $1.3^{+1.8}_{-2.5}\%$. When the ϕ meson is used as input, the difference in the mass of the K^* meson is less than 1%, and that for the mass of the K meson is $1.3 \pm 5.3\%$. The masses of (multi)strange baryons are, within much larger computational error, also in agreement with the experimental values.

Should this relatively low mass for the strange quark be confirmed when dynamic strange quarks are introduced, the speed of production of strangeness at low temperatures $T \gtrsim T_c$ in QGP as perhaps formed at intermediate SPS energies would dramatically increase. Strangeness could develop into a highly sensitive 'low energy' probe of formation of QGP, even when the initial conditions reached are near to the critical temperature. In this context, it is interesting to note the steep rise and threshold of the strangeness-excitation function shown in Fig. 1.5 on page 17.

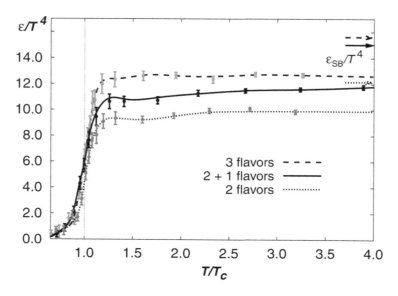

Fig. 15.3. The energy density of hadrons obtained with staggered fermions divided by T^4, as a function of T/T_c. Stefan–Boltzmann limits of a non-interacting gas of quark and gluons are indicated with arrows for each case considered.

In our context, an even more important recent lattice advance has been the extensive study of 2, 2 + 1 and 3 flavors in hot QCD [159–161]. The behavior of the energy density ϵ/T^4 is presented in Fig. 15.3 as a function of temperature T/T_c, obtained with staggered fermions. The Stefan–Boltzmann values expected for asymptotically (high-T) free quarks and gluons are shown by arrows to the right, coded to the shades of the three results presented: two and three light flavors (up and down, respectively), for which the quark mass is $m_q = 0.4T$ and a third case (dark line), in which, in addition to the two light flavors, a heavier flavor $m_s = T$ is introduced. The temperature scale is expressed in units of the critical temperature T_c, as appropriate for each case (T_c changes with the number of flavors). The value of T_c is where one observes a rapid change in the behavior of the quark condensate/susceptibility and, at the same location, one sees also the onset of deconfinement in the Polyakov loop; see Fig. 15.1.

We see that, around $T/T_c = 1$, the number of active degrees of freedom rapidly rises, and the energy density attains as early as $T = 1.2T_c$ the behavior characteristic of an ideal gas of quarks and gluons, but with a somewhat reduced number of active degrees of freedom. The energy density in the deconfined phase can be well approximated by

$$\epsilon_{\text{QGP}} \simeq (11–12)\, T^4. \tag{15.20}$$

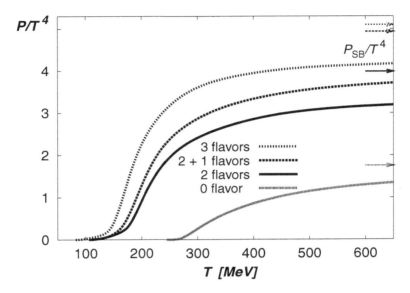

Fig. 15.4.   The pressure of hadrons obtained with staggered fermions in units of T^4 as a function of T.

The critical temperature T_c has the value

$$\frac{T_c}{m_\rho} = 0.20 \pm 0.01, \quad T_c = 154 \pm 8 \text{ MeV}, \quad \text{for } n_f = 3, \tag{15.21}$$

where, in addition to the statistical error quoted above, a systematic error at a similar level, associated with uncertainties in the scaling behavior, is expected. In case of two flavors, the value $T_c = 173 \pm 8$ MeV is found. These results are consistent with calculations performed with clover-shape-improved Wilson action (see section 15.3) by the CP-PACS collaboration [32] for two flavors. In Fig. 15.4, we present the behavior of P/T^4 as a function of temperature, with the temperature scale derived from Eq. (15.21). The expected Stefan–Boltzmann limits are shown by arrows. Apart from the cases of 2, 2 + 1 and 3 flavors, we see also the 'pure gauge' case with zero flavors.

The conclusion we draw from these results is that the lattice-QCD evaluation has matured to the level of being able to offer information directly relevant to the physical properties of hot QCD. These results, in particular, show that there is a rapid phase transformation or even a first-order phase transition at $T_c = 163 \pm 15$ MeV.

We see, in Fig. 15.4, that the phase-transition temperature decreases significantly with increasing number of flavors. However, the shapes of all curves scale similarly. This is shown in Fig. 15.5, in which the temperature scale is expressed in units of T_c and the pressure in terms of the ideal

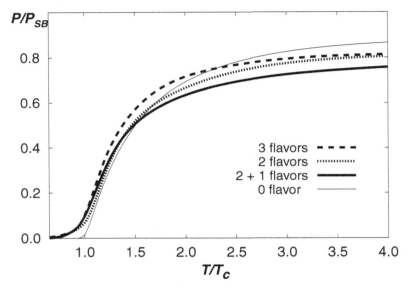

Fig. 15.5.   The pressure of hadrons obtained with staggered fermions divided by the appropriate Stefan–Boltzmann limit, as a function of T/T_c.

Stefan–Boltzmann pressure. The zero-flavor pure gauge case is fastest growing toward the limiting value, but still at a considerably slower rate than that for the energy density we have seen in Fig. 15.3.

An ideal massless relativistic gas should satisfy the relation $\epsilon - 3P = 0$, Eq. (4.64). The difference in behavior, comparing Fig. 15.3 with Fig. 15.5, must originate from the presence of variables with dimensioned scales. We encountered two such (related) variables, the vacuum property $\mathcal{B}$ and the parameter Λ controlling the magnitude of the running variables (α_s and m_i). The deviations of pressure from the Stefan–Boltzmann ideal-gas behavior seen in lattice results, in particular $\epsilon - 3P \neq 0$, are in direct or indirect fashion related to these quantities. We will quantify this in the following section.

We note that, using the Gibbs–Duham relation, Eq. (10.26), we can relate the change in the pressure, seen in Fig. 15.4, to the difference $\epsilon - 3P$. We generalize slightly the argument presented in Eqs. (4.62) and (4.64). We consider the free-energy density

$$f = -\frac{T}{V} \ln Z(T, V). \tag{15.22}$$

$P = -f$ for an infinite system. Moreover, the entropy density $\sigma = \partial/\partial T f$, see Eq. (10.6). We find, employing Eq. (10.26) at zero baryon density,

$$\frac{\epsilon - 3P}{T^4} = T \frac{d}{dT} \left(\frac{P}{T^4} \right). \tag{15.23}$$

As soon as we understand the gentle slope of pressure P in Fig. 15.4, i.e., the right-hand side of Eq. (15.23), we will also understand the difference between the behaviors of energy and pressure, the left-hand side of Eq. (15.23), noted on comparing Fig. 15.3 with Fig. 15.5.

We will show, in section 16.2, that this non-ideal-gas behavior can be interpreted as resulting from perturbative quark–gluon interactions and the presence of the latent heat of the vacuum $\mathcal{B}$. An equivalent explanation invoking the presence of quasi-particles with mass, and quantum numbers of quarks and gluons, will also be considered.

16 Perturbative quark–gluon plasma

16.1 An interacting quark–gluon gas

As explained in section 14.1, the interactions between quarks and gluons are contained in the QCD Lagrangian Eq. (13.79), improved by gauge-fixing and FP-ghost terms Eq. (14.1). Strictly considered, the rules for Feynman diagrams we presented in Eqs. (14.2)–(14.8) are applicable to processes in perturbative vacuum, whereas to compute thermal properties of interacting quark and gluons, we are working in matter at finite temperature T and chemical potential μ. The generalization required is discussed in detail in the textbook by Kapusta [157].

A lot of effort in the past few decades has gone into the development of the perturbative expansion of the partition function. The series expansion, in terms of the QCD coupling constant g, has been carried out to order $[(g/(4\pi)]^5 = (\alpha_s/\pi)^{5/2}/32$ [282]. This series expansion, which was developed using as reference the perturbative vacuum in empty space, does not appear to lead to a convergent result for the range of temperatures of interest to us [36]: the thermodynamic properties vary widely from order to order, oscillating quite strongly around the Stefan–Boltzmann limit. It has therefore been claimed that the perturbative QCD thermal expansion has a zero-range convergence radius in α_s.

Our following considerations will be limited to the lowest-order perturbative term combined with the vacuum energy $\mathcal{B}$ and allow an excellent reproduction of the key features of lattice results. It remains to be understood why this is the case. It is not uncommon to encounter in a perturbative expansion a semi-convergent series. The issue then is how to establish a workable scheme. It is, for example, possible that a different scheme of perturbative approach, in which the QCD parameters (α_s and masses) are made nonperturbative functions of the medium using an in-medium renormalization group, would yield a better converging series in α_s.

Considering the inconclusive and rapidly evolving landscape of thermal

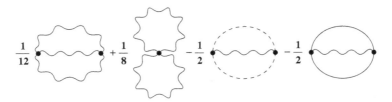

Fig. 16.1. Feynman diagrams contributing to the equation of state of the QGP in order α_s. Wavy lines represent gluons, solid lines represent quarks, and dashed lines denote the ghost subtractions of non-physical degrees of freedom.

QCD, we will in this book not explore the subject beyond the study of the consequences of the lowest-order thermal corrections. The lowest-order contributions are obtained by evaluating the graphs shown in Fig. 16.1. Evaluation of these diagrams at high temperature for massless quarks and gluons is possible analytically [91]. For massless quarks and antiquarks one finds the following two terms in the partition function:

$$\ln \mathcal{Z}_q(\beta, \lambda) = \frac{gV}{6\pi^2} \beta^{-3} \left[\left(1 - \frac{2\alpha_s}{\pi} \right) \left(\frac{1}{4} \ln^4 \lambda_q + \frac{\pi^2}{2} \ln^2 \lambda_q \right) \right.$$
$$\left. + \left(1 - \frac{50\alpha_s}{21\pi} \right) \frac{7\pi^4}{60} \right], \tag{16.1}$$

where $g = n_s n_c n_f = 12$, for $n_s = 2s + 1 = 2$, $n_c = 3$, and $n_f = 3$. The first term in parentheses is c_3, Eq. (4.71c), and the second is c_2, Eq. (4.71b). The quark fugacity λ_q is related to the baryon-number fugacity, as discussed in Eq. (11.3). The glue contribution is

$$\ln \mathcal{Z}_g(\beta, \lambda) = \frac{8\pi^2}{45} \beta^{-3} \left(1 - \frac{15\alpha_s}{4\pi} \right), \tag{16.2}$$

where the last term in parentheses is c_1, Eq. (4.71a). Finally, the vacuum contribution can be added in the form

$$\ln \mathcal{Z}_{\text{vac}}(\beta) = -\beta BV. \tag{16.3}$$

This term insures that the energy density, inside the bag, is positive and simultaneously that the pressure exerted on the surface of the bag is negative. The total grand partition function is

$$\ln \mathcal{Z}_{\text{QGP}} = \ln \mathcal{Z}_q + \ln \mathcal{Z}_g + \ln \mathcal{Z}_{\text{vac}}. \tag{16.4}$$

This equation was presented explicitly in section 4.6, Eq. (4.70). The resulting perturbative interactions between quarks and gluons are obtained by differentiating Eq. (16.4) with respect to V, β, and λ_q; see section 10.1.

Another important consequence of interactions in a conductive color plasma is the change in location of poles of particle propagators. Rather than of particles, one than speaks of quasi-particles, in our instance thermal quarks and gluons. The idea of quasi-particle thermal gluon and thermal quark masses is rooted in the desire to characterize, in a simple way, the behavior of quark and gluon correlators (propagators) evaluated inside the color-conductive medium. The monograph by Kapusta [157] offers an excellent early introduction to this still rapidly evolving subject, and we will restate here a few results of immediate interest.

Allowing for interaction to lowest order in α_s, a relation between the energy E and momentum $p = |\vec{p}|$ (a dispersion relation) of quasi-quarks is fixed by the location of the pole of the quark propagator Eq. (14.2) [270]:

$$E^2 = \vec{p}^2 + (m_q^T)^2 \left[\frac{E+p}{2p} - \frac{E^2 - p^2}{4p^2} \ln\left(\frac{E+p}{E-p}\right) \right]. \tag{16.5}$$

For $p \to \infty$, with $E \to p + m^2/(2p)$, we recognize the quantity

$$\boxed{(m_q^T)^2 = \frac{4\pi}{3}\alpha_s T^2} \tag{16.6}$$

as the mass parameter, establishing the relation with the momentum and controlling the quasi-quark phase space.

Near to $p = 0$, the long-range oscillations described by Eq. (16.6) require more attention. They have the energy

$$E_q(p \to 0) = \frac{m(\vec{p} \to 0, \omega; T)}{\sqrt{2}} + \frac{p}{3}\frac{\sqrt{2}p^2}{3m(\vec{p} \to 0, \omega; T)} + \cdots. \tag{16.7}$$

It is evident from the above that the 'thermal' mass of the quark in a medium may be defined also by considering this zero-momentum limit [270], rather than the high-momentum limit Eq. (16.6) adopted here. However, the domain $p \to 0$ is not important in the counting of states in the phase space, considering the momentum-volume factor $d^3p = 4\pi p^2 \, dp$. We conclude that, in the study of the phase space of light thermal quarks in plasma, we should use for the light-quark energy the high momentum limit

$$E_q \simeq \sqrt{p^2 + (m_q^T)^2}, \quad q = u, d.$$

A similar discussion arises for the behavior of gluons. The collective oscillations in the plasma with the quantum numbers of gluons behave, at high momentum, according to

$$E_g \simeq \sqrt{p^2 + (m_g^T)^2},$$

where

$$\boxed{(m_g^T)^2 = 2\pi\alpha_s T^2\left(1 + \frac{n_f}{6}\right),} \tag{16.8}$$

whereas near $p \to 0$

$$E_{\rm g}(p \to 0) = \sqrt{\frac{2}{3}} m_{\rm g}^{\rm T}. \tag{16.9}$$

The thermal mass of gluons introduced in Eq. (16.8) is not the Debye screening mass limiting the effective range of interactions in plasma. This quantity corresponds to the static limit in the behavior of the longitudinal gluon-like plasma oscillations [269]:

$$\boxed{(m_{\rm g}^{\rm D})^2 = 4\pi\alpha_{\rm s}T^2\left(1 + \frac{n_{\rm f}}{6}\right).} \tag{16.10}$$

16.2 The quark–gluon liquid

We consider the properties of the QGP allowing for the lowest-order inter-actions with a temperature-dependent interaction strength, and vacuum pressure $\mathcal{B}$. We refer to this model as the quark–gluon liquid [147]. Naturally, we hope and expect to reproduce lattice-QCD results [160]. There is considerable sensitivity to the value of $\alpha_{\rm s}(\mu)$ and it is necessary to use its precise form; see section 14.2.

Along with many other authors, we adopt the relation

$$\boxed{\mu \simeq 2\pi T} \tag{16.11}$$

between the scale of the QCD coupling constant and the temperature of the thermal bath. On the one hand, the right-hand side is close to the average collision energy of two massless quanta at T, and on the other, the relation Eq. (16.11) makes the thermal-QCD expansion least sensitive to the renormalization scale [283].

In Fig. 16.2, the 'experimental' values are the numerical lattice simulations [160], see section 15.5, for 2 (diamonds), 3 (triangles), and 2 + 1 (squares) flavors. The non-interacting Stefan–Boltzmann quark–gluon gas, Eq. (4.70), with $c_i = 1$ and with the bag constant

$$\mathcal{B} = 0.19\,{\rm GeVfm}^{-3} \tag{16.12}$$

is shown as thin lines, dotted for the case of three flavors and dashed for the case of two flavors. We see that the effect of vacuum pressure disappears as $T \to 2T_{\rm c}$, and that the lattice results differ significantly from those for the free gas, even at $T = 4T_{\rm c}$.

To obtain agreement with the lattice results, it is necessary to introduce perturbative coefficients c_i, Eqs. (4.71a)–(4.71c), with numerically computed $\alpha_{\rm s}(\mu = 2\pi T)$ [147]. The thick lines seen in Fig. 16.2 were obtained allowing for $\alpha_{\rm s}(2\pi T)$ shown in Fig. 14.3 with $T_{\rm c} = 160\,{\rm MeV}$. To find the behavior near to $T = T_{\rm c}$, the only 'free' choice we can make is

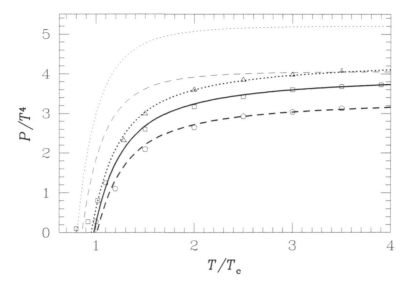

Fig. 16.2. Lattice-QCD results [160] for P/T^4 at $\lambda_q = 1$ (for 2 (diamonds), 3 (triangles), and $2 + 1$ (squares) flavors) compared with free quark–gluon gas with bag pressure $\mathcal{B} = 0.19\,\text{GeV fm}^{-3}$, thin dotted (three flavors) and dashed (two flavors) lines. Thick lines (agreeing with lattice data) are derived from the quark–gluon liquid model: dotted line, 3 flavors; solid line, $2 + 1$ flavors; and dashed line, 2 flavors.

the value of $\mathcal{B}$, and this is the reason why that particular number was selected in Eq. (16.12). To achieve the agreement with lattice results seen in Fig. 16.2, the relevant relation is

$$\frac{\mathcal{B}}{T_c^4} = 2.2, \quad \mathcal{B}^{1/4} = 1.22 T_c. \qquad (16.13)$$

The precise relationship between the scale μ and T has negligible impact on the result shown, as long as the natural order of magnitude seen in Eq. (16.11) is maintained. Within the simple model we introduced in section 1.3 to describe the phase transition, Eq. (16.13) implies nearly the correct number of degrees of freedom freezing in the transition, $\Delta g \simeq 20$.

It has been shown that it is also possible to reproduce the lattice results using fine-tuned thermal masses (see table I in [204]). In Fig. 16.3, we show the light-quark (solid thick line) and gluon (dashed thick line) thermal masses which were used to fit the lattice data. The actual thermal quark and gluon masses, defined in Eqs. (16.6) and (16.8), are also shown in Fig. 16.3, as thin lines (dashed for gluons) obtained using $\alpha_s(\mu = 2\pi T)$, from Fig. 14.3. We conclude that the thermal masses required to describe the reduction of the number of degrees of freedom for $T > 2T_c$ are

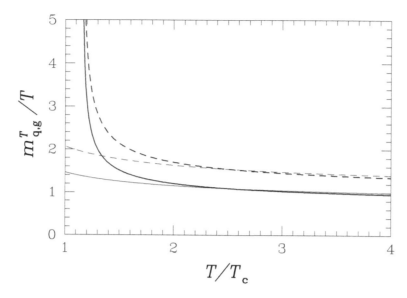

Fig. 16.3.   Thermal masses fitted to reproduce lattice-QCD results for $n_f = 2$ [204], thick solid line for quarks, and thick dashed line for gluons. Thin lines, perturbative QCD result for $\alpha_s(\mu = 2\pi T)$.

just the perturbative QCD result. Importantly, this means that thermal masses express, in a different way, the effect of perturbative quantum-chromodynamics, and thus, for $T > 2T_c$, we have the option of using Eq. (16.1), or the more complex thermal-mass approach.

However, in the temperature domain $T < 2T_c$, in which the vacuum pressure $\mathcal{B}$ is relevant, see Fig. 16.2, the thermal mass required to fit lattice results, as can be seen in Fig. 16.3, is quite different from the perturbative QCD result, and we believe that the interpretation of lattice results in this phenomenologically important domain is much less natural than the quark–gluon-liquid approach. The introduction of thermal masses, in order to describe the behavior seen in Fig. 15.5, is expressing just the same fact that the pressure must be some function of the parameter Λ which is controlling the magnitude of the running α_s, and that additional physics, such as the vacuum pressure $\mathcal{B}$, is required in order to understand the behavior of the QGP, for $T < 2T_c$.

We have seen that the suppression of the number of degrees of freedom, seen in the QGP pressure can be well described by a first-order thermal QCD result, either using thermal masses, or more directly using the first-order corrections seen in Eqs. (16.1) and (16.2). However, to describe the behavior for $T \to T_c$, we should invoke nonperturbative properties of the vacuum. As discussed at the end of section 15.5, once the variation of

P/T^4 with T has been described, the deviation of lattice results from the ideal-gas law, $\epsilon - 3P \rightarrow 0$, is also understood.

16.3 Finite baryon density

It is in the consideration of the finite baryon density that the issues regarding how to model the lattice results we raised in the last section are most relevant. We believe that it is appropriate to obtain the properties of QGP in a manner allowing the magnitude of the color interaction to be controlled by the energy scale which depends on the baryon-chemical potential. The dependence of the scale μ of α_s on the fugacity we adopt is [264]

$$\mu = 2\sqrt{(\pi T)^2 + \mu_q^2} = 2\pi\beta^{-1}\sqrt{1 + \frac{1}{\pi^2}\ln^2\lambda_q}. \tag{16.14}$$

Like with Eq. (16.11), there is no exact mathematical rationale for this expression; it is entirely based on intuition and the particle-energy behavior seen in studies of thermal QCD.

We note that Eq. (16.14) implies that

$$-\beta\frac{\partial\alpha_s(\beta,\lambda_q)}{\partial\beta} = \mu\frac{\partial\alpha_s}{\partial\mu},$$

$$\lambda_q\frac{\partial\alpha_s(\beta,\lambda_q)}{\partial\lambda_q} = \frac{\ln\lambda_q}{\pi^2 + \ln^2\lambda_q}\mu\frac{\partial\alpha_s}{\partial\mu}, \tag{16.15}$$

$$T\frac{\partial\alpha_s(T,\mu_q)}{\partial T} = \frac{\pi^2 T^2}{(\pi T)^2 + \mu_q^2}\mu\frac{\partial\alpha_s}{\partial\mu}.$$

The derivative of the QCD coupling constant can be expressed as, Eqs. (14.18) and (14.21),

$$\mu\frac{\partial\alpha_s}{\partial\mu} = -b_0\alpha_s^2 - b_1\alpha_s^3 + \cdots \equiv \beta_2^{\text{pert}}. \tag{16.16}$$

β_2^{pert} is the beta-function of the renormalization group in the two-loop approximation, with b_i defined in Eq. (14.23). β_2^{pert} does not depend on the renormalization scheme, and solutions of Eq. (16.16) differ from higher-order results by less than the error introduced by the experimental uncertainty in the measured value of $\alpha_s(M_Z)$; see section 14.3.

We are now prepared to study physical properties of the quark–gluon liquid. The energy density is obtained from Eq. (4.70):

$$\epsilon_{\mathrm{QGP}} = -\frac{\partial \ln \mathcal{Z}_{\mathrm{QGP}}(\beta, \lambda)}{V \, \partial \beta},$$

$$= 4B + 3P_{\mathrm{QGP}} - \beta \frac{\partial \alpha_{\mathrm{s}}}{\partial \beta} \sum_{i=1}^{3} \frac{\partial c_i}{\partial \alpha} \frac{P_{\mathrm{QGP}}}{\partial c_i}. \tag{16.17}$$

We find that

$$\boxed{\epsilon_{\mathrm{QGP}} - 3P_{\mathrm{QGP}} = \mathcal{A} + 4B,} \tag{16.18}$$

where

$$\mathcal{A} = (b_0 \alpha_{\mathrm{s}}^2 + b_1 \alpha_{\mathrm{s}}^3) \left[\frac{2\pi}{3} T^4 + \frac{n_{\mathrm{f}} 5\pi}{18} T^4 + \frac{n_{\mathrm{f}}}{\pi} \left(\mu_{\mathrm{q}}^2 T^2 + \frac{1}{2\pi^2} \mu_{\mathrm{q}}^4 \right) \right], \tag{16.19}$$

and P_{QGP} is stated explicitly in Eq. (4.70). We see in Eq. (16.18) the interesting property

$$\frac{\epsilon_{\mathrm{QGP}} - 3P_{\mathrm{QGP}}}{T^4} \rightarrow \frac{\pi}{18}(12 + 5n_{\mathrm{f}})(b_0 \alpha_{\mathrm{s}}^2 + b_1 \alpha_{\mathrm{s}}^3) + 4\frac{B}{T^4}, \quad \mu_{\mathrm{q}} \rightarrow 0, \tag{16.20}$$

where the thermal interaction (the first term) is determining the behavior at $T \gtrsim 2T_{\mathrm{c}}$, Fig. 16.2.

A convenient way to obtain the entropy and baryon density uses the thermodynamic potential $\mathcal{F}$; see Eq. (4.70) and chapter 10. For the quark–gluon liquid, we have

$$\frac{\mathcal{F}(T, \mu_{\mathrm{q}}, V)}{V} = -\frac{T}{V} \ln \mathcal{Z}(\beta, \lambda_{\mathrm{q}}, V)_{\mathrm{QGP}} = -P_{\mathrm{QGP}}(T, \mu_{\mathrm{q}}), \tag{16.21}$$

with entropy density s_{QGP} and baryon density ρ_{b}, which is a third of the quark density:

$$s_{\mathrm{QGP}} = -\frac{d\mathcal{F}}{V \, dT}, \quad \rho_{\mathrm{b}} = -\frac{1}{3} \frac{d\mathcal{F}}{V \, d\mu_{\mathrm{q}}}. \tag{16.22}$$

The entropy density is

$$\boxed{s_{\mathrm{QGP}} = \frac{32\pi^2}{45} c_1 T^3 + n_{\mathrm{f}} \left(\frac{7\pi^2}{15} c_2 T^3 + c_3 \mu_{\mathrm{q}}^2 T \right) + \mathcal{A} \frac{\pi^2 T}{\pi^2 T^2 + \mu_{\mathrm{q}}^2}.} \tag{16.23}$$

The coefficients c_i are defined in Eq. (4.71a) and are the same as in Eqs. (16.1), (16.2), and (4.70). The baryon density is

$$\boxed{\rho_{\mathrm{b}} = \frac{n_{\mathrm{f}}}{3} c_3 \left(\mu_{\mathrm{q}} T^2 + \frac{1}{\pi^2} \mu_{\mathrm{q}}^3 \right) + \frac{1}{3} \mathcal{A} \frac{\mu_{\mathrm{q}}}{\pi^2 T^2 + \mu_{\mathrm{q}}^2}.} \tag{16.24}$$

16.4 Properties of a quark–gluon liquid

The behavior of the quark–gluon liquid is of particular interest in the study of

- initial conditions corresponding to an early instant in time during the heavy-ion collision when the light quarks and gluons are in approximate chemical equilibrium, but strange quarks have not yet been produced, thus $n_f = 2$; and
- properties of the fireball of quarks and gluons at the time of its breakup near to $T = T_c$, at which strangeness is practically equilibrated, and $n_f = 2.5$.

Given the good agreement with the lattice results for vanishing chemical potential, the following study offers a quantitative description of the relationship between the temperature and physical properties reached in the deconfined phase, but with an unknown systematic error when properties involving baryon density are physically relevant (e.g., in the AGS energy range).

In the left-hand panels of Fig. 16.4, we show the physical properties at fixed energy per baryon, in the range $2\,\text{GeV} \le E/b \le 15\,\text{GeV}$, as functions of temperature, while in the right-hand panels we study the behavior at fixed value of the (dimensionless) entropy per baryon $10 \le S/b \le 60$. In panels (a) and (b) of Fig. 16.4, as we step from line to line from left to right, the energy per baryon is incremented by 1 GeV; in panels (d) and (e) the entropy per baryon is incremented by 5 units; in panel (c) we step from bottom to top incrementing by 1 GeV; and in panel (f) from bottom to top by 5 entropy units. The light-dashed boundaries are obtained from the conditions

- on energy density $\epsilon_{q,g} \ge 0.5\,\text{GeV fm}^{-3}$ (excluding here the latent heat of the vacuum $\mathcal{B} \simeq 0.19\,\text{GeV fm}^{-3}$), and/or
- baryochemical potential, $\mu_b \le 1\,\text{GeV}$.

We highlight the result for $E/b = 8.5\,\text{GeV}$ by using a thick dashed line on the left in Fig. 16.4, and that for $S/b = 40$ by using a thick solid line on the right.

This systematic exploration should allow one to assess the behavior of the quark–gluon liquid possibly created in collisions performed in the energy range between those of the AGS and SPS accelerators, and comprising chemically equilibrated u and d quarks and gluons. The particle multiplicity in the final state tells us that the entropy per baryon is at the level of 40 units for the high-energy range of the SPS; see section 7.4. The corresponding temperature which we read from panel (c) for $E/b < 8.5$ GeV is $T < 280$ MeV. Evaluation of properties of the final abundances of particles, section 19.3, shows that it is easier to deposit baryon number

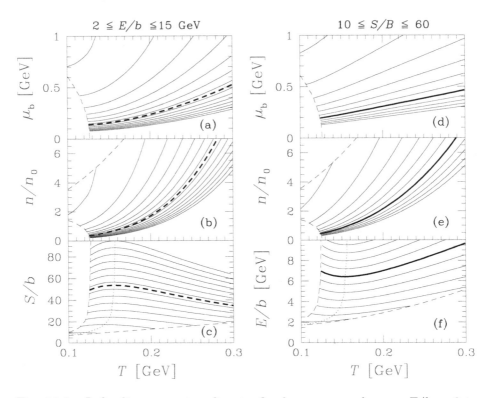

Fig. 16.4. Left: lines corresponding to fixed energy per baryon $E/b = 2$ to 15 GeV in steps of 1 GeV: (a) baryo-chemical potential μ_b (highest E/b at the bottom), (b) baryon density n/n_0 in units of equilibrium nuclear density (highest E/b at the bottom), and (c) S/b, the entropy per baryon (highest E/b at the top). Right: lines corresponding to fixed entropy per baryon $S/b = 10$ to 60 in steps of 5, from top to bottom: (d) μ_b, (e) n/n_0, and (f) E/b; see the text for further details.

than energy in the fireball, and, in general, the initial energy per baryon is smaller than the collisional kinematic limit. Taking $E/b \simeq 7$ GeV, we obtain $T_{ch} \simeq 220$. Before the light-quark flavor has been equilibrated, the temperature of gluons could be as high as $T_{th} \simeq 250$ MeV, for the SPS top energy.

We see in Fig. 16.4 panels (a) and (d) the appropriate ranges of the baryo-chemical potential; and in panels (b) and (e) the baryon density in units of equilibrium nuclear density, $n_0 = 0.16$ fm^{-3}. The dotted lines in panels (c) and (d) show where $P_{q,g} - \mathcal{B} = 0$: there, the entropy per baryon at fixed energy per baryon reaches its maximum as a function of temperature, and the energy per baryon at fixed entropy per baryon reaches its minimum. In an equilibrium transition, the QGP transforms

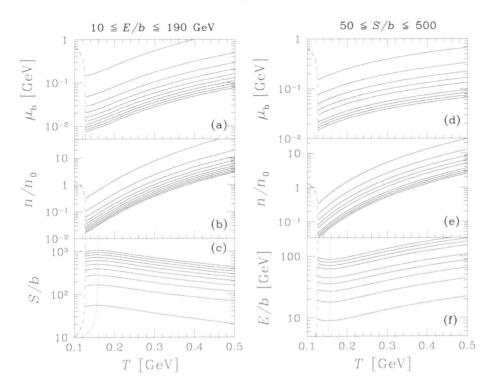

Fig. 16.5.   The analog of Fig. 16.4 for the RHIC energy domain. Left, lines at fixed energy per baryon $E/b = 10$ to $190\,\mathrm{GeV}$ in steps of $20\,\mathrm{GeV}$. Right, lines at fixed entropy per baryon $S/b = 50$ to 500 in steps of 50; see the text for more details.

for T above this condition, since the 'external' hadron pressure needs to be balanced. In an exploding system, the breakup occurs at T below this condition, since the flow of the quark–gluon liquid adds to the pressure working against the vacuum; see section 3.5.

In Fig. 16.5, a similar discussion of the RHIC physical conditions is shown, following the same pattern as Fig. 16.4, except for the use of logarithmic scales. On the left-hand side, the energy range is now 10 GeV $\leq E/b \leq$ 190 GeV, and lines are in steps of $20\,\mathrm{GeV}$. On the right-hand side, the specific entropy range is $50 \leq S/b \leq 500$, in step of 50 units. The lines become denser toward higher energy or entropy. The dotted lines in panels (c) and (f) indicate where the particle pressure is balanced by the vacuum pressure.

In the RHIC run at $\sqrt{s_{\mathrm{NN}}} = 130$ GeV, the final hadron phase space at central rapidity, the intrinsic local-rest-frame energy per baryon is $E/b \simeq$ 25 GeV, while the entropy content is $S/b \simeq 150$, the uncertainties in both values are of 10%–15%, compare with table 19.4 on page 367. The exact

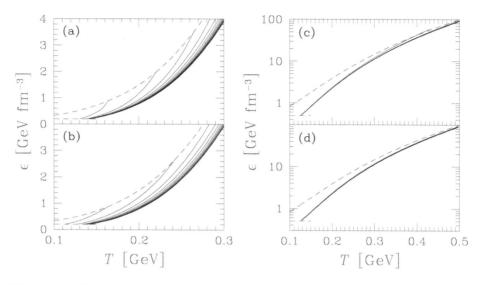

Fig. 16.6.   The energy density as a function of temperature for a quark–gluon liquid. Top, at fixed E/b; bottom, at fixed S/b; left for the SPS and right for the RHIC energy domain. Limits are energy density $\epsilon = 0.5$ GeV fm^{-3} and baryo-chemical potential $\mu_b = 1$ GeV. Lines: left top for fixed $E/b = 2$ to 15 GeV in steps of 1 GeV; left bottom, at fixed $S/b = 10$ to 60 in steps of 5; right top, at fixed $E/b = 10$ to 190 GeV in steps of 20 GeV; and right bottom, at fixed $S/b = 50$ to 500 in steps of 50.

values of E/b and S/b depend on the way one accounts for the influence on the yield analysis of unresolved (at the time of writing) weak decays of hyperons and assumptions made about chemical equilibria. Despite this substantial increase compared to the energy and entropy content seen at the SPS, the value of T_{ch} consistent with this result is only 30–40 MeV higher than that at the SPS. For this reason, the increases in the particle multiplicity we have discussed in section 9.5 are relatively modest.

The rise of energy density with temperature at fixed E/b and S/b is shown in Fig. 16.6, for the expected domain of parameters at the SPS on the left and for the domain of RHIC parameters on the right; the lines follow the same key as that used in Figs. 16.4 and 16.5. We see the expected rise with T^4, but the narrow band of values associated with baryon content is quite striking. In fact, in the right-hand panel the different lines coalesce, energy density is effectively solely a function of T for $E/b > 10$ GeV and $S/b > 50$.

These results indicate clearly how the presence of baryon-rich quark matter possibly formed at lower collision energies augments the entropy and energy content. At large baryon density these results depend strongly

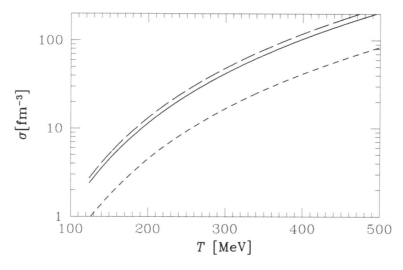

Fig. 16.7. The entropy density in chemically equilibrated QGP at $\lambda_q = 1$ as a function of temperature: solid line, $n_f = 2$; long-dashed line, $n_f = 2.5$; and short-dashed line, 'pure glue' $n_f = 0$.

on the validity of the extrapolation made from $\mu_b = 0$ of the quark–gluon liquid and there is no way to estimate the associated systematic error until lattice results of comparable quality for $\mu_b \neq 0$ become available. However, it would appear that at above $E/b > 3$ GeV, conditions suitable for formation of deconfined state are present.

Entropy plays a very important role in the study of scaling one dimensional hydrodynamic expansion, and the hadron yield in the final state offers a reliable measure of the product $\sigma_0 \tau_0$, Eq. (7.27). In Fig. 16.7, the entropy density σ is shown as a function of temperature. The solid line is for the case of an equilibrated light-quark–glue system in the limit of vanishing chemical potential. We note that initially the entropy rises faster than the asymptotic T^3 behavior, since the quantum chromodynamics interactions weaken, and there is an increase in the effective number of acting quark and gluon degrees of freedom. Thus, the drop in entropy density on going toward the hadronization condition is considerable. In order to preserve the entropy content in the fireball when the QGP fireball expands, from $T \simeq 300$ MeV toward 150 MeV, a volume growth by a factor of nine must occur.

The 'pure-glue' case (short-dashed line) contains as expected about half of the entropy when one makes a comparison at equal temperature. The addition of strangeness expressed by increasing $n_f = 2$ to 2.5 adds about 10% to the entropy content.

VI

Strangeness

17 Thermal production of flavor in a deconfined phase

17.1 The kinetic theory of chemical equilibration

Strangeness, and more generally heavy-flavor quarks, can be produced either in the first interactions of colliding matter, or in the many ensuing less-energetic collisions. The mass of the strange quark m_s is comparable in magnitude to the typical temperatures reached in heavy-ion interactions, and the numerous 'soft' collisions of secondary partons dominate the production of strangeness, and naturally, of the light flavors u and d.

The masses of charm and bottom quarks are well above typical temperatures; these quarks are predominantly produced in the hard initial scattering. This process remains today a topic of current intense study both for the elementary and for the nuclear collisions [124]. We will not discuss it further in this book.

At the time at which the strange flavor approaches chemical equilibrium in soft collisions, the back reaction is also relevant. The quantum-mechanical matrix element driving a two-body reaction must be, channel by channel, the same for forward- and backward-going reactions. The actual rates of reaction differ since there are usually considerable differences in statistical and phase-space factors. However, the forward and backward reactions will balance when equilibrium yields of particles are established. This principle of detailed balance can sometimes be used to evaluate reaction rates.

The net change in yield of flavors f and $\bar{\text{f}}$ is given by the difference between the rates of production and annihilation. The evolution in the density of heavy quarks in QGP can be described by the master equation

$$\frac{dN_{\text{f}}(t)}{d^3x\,dt} = \frac{dN(\text{gg}, \text{q}\bar{\text{q}} \to \text{f}\bar{\text{f}})}{d^3x\,dt} - \frac{dN(\text{f}\bar{\text{f}} \to \text{gg}, \text{q}\bar{\text{q}})}{d^3x\,dt}. \tag{17.1}$$

When locally at a point in space there is exact balance between the two terms on the right-hand side of Eq. (17.1), chemical equilibrium has been established, which, as we recall, is the state of maximum entropy; see section 7.1.

Each of the terms in Eq. (17.1) expresses a change in number of particles per unit of 4-volume and it is a Lorentz-invariant quantity. We take advantage of this to write

$$
\partial_\mu j_f^\mu \equiv \frac{\partial \rho_f}{\partial t} + \frac{\partial \vec{v} \, \rho_f}{\partial \vec{x}} = \rho_g^2(t) \langle \sigma v \rangle_p^{gg \to f\bar{f}} + \rho_q(t) \rho_{\bar{q}}(t) \langle \sigma v \rangle_p^{q\bar{q} \to f\bar{f}}
$$
$$
- \rho_f(t) \rho_{\bar{f}}(t) \langle \sigma v \rangle_p^{f\bar{f} \to gg, q\bar{q}}. \tag{17.2}
$$

The left-hand side describes the change in the local particle density including the effect of flow; the right-hand side is another way to express the change in number of particles in terms of individual reactions, as we shall show, see Eq. (17.7).

The momentum-averaged cross section of reacting particles is

$$
\boxed{ \langle \sigma v_{\rm rel} \rangle_p \equiv \frac{1}{1 + I_{12}} \frac{\int d^3 p_1 \int d^3 p_2 \, \sigma_{12} v_{12} f(\vec{p}_1) f(\vec{p}_2)}{\int d^3 p_1 \int d^3 p_2 \, f(\vec{p}_1) f(\vec{p}_2)}. } \tag{17.3}
$$

The factor $1/(1 + I_{12})$ is introduced to avoid double counting of indistinguishable pairs of particles, $I_{12} = 1$ for an identical pair of bosons (gluons, pions), otherwise $I_{12} = 0$. Some authors introduce this factor into the kinetic equation Eq. (17.2). Considering that the cross section is obtained as an average over all reaction channels, the implicit sums over spin, color, and any other discrete quantum numbers can be combined in the particle density,

$$
\boxed{ \rho = \int \frac{d^3 p}{(2\pi)^3} f(\vec{p}) = \int \frac{d^3 p}{(2\pi)^3} \sum_{i=s,c,\dots} f_i(\vec{p}). } \tag{17.4}
$$

We have suppressed, in the above discussion, the dependence of the phase-space distributions $f(\vec{p}, \vec{x}, t)$ on the spatial coordinates, as well as their evolution with time.

In general terms, we need to obtain $f(\vec{p}, \vec{x}, t)$ for gluons and light quarks from a solution of a transport master equation such as the Boltzmann equation. However, this introduces a large uncertainty due to our great ignorance of the early collision (quantum) dynamics. Moreover, a seven dimensional evolution equation for $f(\vec{p}, \vec{x}, t)$ cannot yet be handled with the available computing power without simplifying assumptions invoking spherical symmetry. Moreover, the uncertainty about the initial temperature, initial yield of strangeness from pre-equilibrium reactions, and the poorly known mass of the strange quark m_s introduce significant uncer-

tainties into calculations of the yield of strangeness, and limit the need
for very precise methods.

We proceed to simplify by the use of two assumptions, which follow
from the discussion we presented in section 5.5.

- The kinetic (momentum-distribution) equilibrium is approached faster
  than the chemical (abundance) equilibrium [23, 231, 246]. This allows
  us to study only the chemical abundances, rather than the full momentum distribution of the (strange) quark flavor.

- Gluons equilibrate chemically significantly faster than does strangeness
  [276]. We consider the evolution of the population of strangeness only
  after gluons have (nearly) reached chemical equilibrium.

In view of these assumptions, the phase-space distribution f_s can be
characterized by a local temperature $T(\vec{x}, t)$ of a (Boltzmann) equilibrium
distribution reached for $t \to \infty$, f_s^∞, with normalization set by a phase-
space-occupancy factor:

$$f_s(\vec{p}, \vec{x}; t) \simeq \gamma_s(T) f_s^T(p), \quad f_s^T(p) = e^{-\sqrt{m_s^2 + p^2}/T}, \tag{17.5}$$

where f_s^T is the equilibrium Boltzmann momentum distribution. Equation (17.5) invokes in the momentum independence of γ_s the first assumption. The factor γ_s allows the local density of strange quarks to evolve
independently of the local temperature.

Using the Boltzmann momentum distribution Eq. (17.5) in Eq. (17.3),
we are performing a thermal average of the cross section and relative
velocity, and the result is a thermally averaged cross section, a function
that depends on T instead of $\sqrt{s}$. Some books refer to this as thermal
reactivity; we will often call it the thermal cross section:

$$\langle \sigma v_{\text{rel}} \rangle_T \equiv \frac{1}{1 + I_{12}} \frac{\int d^3 p_1 \int d^3 p_2 \, \sigma_{12} v_{12} f_1^T(p) f_2^T(p)}{\int d^3 p_1 \int d^3 p_2 \, f_1^T(p) f_2^T(p)}. \tag{17.6}$$

This thermal cross section is dependent on T, and on the masses of reacting particles, and its physical dimension is volume per time. We will
often drop the subscripts T and rel, since the only average to which we
refer in a cross section is 'thermal', and, in this context, the velocity is
always relative.

The thermal reaction rate per unit time and volume, $R_{12}(T)$, is obtained as follows: consider that a single particle '1' enters at velocity v_{12}
a medium of particles '2'; the rate of reactions is $\langle \sigma v_{\text{rel}} \rangle_T \rho_2$. If per unit
volume there are N_1 particles, i.e., we have a density ρ_1, then

$$\boxed{R_{12}(T) \equiv \langle \sigma v_{\text{rel}} \rangle_T \rho_1 \rho_2.} \tag{17.7}$$

The densities ρ_1 and ρ_2 arise from the momentum integration of the Boltzmann distributions f_1 and f_2, Eq. (17.4), and contain the degeneracy factors g_1 and g_2. This rate Eq. (17.7) is Lorentz invariant, i.e., all observers agree by how much the number of particles changes per (invariant) unit volume in space–time.

The following evaluation of R_{12} applies to reactions occurring in confined and deconfined matter. However, for almost all particles (except pions) in the hadronic gas, it is sufficient to use the Boltzmann momentum distribution function, since the phase-space cells are nearly empty, while the density of particles arises from the numerous resonances encountered (see section 12.1). In QGP, we must in general use the Bose and Fermi distributions, as appropriate, which will complicate the results slightly.

We recall here the (Mandelstam) variables s, t, and u characterizing, in an invariant way, the two-particle reaction $1 + 2 \rightarrow 3 + 4$,

$$s = (p_1 + p_2)^2 = (p_3 + p_4)^2, \tag{17.8a}$$

$$t = (p_1 - p_3)^2, \tag{17.8b}$$

$$u = (p_2 - p_3)^2, \tag{17.8c}$$

$$s + t + u = \sum_{i=1}^{4} m_i^2, \tag{17.8d}$$

where the 4-momenta $p^\mu = (E_p, \vec{p})$ are used with $E_p = \sqrt{\vec{p}^2 + m^2}$. $\sqrt{s}$ is as usual the total CM energy and t is the invariant generalization of the scattering angle.

The cross section for reaction of two particles to give n final-state particles is computed according to

$$\sigma_{12} v_{12} E_1 E_2 = \int \prod_{i=3}^{n+2} d^4 p_i \, \delta(p_i^2 - m_i^2) \Theta(p_i^0)$$

$$\times \delta^4 \left(p_1 + p_2 - \sum_{i=3}^{n+2} p_i \right) |\mathcal{M}|^2. \tag{17.9}$$

$|\mathcal{M}(s,t)|^2$ is the reaction-matrix element obtainable, for perturbative processes, using the Feynman rules described in section 14.1. The relative velocity of two collinear particles, which is used in the definition of the cross section, is*

$$v_{12} 2E_1 2E_2 \equiv 2\lambda_2^{1/2}(s),$$

$$= 2\sqrt{s - (m_1 + m_2)^2} \sqrt{s - (m_1 - m_2)^2}. \tag{17.10}$$

---

* $\lambda_2^{1/2}(s)$ has nothing to do with a fugacity.

The invariant reaction rate, Eq. (17.7), thus is

$$R_{12} = \int \frac{d^3p_1}{(2\pi)^3 2E_1} \frac{d^3p_2}{(2\pi)^3 2E_2} \frac{f_1\, f_2}{1 + I_{12}} \sigma_{12} v_{12} \times 2E_1 \times 2E_2 \qquad (17.11)$$

$$= \frac{g_1 g_2}{(2\pi)^6} \int_{s_{\rm th}}^{\infty} ds\, \frac{2\lambda_2^{1/2} \sigma_{12}}{1 + I_{12}} \left(\int \frac{d^3p_1}{2E_1} \frac{d^3p_2}{2E_2} e^{-E_1/T} e^{-E_2/T} \delta(s - (p_1 + p_2)^2) \right),$$

where we have inserted a (dummy) integration over s. The lower limit $s_{\rm th}$ of the integration over s is the threshold for the reaction, usually $s_{\rm th} = (\sum_i m_i)^2$, the sum of masses of the final state created.

We are also interested in understanding at which values of $\sqrt{s}$ the production processes occur. We will evaluate the p_1 and p_2 momentum integrals first, which will leave us with a final integral over $\sqrt{s}$ in Eq. (17.11). We can present the rate of production as an integral over the differential rate dR_i/ds, where i refers to the reaction channel considered:

$$R \equiv \sum_i \int_{s_{\rm th}}^{\infty} ds\, \frac{dR_i}{ds} \equiv \sum_i \int_{s_{\rm th}}^{\infty} ds\, \sigma_i(s) P_i(s). \qquad (17.12)$$

$\sigma_i(s)$ is the cross section of the channel. The factor $P_i(s)$, which has the same dimension as the invariant rate R, is interpreted as the number of collisions per unit volume and time, and corresponds to the expression in the second line in Eq. (17.11), with the channel i corresponding to the collision of particles $\{1, 2\}$.

In order to evaluate the p_1 and p_2 momentum integrals in Eq. (17.11), it is convenient to introduce, for the Boltzmann distributions, the 4-vector of temperature, Eq. (12.39), in the local restframe $\beta = (1/T, 0)$ and to write the (invariant) factor in large brackets in Eq. (17.11),

$$[\cdots] = \int d^4p_1\, d^4p_2\, \delta^4(p - p_1 - p_2)\delta_0(p_1^2 - m_1^2)\delta_0(p_2^2 - m_2^2)$$

$$\times \int d^4p\, e^{-\beta \cdot p}\delta(p^2 - s), \qquad (17.13)$$

with δ_0 being the δ-function restricted to positive roots of the argument only (compare with Eq. (12.45)). A dummy integration over $p = p_A + p_B$ allows one to rearrange the terms in a way that separates the expression into the two factor integrals. The first is known as the two particle invariant phase-space integral 'IMS2' [86]:

$$\int d^4p_1\, d^4p_2\, \delta^4(p - p_1 - p_2)\delta_0(p_1^2 - m_1^2)\delta_0(p_2^2 - m_2^2) = \frac{\pi}{2} \frac{\sqrt{\lambda_2}}{s}. \qquad (17.14)$$

$\lambda_2^{1/2}(s)$ is as defined in Eq. (17.10). The second integral can be obtained by evaluating Eq. (10.43), it appeared previously in Eq. (12.22):

$$\int d^4p\, e^{-\beta p}\delta_0(p^2 - s) = \frac{2\pi}{\beta}\sqrt{s}K_1(\beta\sqrt{s}).\tag{17.15}$$

The invariant reaction rate, Eq. (17.7), is

$$R_{12} = \frac{g_1 g_2}{32\pi^4}\frac{T}{1 + I_{12}}\int_{s_{\mathrm{th}}}^{\infty}ds\,\sigma(s)\frac{\lambda_2(s)}{\sqrt{s}}K_1(\sqrt{s}/T).\tag{17.16}$$

So far, we have not addressed the quantum nature of colliding particles. The difficult case is that of a pair of light quarks reacting at finite baryon density, for which one of the distributions cannot be expanded. Since, in this case, the mass of the light quark is negligible, one of the integrals can be done analytically. The integral of interest, which is obtained after performing the angular integrals in Eq. (17.11), is

$$K_1(\sqrt{s}/T) \to \frac{1}{T\sqrt{s}}\int_0^{\infty}dp_1\int_0^{\infty}dp_2\,\Theta(4p_1p_2 - s)f_q(p_1)f_{\bar{q}}(p_2),\tag{17.17}$$

with (compare with Eqs. (10.34a) and (10.34b))

$$f_q(p) = \frac{1}{\gamma_q^{-1}\lambda_q^{-1}e^{p/T} + 1}, \qquad f_{\bar{q}}(p) = \frac{1}{\gamma_q^{-1}\lambda_q e^{p/T} + 1}.$$

Assuming that $\gamma_q/\lambda_q < 1$ (baryon-rich matter), we can expand the distributions for antiquarks and obtain the generalization of K_1 in Eq. (17.16):

$$K_1(\sqrt{s}/T) \to \sum_{l=1}^{\infty}(-)^{l+1}\frac{\gamma_q^l}{l\lambda_q^l}\int_0^{\infty}\frac{dp_1}{\sqrt{s}}\frac{\exp\left(-l\dfrac{s}{4Tp_1}\right)}{\gamma_q^{-1}\lambda_q^{-1}e^{p_1/T} + 1},\tag{17.18}$$

which has to be evaluated numerically.

In the special case that all chemical factors are unity (or otherwise allow the expansion), we expand Eq. (17.18) again to obtain

$$K_1(\sqrt{s}/T) \to \sum_{l,n=1}^{\infty}\frac{(\pm)^{l+n}}{\sqrt{s}\,l}\int_0^{\infty}dp_1\exp\left(-l\frac{s}{4Tp_1} - \frac{np_1}{T}\right).\tag{17.19}$$

We have allowed for Fermi and Bose distributions, recalling the expansion

$$\frac{1}{e^{E/T}\mp 1} = \pm\sum_{n=1}^{\infty}(\pm)^n e^{-nE/T}.$$

The upper sign in Eq. (17.19) is for bosons and the lower for fermions. We now use

$$\int_0^\infty dx\, e^{-a/(4x)-bx} = \sqrt{\frac{a}{b}} K_1(\sqrt{ab}), \tag{17.20}$$

and obtain in the case $\gamma_q = \lambda_q = 1$ the generalization of K_1 in Eq. (17.16):

$$K_1(\sqrt{s}/T) \rightarrow \sum_{l,n=1}^\infty (\pm)^{l+n} \frac{K_1(\sqrt{lns}/T)}{\sqrt{ln}}. \tag{17.21}$$

We see how the powers of $\sqrt{s}$ cancel out, leaving only the slowly converging pre-factor. It turns out that many terms in the sum of l and n are required in order to arrive at a precise result.

17.2 Evolution toward chemical equilibrium in QGP

The conservation of current used in Eq. (17.2) applies to the laboratory 'Eulerian' formulation. This can also be written with reference to the individual particle dynamics in the so-called 'Lagrangian' description: consider ρ_s as the inverse of the small volume available to each particle. Such a volume is defined in the local frame of reference (subscript 'l') for which the local flow vector vanishes, $\vec{v}(\vec{x}, t)|_{\text{local}} = 0$. For the considered volume δV_l being occupied by a small number of particles δN (e.g., $\delta N = 1$), we have

$$\delta N_s \equiv \rho_s\, \delta V_l. \tag{17.22}$$

The left-hand side of Eq. (17.2) can be now written as

$$\frac{\partial \rho_s}{\partial t} + \frac{\partial \vec{v}\, \rho_s}{\partial \vec{x}} \equiv \frac{1}{\delta V_l} \frac{d\delta N_s}{dt} = \frac{d\rho_s}{dt} + \rho_s \frac{1}{\delta V_l} \frac{d\delta V_l}{dt}. \tag{17.23}$$

Since δN and $\delta V_l\, dt$ are Lorentz-invariant quantities, the actual choice of the frame of reference in which the right-hand side of Eq. (17.23) is studied is irrelevant and, in particular, it can be considered in the local rest frame. The last term in Eq. (17.23) describes the effect of volume dilution due to the dynamic expansion of matter. The other term on the right-hand side is then interpreted as the evolution of the local density in proper time of the volume element.

We continue to use the first form of Eq. (17.23) and evaluate the local change in number of particles. We introduce $\rho_s^\infty(T)$ as the (local) chemical-equilibrium abundance of strange quarks which arises at $t \rightarrow \infty$, thus $\rho_s = \gamma_s(t)\rho_s^\infty$. We use the Boltzmann equilibrium abundance, section 10.4,

$$\delta N_s = \delta V\, \gamma_s \rho_s^\infty = [T^3\, \delta V]\gamma_s \frac{3}{\pi^2} z^2 K_2(z). \quad z = \frac{m_s}{T}, \tag{17.24}$$

In an entropy-conserving evolution, e.g., subject to (ideal) hydrodynamic flow, section 6.2, the first factor on the right-hand side in Eq. (17.24) (in square brackets) is a constant in time, $\delta V \, T^3 = \delta V_0 \, T_0^3 = $ constant. We now substitute in Eq. (17.23) and obtain, using Eq. (10.54b),

$$\frac{\partial \rho_s}{\partial t} + \frac{\partial \vec{v} \, \rho_s}{\partial \vec{x}} = \dot{T} \rho_s^\infty \left(\frac{d\gamma_s}{dT} + \frac{\gamma_s}{T} z \frac{K_1(z)}{K_2(z)} \right), \tag{17.25}$$

where $\dot{T} = dT/dt$. Only a part of the usual flow-dilution term is left, since we implemented the adiabatic volume expansion, and study the evolution of the phase-space occupancy in lieu of the particle density.

We include the collision term seen in Eq. (17.2) and two channels, the fusion of gluons and light-quark–antiquark fusion into a pair of strange (or equivalently charm) quarks:

$$\boxed{\dot{T} \rho_s^\infty \left(\frac{d\gamma_s}{dT} + \frac{\gamma_s}{T} z \frac{K_1(z)}{K_2(z)} \right) = \gamma_g^2(\tau) R^{gg \to s\bar{s}} + \gamma_q(\tau) \gamma_{\bar{q}}(\tau) R^{q\bar{q} \to s\bar{s}}}$$

$$-\gamma_s(\tau)\gamma_{\bar{s}}(\tau)(R^{s\bar{s} \to gg} + R^{s\bar{s} \to q\bar{q}}). \tag{17.26}$$

Similar equations can be formulated for the evolution of γ_g and γ_q. Knowledge of the dynamics of the local temperature, along with the required invariant rate of production $R(T)$, allows one to evaluate the dynamic behavior of occupancy fugacities $\gamma_i(t)$.

Since only weak interactions convert quark flavors, on the hadronic time scale we have $\gamma_{s,q}(\tau) = \gamma_{\bar{s},\bar{q}}(\tau)$. Moreover, detailed balance, arising from the time-reversal symmetry of the microscopic reactions, assures that the invariant rates for forward/backward reactions are the same, specifically

$$R^{12 \to 34} = R^{34 \to 12}, \tag{17.27}$$

and thus

$$\dot{T} \rho_s^\infty \left(\frac{d\gamma_s}{dT} + \frac{\gamma_s}{T} z \frac{K_1(z)}{K_2(z)} \right) = \gamma_g^2(\tau) R^{gg \to s\bar{s}} \left(1 - \frac{\gamma_s^2(\tau)}{\gamma_g^2(\tau)} \right)$$

$$+ \gamma_q^2(\tau) R^{q\bar{q} \to s\bar{s}} \left(1 - \frac{\gamma_s^2(\tau)}{\gamma_q^2(\tau)} \right). \tag{17.28}$$

When all $\gamma_i \to 1$, the right-hand side vanishes; chemical equilibrium is established.

In order to be able to evolve the population of (strange) quarks we need to understand the population of gluons, i.e., γ_g. Several workers have considered the glue approach to equilibrium in perturbative processes such as gluon splitting, e.g., $gg \to ggg$ [71, 252, 253, 275]. They find that, in the thermal environment, there is not enough production of gluons to reach chemical equilibrium. Accordingly, there are too few gluons to

derive the approach to chemical equilibrium of light quarks, and, even
more so, of strange quarks. The initial conditions for such a full-kinetic-
evolution study are relatively unreliable, but can not alter this conclusion.
We pursue an entirely different point of view in this book, based on the
belief that strong multigluon processes gg → ng, $n > 3$ can equilibrate
the abundance of gluons faster [278]. Therefore, we assume that gluon
chemical equilibrium is reached rapidly, $\mathcal{O}(1\,\mathrm{fm})$, on a scale not much
longer than the time required to reach thermal equilibrium. Our choice
of an initial condition for consideration of strange flavor equilibration is
$\gamma_g = 1$. In the study of evolution of strange (and charm) quarks we also
take $\gamma_q = 1$. A large error, in this last assumption, is without significance
for what follows, since gluons dominate the thermal production of strange
quarks.

Given the chemical equilibrium of gluons and light quarks, we obtain
the dynamic equation describing the evolution of the local phase-space
occupancy of strangeness (and, in analogy, charm),

$$2\tau_s \dot{T}\left(\frac{d\gamma_s}{dT} + \frac{\gamma_s}{T}z\frac{K_1(z)}{K_2(z)}\right) = 1 - \gamma_s^2. \tag{17.29}$$

As discussed at the end of section 5.7, we introduce the relaxation time
τ_s of chemical (strangeness) equilibration as the ratio of the equilibrium
density that is being approached and the rate at which this occurs,

$$\tau_s \equiv \frac{1}{2}\frac{\rho_s^\infty}{(R^{gg\to s\bar{s}} + R^{q\bar{q}\to s\bar{s}} + \cdots)}. \tag{17.30}$$

The factor $\frac{1}{2}$ is introduced by convention, in order for the quantity $\tau_s(T)$
to describe an exponential approach to equilibrium, Eq. (5.41).

One generally expects that $\gamma_s \to 1$ monotonically as a function of time.
However, Eq. (17.29) allows the range $\gamma_s > 1$, for it incorporates the
physics of a rapidly expanding high yield of strangeness created in the
early stage at high T. At a high background temperature, the evolution
$\gamma_s(t) \to 1$ produces a high yield of particles, which corresponds, at the
lower temperature established after expansion of the system, to $\gamma_s > 1$.
One finds that thermal annihilation of flavor cannot keep up with the
rapid evolution of a fireball of QGP, and an overabundance will generally
result. Annihilation is slow, since the density of strange and antistrange
quarks is about four times smaller than the density of gluons (an effect of
color and mass) and the rate of annihilation for strange quarks is 16 times
slower. With a relaxation time for the production of strangeness of 1.5–10
fm (depending on temperature), see Fig. 17.11 below, the relaxation time
for s̄s annihilation is 20–150 fm, so practically all strangeness is preserved
on the time scale of 5 fm of a QGP fireball. For charm this argument
is much stronger, given the greater effect of mass. Once it has been

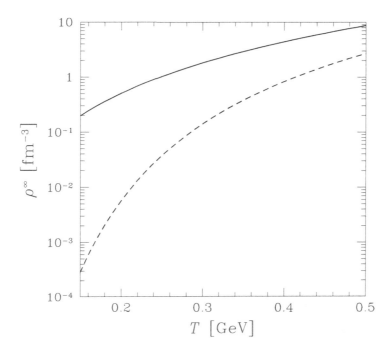

Fig. 17.1.   The statistical equilibrium densities of strange or antistrange quarks with $m_\mathrm{s} = 160\,\mathrm{MeV}$ (solid line) and of charmed or anticharmed quarks with $m_\mathrm{c} = 1500\,\mathrm{MeV}$ (dashed line), as functions of temperature T.

produced, heavy flavor has no time to annihilate and reequilibrate. This is a very important feature that makes the yield of strangeness a 'deep' probe of the deconfined phase.

Said differently, the high abundance of strangeness (or charm) formed in the high-temperature QGP stage over-populates the available phase space at lower temperature, when the equilibration rate cannot keep up with the cooling due to expansion. We will quantify this effect in more detail in section 17.5. In the kinetic equation Eq. (17.29), this is seen most clearly by considering the case $T < m_\mathrm{s}$. In this limit, $1/\tau_\mathrm{s}$ becomes small, the dilution term (second term on the left-hand side in Eq. (17.29)) dominates the evolution of γ_s. For the massive charm quarks $T \ll m_\mathrm{c}$, so expansion dilution can generate a very large phase-space overabundance, compared with the equilibrium yields expected in hadronization.

To grasp the sensitivity of these remarks to the early QGP stage, we look at the equilibrium densities of strange ($m_\mathrm{s} = 160\,\mathrm{MeV}$) and charmed ($m_\mathrm{c} = 1500\,\mathrm{MeV}$) quarks, shown in Fig. 17.1. We note that, for $T \simeq 250\,\mathrm{MeV}$, the equilibrium abundance of strangeness exceeds one $\bar{\mathrm{s}}$ quark for each fm³ of matter, and that charm reaches this for $T \geq 450\,\mathrm{MeV}$.

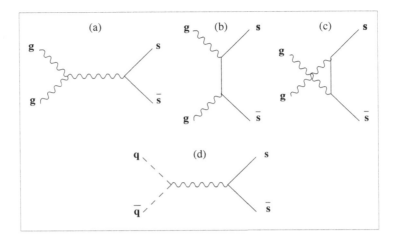

Fig. 17.2. Leading order Feynman diagrams for production of s s̄ (and similarly
c c̄) by fusion of gluons and pairs of quarks (q = u, d).

Looking at charm, we note that, when the system expands and cools
rapidly from $T = 450$ to 150 MeV, the volume grows by less than a factor
of 70, see the entropy density in Fig. 16.7, while the equilibrium density
of charm declines by factor 10 000 and the charm saturation factor γ_c increases by the factor 150, to preserve the yield of charm. This abundance
of charm is a product of the first interactions, not of thermal processes,
except possibly at the LHC in a very-high-T scenario.

17.3 Production cross sections for strangeness and charm

The production processes involving quark and gluon degrees of freedom
in QGP are

$$u + \bar{u} \to s + \bar{s}, \qquad d + \bar{d} \to s + \bar{s}, \tag{17.31a}$$
$$g + g \to s + \bar{s}. \tag{17.31b}$$

These three processes describing perturbative production of pairs of
strange quarks are represented to lowest order in Fig. 17.2, and have to
be summed incoherently. These lowest-order diagrams were studied in
the early eighties, for the quark process [68] and for the gluon process
[226], employing fixed values of $\alpha_s = 0.6$ and $m_s = 160$–180 MeV.
    The evaluation of the lowest-order Feynman diagrams shown in Fig. 17.2
yields the cross sections [95]:

$$\sigma_{q\bar{q}\to f\bar{f}}(s) = \frac{8\pi\alpha_s^2}{27s}\left(1 + \frac{2m_f^2}{s}\right)w(s), \quad w(s) = \sqrt{1 - \frac{4m_f^2}{s}}, \tag{17.32a}$$

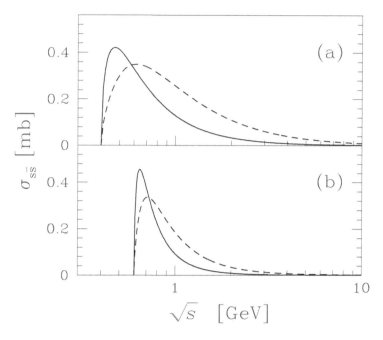

Fig. 17.3. Production cross sections for strangeness in leading order: (a) for $\alpha_s = 0.6$ and $m_s = 200\,\mathrm{MeV}$; (b) for running $\alpha_s(\sqrt{s})$ and $m_s(\sqrt{s})$, with $\alpha_s = 0.118$. Solid lines, $q\bar{q} \to s\bar{s}$; dashed lines, $gg \to s\bar{s}$.

$$\sigma_{gg\to f\bar{f}}(s) = \frac{\pi\alpha_s^2}{3s}\left[\left(1 + \frac{4m_f^2}{s} + \frac{m_f^4}{s^2}\right)\ln\left(\frac{1+w(s)}{1-w(s)}\right)\right.$$

$$\left.-\left(\frac{7}{4} + \frac{31m_f^2}{4s}\right)w(s)\right]. \tag{17.32b}$$

Inspecting Fig. 17.3(a), we see that the magnitudes (up to 0.4 mb) of both types of reactions considered, quark fusion and gluon fusion, are similar. At this stage, it is not immediately apparent that gluons dominate the production of flavor.

The magnitude of the cross section of interest is normalized by α_s. To obtain Fig. 17.3(a), we took $\alpha_s = 0.6$. While the value seems reasonable, a value of $\alpha_s = 0.3$ would lengthen the relaxation time of strangeness, $\tau_s \propto \alpha_s^{-2}$, by a factor of four, nearly beyond the expected life span of the QGP fireball. Thus, we must improve the determination of α_s. There are two natural ways to do this; the easier one is to adopt the functional $\alpha_s(T)$ seen in Fig. 14.3. However, in such an approach, two-body collisions occurring at very different $\sqrt{s}$ but in a thermal bath at the same temperature T are evaluated with the same value of α_s. Only for the thermal production of

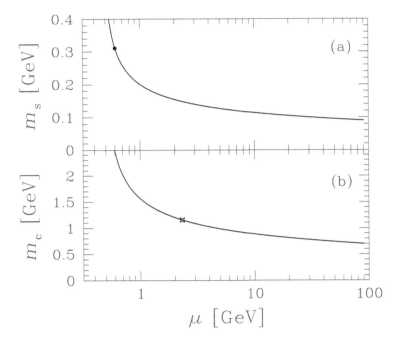

Fig. 17.4.   Running masses for $\alpha_s = 0.118$: (a) the mass of the strange quark m_s, for which the dot indicates the production thresholds for pairs of strange quarks for $m_s(M_Z) = 90$ MeV; (b) the mass of the charmed quark m_c, for which the cross indicates production thresholds for pairs of charmed quarks for $m_c(M_Z) = 700$ MeV.

charm does this approach turn out to yield the same result as does the more complex, but more precise, consideration of an appropriate value of α_s for each collision, governed by the applicable $\alpha_s(\mu)$, Fig. 14.1, with $\mu \simeq \sqrt{s}$.

This second method, in which for each collision in the thermal bath an appropriate coupling strength is selected, is necessary for studying the production of strangeness in order to account for the growth of the cross section for soft scattering. The increase of cross section in soft collisions is, however, largely balanced by the concurrent suppression of the cross section due to the increase in mass of the strange quark m_s on the soft momentum scale. We adopt the running-mass and coupling-constant results presented in chapter 14, for $\alpha_s(M_Z) = 0.118$. In Fig. 17.4, the running masses of the strange and charm quarks $m_i(\mu)$, $i = s, c$, for $m_s(M_Z) = 90$ MeV and $m_c(M_Z) = 700$ MeV, derived from Fig. 14.1, are shown. These values imply that $m_s(1\,\mathrm{GeV}) \simeq 200\,\mathrm{MeV}$ and $m_c(1\,\mathrm{GeV}) \simeq 1.55$ GeV.

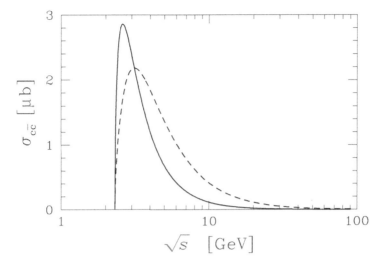

Fig. 17.5.   Cross sections for leading order production of charm for the running α_s and running mass of the charmed quark with $m_c(M_Z) = 0.7\,\mathrm{GeV}$: the solid line is for fusion of pairs of light quarks, whereas the dashed line is for gg $\rightarrow$ c$\bar{\mathrm{c}}$.

The energy scale of greatest interest for studying the production of strangeness is certainly $\mu = 2\pi T \simeq 1\text{–}2$ GeV, which is precisely the region of rapid change of the value of m_s. Below $\sqrt{s} = 1$ GeV, the mass of the strange quark increases rapidly and the threshold for producing pairs of strange quarks increases to above $2m_s(1\,\mathrm{GeV})$. Half of the threshold energy is indicated by the black dot in Fig. 17.4(a). The pair-production threshold is, section 14.4,

$$2m_{\mathrm{s}}^{\mathrm{th}}(\mu = 2m_{\mathrm{s}}^{\mathrm{th}}) = 611\,\mathrm{MeV}, \ m_{\mathrm{s}}(M_Z) = 90\,\mathrm{MeV}, \ \alpha_{\mathrm{s}}(M_Z) = 0.118.$$

For charm, the running-mass effect plays differently, since the naive threshold for production of charmed quarks $2m_c(2\,\mathrm{GeV}) > 2\,\mathrm{GeV}$. The running of the mass has the effect of reducing the effective threshold. For $m_c(M_Z) = 700$ MeV, the production threshold is found, rather than at 3.1 GeV, at

$$2m_{\mathrm{c}}^{\mathrm{th}}(\mu = 2m_{\mathrm{c}}^{\mathrm{th}}) = 2.3\,\mathrm{GeV}, \ m_{\mathrm{c}}(M_Z) = 700\,\mathrm{MeV}, \ \alpha_{\mathrm{s}}(M_Z) = 0.118.$$

The cross, in Fig. 17.4(b), indicates the position of half of the threshold energy. Even this small reduction in threshold enhances the production of charm at low energy and especially so in the thermal environment we are considering.

In Fig. 17.3(b), we have presented the cross sections for production of strangeness Eqs. (17.32a) and (17.32b), evaluated using the running QCD parameters obtained in sections 14.3 and 14.4, identifying $\mu \rightarrow \sqrt{s}$.

On comparing this with the 'conventional' result seen in Fig. 17.3(a), a greater threshold and more rapid decline of the cross sections are noted. Near $\sqrt{s} = 1$ GeV, in both cases, the fusion of gluons (dashed line) dominates, even though at lower energy the quark-pair fusion reaction (solid line) has a stronger peak. Similarly, for production of charm, we see in Fig. 17.5 the cross sections for fusion of gluons (dashed line) and pairs of quarks (solid line), for production cross sections which were computed for the running $\alpha_s(M_Z) = 0.118$ and running charmed-quark mass with $m_c(M_Z) = 700$ MeV. These cross sections are a factor 100 smaller than those for strangeness, at the level 1–2 μb. This is due to the fact that an eight-fold greater $\sqrt{s}$ is required, given that $\sigma \propto 1/s$, and a reduction in the effective coupling strength. The smallness of the cross section for production of charm is the reason why thermal production of charm becomes relevant only at $T \to 1$ GeV. Inspecting Fig. 17.5, we can also clearly understand the great sensitivity of the direct production of charm in non-thermal parton collisions to the value, and running, of the mass of the charmed quark: using a production threshold at 3 GeV, we cut 40% of the available strength of the cross section.

The use of scale dependent QCD parameters, α_s and m_f, f $=$ s, c, with $\mu \propto \sqrt{s}$ amounts to a re-summation of many QCD diagrams comprising vertex, and self energy corrections. A remaining shortcoming of thermal production evaluation is that up to day, there has not been a study of the next to leading order final state accompanying gluon emission in thermal processes, e.g., gg $\to$ s$\bar{\text{s}}$ + g. In direct parton induced reactions, this next to leading order effect enhances the production rate by a factor $K = 1.5$–3. This causes a corresponding increase in the rate of production, and a reduction in the thermally computed chemical equilibration time of strangeness and charm.

17.4 Thermal production of flavor

The thermal production processes occur over a wide range of $\sqrt{s}$. There are two factors determining this. Aside from the cross section, the collision frequency is the determining factor. We have introduced the thermal collision frequency per unit time and volume $P_i(s)$ in channel i, Eq. (17.12). Employing the result Eq. (17.16) and the discussion of quantum corrections which followed, we obtain, setting $g_1 = g_2 = 16$, $I_{12} = 1$ and $\lambda_2 = s^2$ (for massless gluons)

$$P_{\text{g}} = \frac{4Ts^{3/2}}{\pi^4} \sum_{l,n=1}^{\infty} \frac{1}{\sqrt{nl}} K_1\left(\frac{\sqrt{nl}\,s}{T}\right). \tag{17.33}$$

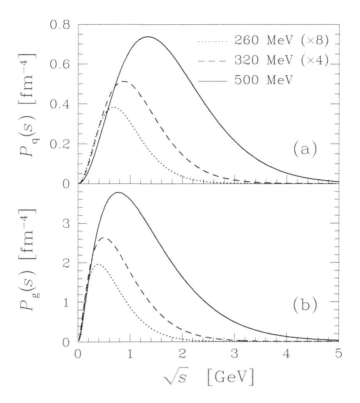

Fig. 17.6.   The collision distribution functions as functions of $\sqrt{s}$: (a) for quarks and (b) for gluons, computed for temperature $T = 260\,\mathrm{MeV}$, $\lambda_\mathrm{q} = 1.5$ (dotted lines, amplified by a factor of eight); $T = 320\,\mathrm{MeV}$, $\lambda_\mathrm{q} = 1.6$ (dashed lines, amplified by a factor of four); and $T = 500\,\mathrm{MeV}$, $\lambda_\mathrm{q} = 1.05$ (solid lines). In all cases $\gamma_\mathrm{q}, \gamma_\mathrm{g} = 1$.

For quark processes, using Eq. (17.18) for $\gamma_\mathrm{q}/\lambda_\mathrm{q} < 1$, and setting $g_1 = g_2 = 6$, $I_{12} = 0$ and $\lambda_2 = s^2$ (for massless quarks), and taking the result twice to allow for the two quark flavors which can undergo incoherent reactions,

$$P_\mathrm{q}|_{\mu_\mathrm{q}=0} = \frac{9T s^{3/2}}{4\pi^4} \sum_{l=1}^{\infty} (-)^{l+1} \frac{\gamma_\mathrm{q}^l}{l\lambda_\mathrm{q}^l} \int_0^\infty \frac{dp_1}{\sqrt{s}} \frac{\exp\left(-l\dfrac{s}{4Tp_1}\right)}{\gamma_\mathrm{q}^{-1}\lambda_\mathrm{q}^{-1} e^{p_1/T} + 1}. \quad (17.34)$$

In Fig. 17.6, the distribution functions for the collision frequency Eqs. (17.33) and (17.34) are presented as functions of $\sqrt{s}$ for gluons (b), and a q$\bar{\mathrm{q}}$ pair of light quarks (a). The temperatures correspond to a range of possible initial fireball temperatures at the SPS and RHIC: $T = 260$ MeV (dotted, $\lambda_\mathrm{q} = 1.5$), 320 MeV (dashed, $\lambda_\mathrm{q} = 1.6$), and 500 MeV

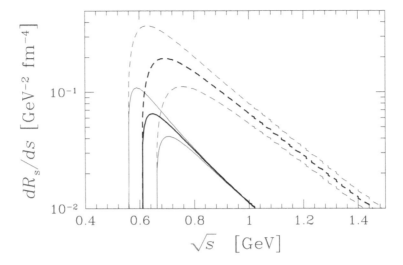

Fig. 17.7. The differential thermal production rate for strangeness dR_s/ds, with $T = 250$, $\lambda_q = 1.5$ for gluons (dashed line), and $q\bar{q} \to s\bar{s}$ (solid line, includes two interacting flavors), for the running $\alpha_s(M_Z) = 0.118$ and running mass of the strange quark $m_s(M_Z) = 90$ MeV $\pm 20\%$ (thin lines).

(solid $\lambda_q = 1.05$). Gluons and light quarks are assumed to be in chemical equilibrium, $\gamma_q, \gamma_g = 1$.

There is a shift in the maximum of the distribution of $P_{q,g}$ to higher $\sqrt{s}$ with increasing temperature. The collision frequency for gluons is about five times greater, $P_g \simeq 5P_q$, than that for a pair of quarks, and this is the origin of the glue dominance of thermal production of strangeness. Moreover, the peak in the gluon collision frequency P_g is more coincident with the peak in the cross section, as a function of $\sqrt{s}$, Fig. 17.3(b). This further amplifies the gluon dominance. This combined enhancement effect can be seen in the thermal differential production rates,

$$\frac{dR_f}{ds} = \sum_{i=q,g} \sigma_f^i(s) P_i(s), \qquad f = s, c, \qquad (17.35)$$

shown for thermal production of strangeness, in Fig. 17.7 for $T = 250$ MeV, and for thermal production of charm, in Fig. 17.8 for $T = 500$ MeV. Gluons (dashed lines) dominate quark-pair processes (solid lines), which are contributing at the level of 15%. The uncertainty in mass of the strange quark has significant impact; thin lines bordering thick lines show the effect of 20% variation in the value of the quark mass considered.

Since formation of charm occurs in the domain $T \ll m_c$, near to the threshold, where the cross section for fusion of a pair of quarks dominates gluon fusion, the gluon dominance of the production rate is not

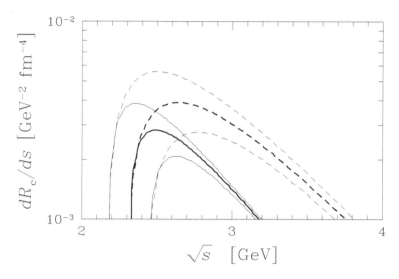

Fig. 17.8.   The differential thermal production rate for charm dR_c/ds, with T = 500 MeV, $\lambda_q = 1$ for gluons (dashed lines) and $q\bar{q} \to c\bar{c}$ (solid line, includes three interacting flavors), for the running $\alpha_s(M_Z) = 0.118$ and running mass of the charmed quark $m_c(M_Z) = 700$ MeV $\pm$ 7% (thin lines).

as pronounced as it is for strangeness. Only for $T \geq 500$ MeV does the glue fusion pick up strength and clearly dominate the thermal production of charm. Yet, even at $T = 500$ MeV, the rates for charm are a factor 100 smaller than those for strangeness at $T = 250$ MeV, and thermal production of charm is expected to be irrelevant at the RHIC.

The differential production rate can easily be integrated, and we show the results in Fig. 17.9 for strangeness, and in Fig. 17.10 for charm. In Fig. 17.9, we see that the early results (dotted line) [226] are found within the uncertainty in mass of the strange quark (a smaller mass leads to a bigger value of R). A yet greater value of R should result after the K-factor has been introduced, describing the next-to-leading-order effects. The rate of production of strangeness per unit volume and time is at the level of unity at temperatures reached at the RHIC, and production of strangeness is very abundant.

The thermal production of charm could be significant at the LHC relative to the direct first-parton-collision production, if temperatures well above $T = 500$ MeV are reached. We see, in Fig. 17.10, that the production rate for charm changes by six orders of magnitude as the temperature varies between 200 and 600 MeV. This sensitivity to the initial temperature is due to the exponential suppression with $m/T > 1$. In turn, this implies that the thermal production of charm can become important at sufficiently high temperature.

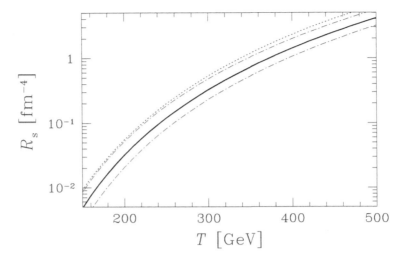

Fig. 17.9.   Thermal production rates for strangeness R_s in QGP (thick solid line), calculated for $\lambda_q = 1.5$, $\alpha_s(M_Z) = 0.118$, and $m_s(M_Z) = 90$ MeV, as a function of temperature. Chain lines show the effect of variation of the mass of the strange quark by 20%. The dotted line shows comparison results for fixed $\alpha_s = 0.6$ and $m_s = 200$ MeV.

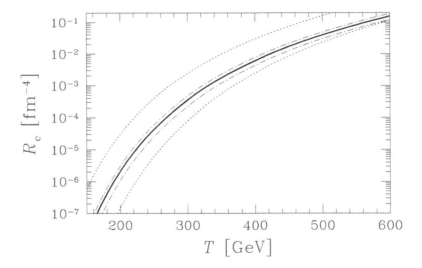

Fig. 17.10.   Thermal production rates for charm R_c in QGP (solid line), calculated for $\lambda_q = 1.05$, $\alpha_s(M_Z) = 0.118$, and $m_c(M_Z) = 0.7$ GeV, as a function of temperature. Chain lines show the effect of variation of mass of the charmed quark by 7%. Dotted lines are comparison results for fixed $\alpha_s = 0.35$ and $m_c = 0.9$ GeV (upper) and $\alpha_s = 0.4$ and $m_c = 1.5$ GeV (lower).

Insertion of the rates R_i into Eq. (17.30) allows us to obtain the time constants for chemical relaxation τ_s and τ_c. It should be stressed that, in an actual kinetic evaluation of the production of flavor, the relaxation time τ_f enters only when we relate the actual yield of flavor to the expected yield ρ_f^∞. Namely, the back reaction, $f\bar{f} \rightarrow gg$, $q\bar{q}$, is driven by the actual density of strangeness, whereas the forward rate, ignoring Pauli blocking, is not affected by the equilibrium yield at all. Which m_f is used in Eq. (17.30) to define ρ_f^∞ is physically irrelevant, as long as the same values of m_f and ρ_f^∞ are used both in the definition of τ_f, Eq. (17.30), and in $\gamma_f(t) = \rho_f/\rho_f^\infty$. In Fig. 17.11, we see τ_s evaluated with $m_s(1\text{GeV}) = 200$ MeV. The range of the assumed 20% uncertainty in $m_s(M_Z)$ is indicated by the hatched areas. The initial predictions obtained 20 years ago [226] at fixed values $\alpha_s = 0.6$ and $m_s = 200$ MeV (the dotted line in Fig. 17.11) are well within the band of values related to the uncertainty in mass of the strange quark. The approximate formula obtained in [226],

$$\tau_f = \frac{1.6}{\alpha_s^2 \gamma_g^2 T} \frac{m_f/T \, e^{m_f/T}}{[1 + (99/56)T/m_f + \cdots]}, \tag{17.36}$$

allows a quick estimate of the expected relaxation time in all the environments discussed in this subsection. We have added the pre-factor γ_g^{-2} relevant in case the dominant source of heavy flavor, gluons, is not in chemical equilibrium. We see that the equilibration time lengthens accordingly.

Thermal nonperturbative effects on the relaxation of strangeness were studied by introducing thermal, temperature-dependent, particle masses [70]. After the new production rates, including the now possible gluon decay, were added up, the total rate of production of strangeness was found to be little changed compared with the free-space rate. This finding was challenged [34], but a further reevaluation [66] confirmed that the rates obtained with perturbative glue-fusion processes are describing precisely the rates of production of strangeness in QGP, for the relevant temperature range $T > 200$ MeV. We can thus assume today that the 'prototype' strangeness-production processes seen in Fig. 17.2, re-summed using the renormalization-group method, are dominating the rates of production of strangeness in QGP.

The poor knowledge about the mass of the strange quark makes it possible that the actual relaxation time for strangeness is even smaller. In quenched-lattice calculations, see section 15.5, a much smaller value $m_s(M_Z) \lesssim 50$ MeV is found. The thin-dotted line in Fig. 17.11 gives the corresponding result for $m_s(M_Z) = 50$ MeV. We see that the relaxation time is already small enough to allow chemical equilibration at $T < 200$ MeV. Moreover, next-to-leading-order effects (the K-factor) should further reduce the chemical relaxation constant.

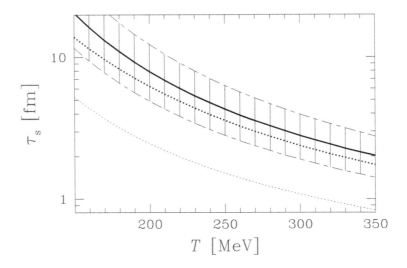

Fig. 17.11. The QGP chemical relaxation time for strangeness τ_s, for $\alpha_s(M_Z) = 0.118$ with $m_s(M_Z) = 90$ MeV and $\rho_s^{\infty}(m_s \simeq 200\,\text{MeV})$ (thick line). Hatched areas show the effect of variation of the mass of the strange quark by 20%. The fat-dotted line shows comparison results for fixed $\alpha_s = 0.6$ and $m_s = 200\,\text{MeV}$. The thin-dotted line shows the result for $m_s(M_Z) = 50$ MeV.

While precise evaluation of the production of strangeness at temperatures as low as $T_c \simeq 160$ MeV is not reliable within the scheme we have presented, it is highly probable that the combined effect of low m_s and the K-factor would ensure that near-chemical equilibrium for strangeness can develop as soon as the QGP phase can be formed. As a result, the strangeness energy excitation function, seen in Fig. 1.5 on page 17, can then be interpreted as due to the onset of deconfinement already in collisions below SPS energies. We see that, despite 20 years' work on strangeness, we still have many new, interesting insights to gain.

In Fig. 17.12, the chemical relaxation time for charm is shown in the extended interval through $T = 1000$ MeV. Since charm is considerably more massive than strangeness, there is less uncertainty in the extrapolation of the running QCD coupling constant. There is also less relative uncertainty in the value of the mass of the charmed quark, shown by the hatched area. We also see (dotted lines) the results for fixed m_c and α_s with parameters selected to border high- and low-T limits of the results presented. In the high-T limit, the choice (upper dotted line) $m_c = 1.5\,\text{GeV}$, $\alpha_s = 0.4$ is appropriate, whereas to follow the result at small T (lower dotted line), we take a much smaller mass $m_c = 0.9\,\text{GeV}$, with $\alpha_s = 0.35$.

The important result, see Fig. 17.12, is that, above $T = 700$ MeV, the relaxation time drops below to 10 fm; the curves flatten. At gluon

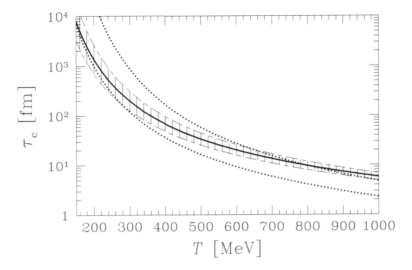

Fig. 17.12.   Solid lines show the thermal relaxation constant for charm in QGP, calculated for running $\alpha_s(M_Z) = 0.118$, $m_c(M_Z) = 0.7\,\text{GeV}$ and $\rho_c^\infty(m_c \simeq 1.5\,\text{GeV})$. Lower dotted line, for fixed $m_c = 0.9\,\text{GeV}$ and $\alpha_s = 0.35$; upper dotted line, for fixed $m_c = 1.5\,\text{GeV}$ and $\alpha_s = 0.4$. The hatched area shows the effect of variation $m_c(M_Z) = 0.7\,\text{GeV} \pm 7\%$.

collision energies of several GeV, it is quite natural to expect that the next-to-leading-order effects enhance the cross section for production of charm by a factor of two, and this reevaluated relaxation time would correspond to a true value of a few fermis only. At this juncture in time, it is quite impossible to be sure how important the thermal component is in the production of charm at the LHC. On the other hand, even the first parton collisions are expected to produce $200 \pm 50\%$ $c\bar{c}$ pairs, and thus, in one way or another, charm certainly will be the novel-physics frontier at LHC energies, replacing strangeness as the flavor signature of new physics.

17.5 Equilibration of strangeness at the RHIC and SPS

Given the relaxation constant τ_s, we evaluate the thermal yield of strangeness in the QGP which arises on integrating the kinetic equation Eq. (17.29). Since this requires as input initial conditions the temporal evolution of the fireball, results are somewhat model-dependent. Indeed, there is considerable difference of opinion among groups regarding the well-studied RHIC system [71, 221, 252, 253, 275], since the experimental data which would narrow down the models is only now being obtained. The most important issue on which the various groups differ is the directly (or indirectly) assumed gluon content. In this book, we assume rapid chemical equilibration of gluons, which is not reached in studies relying on kinetic evolution by implementing the lowest-order gg → ggg gluon fragmentation.

For the RHIC conditions, we present a qualitative model with a cylindrical longitudinal flow, and transverse expansion [221]. We assume that the transverse flow of matter occurs at the velocity of sound for relativistic matter $v_\perp \simeq c/\sqrt{3} = 0.58c$. For a purely longitudinal expansion, the local entropy density scales according to $S \propto T^3 \propto 1/\tau$; see Eqs. (6.35) and (7.22). The transverse flow of matter accelerates the drop in entropy density. To model this behavior without too great a numerical effort, considering the other uncertainties, the following temporal-evolution function of the temperature was proposed:

$$T(\tau) = T_0 \left(\frac{1}{(1 + 2\tau c/d)(1 + \tau v_\perp/R_\perp)^2} \right)^{1/3}. \tag{17.37}$$

Considering various values of T_0, the temperature at which the gluon equilibrium is reached, the longitudinal dimension is scaled according to

$$d(T_0) = (0.5\,\mathrm{GeV}/T_0)^3 \times 1.5\,\mathrm{fm}. \tag{17.38}$$

This adjustment of the initial volume V_0 assures that the different evolution cases refer to a fireball with a similar entropy content. The following results are thus a study of one and the same collision system, and the curves reflect the uncertainty associated with unknown initial conditions of a fireball of QGP with identical, (but large by current standards) entropy content.

The numerical integration of Eq. (17.29) is started at $\tau_0 = 1$ fm/c, the time at which thermal initial conditions are reached. A range of initial temperatures 300 MeV $\leq T_0 \leq$ 600 MeV, varying in steps of 50 MeV, is considered. Since the initial p–p collisions also produce strangeness, to estimate the initial abundance a common initial value $\gamma_s(T_0) = 0.15$ is used. For $T_0 = 0.5$ GeV, the thickness of the initial collision region is $d(T_0 = 0.5)/2 = 0.75$ fm. The initial transverse dimension in nearly central Au–Au collisions is taken to be $R_\perp = 4.5$ fm. The initial volume of QGP is 190 fm^3, which, at the temperature of $T_0 = 0.5$ GeV, implies, according to results seen in Fig. 16.7, a total entropy content of $S = 38\,000$. We divide this by the specific entropy content per hadron in the final state, $S/N = 4$; see section 10.6. We see that the primary final-state hadron multiplicity has implicitly been assumed to be 9500. This is somewhat above results seen even during the RHIC run at $\sqrt{s_N} = 200$ GeV, for which we estimate, for the 3% most central events, a total hadron multiplicity, after resonance cascading, of 8000.

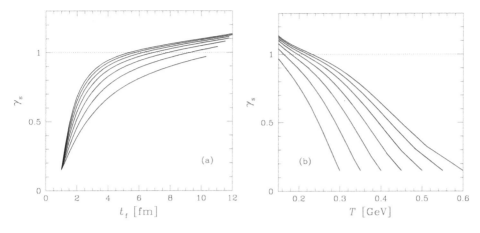

Fig. 17.13.   The evolution of QGP phase-space occupancy for strangeness γ_s: (a) as a function of time and (b) as a function of temperature for $m_s(1\,\mathrm{GeV}) = 200\,\mathrm{MeV}$; see the text for details.

The evolution with time in the plasma phase is followed up to the breakup of the QGP at a temperature $T_f^{\mathrm{RHIC}} \simeq 150 \pm 5$ MeV. The numerical solution of Eq. (17.29) for γ_s is shown as a function both of time t, in Fig. 17.13(a), and of temperature T, in Fig. 17.13(b). This evolution is physically meaningful until it reaches the QGP breakup condition. Since the results for higher temperatures are also displayed, the reader who prefers hadronization at $T = 170 \pm 5$ MeV can easily draw his own conclusions.

We see in Fig. 17.13 the following phenomena.

- A steep rise at early times, showing actual production of strangeness, which is followed by a dilution-driven increase of γ_s near the breakup temperature.
- Widely different initial conditions (but with similar initial entropy contents) lead to rather similar chemical conditions at chemical freeze-out of strangeness.
- Despite the use of a high mass of $m_s = 200$ MeV, we find that strangeness nearly equilibrates chemically, and that the dilution effect allows in certain cases a small over-population of the strange-quark phase space even in the strangeness-dense QGP.
- For a wide range of initial conditions and final freeze-out temperature a narrow band of final result is seen, $1.10 > \gamma_s(T_f) > 0.9$.

Since strangeness is more easily made in the 'hot' era in glue–glue interactions, we can estimate that, if the abundance of glue had been at the time 70% chemically equilibrated, then $\gamma_s \simeq 0.5$. This high sensitivity to the glue density is at the origin of the claim that measuring the yield of strangeness probes the presence and abundance of glue, which is a specific property of QGP.

In the model calculations presented, the fireball begins to expand in the transverse direction instantly at the full velocity. For this reason, the initial drop in temperature is very rapid. This defect also makes the transverse size at the end of the expansion too large, $R_\perp \simeq R_0 + t_f/\sqrt{3} \simeq 9$ fm, compared with the results of HBT analysis, Fig. 9.11. This can easily be fixed by introducing a more refined model of the transverse velocity, which needs time to build up. The yield of strangeness may slightly increase in such a refinement, since the fireball will spend more time near to the high initial temperature.

The RHIC results presented are typical for all collision systems. In the top SPS energy range, the initial temperature reached is certainly less (by 10%–20%) than that in the RHIC 130-GeV run, and the baryon number in the fireball is considerably greater; however, the latter difference matters little for production of strangeness, which is driven by gluons. A model similar to the above yields $\gamma_s^{\rm QGP} \simeq 0.6$–$0.7$, the upper index reminds us that in this section the strangeness occupancy factor γ_s refers to the property of the deconfined phase. The experimental observable directly related to $\gamma_s^{\rm QGP}$ is the total yield of strangeness per participating baryon. We will return to discuss the significance of these results in section 19.4.

18 The strangeness background

18.1 The suppression of strange hadrons

Since the matter around us does not contain valence strange quarks, all strange hadrons produced must contain newly made strange and antistrange quarks. If strangeness is to be used as a diagnostic tool for investigating QGP, we need to understand this background rate of production of strange hadrons. In that context, we are interested in measuring how often, compared with pairs of light quarks, strange quarks are made. One defines for this purpose the strangeness-suppression factor[†]

$$W_s = \frac{2\langle s\bar{s}\rangle}{\langle u\bar{u}\rangle + \langle d\bar{d}\rangle}. \tag{18.1}$$

In W_s, all newly made $s\bar{s}$, $u\bar{u}$, and $d\bar{d}$ quark pairs are counted.

If strangeness were to be as easily produced as light u and d quarks, we would find $W_s \to 1$. To obtain the experimental value for W_s, a careful study of produced hadron yields is required [277]. Results shown in

---

[†] We chose W_s in lieu of the usual symbol λ_s, which clashes with the strangeness fugacity.

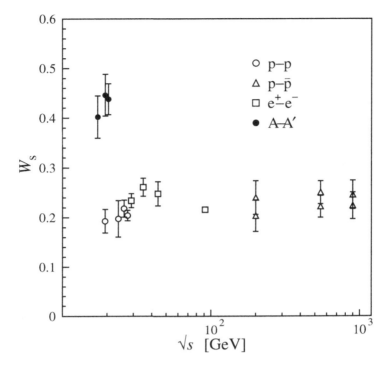

Fig. 18.1. W_s as a function of the center-of-mass energy $\sqrt{s}$ [GeV] (for heavy ion collisions the nucleon–nucleon center of mass energy $\sqrt{s_{NN}}$) [61].

Fig. 18.1 were obtained using a semi-theoretical method [61], in which numerous particle yields are described within the framework of a statistical model, and the computed hadron yields are analyzed in terms of the Wróblewski procedure [277].

We see in Fig. 18.1 that, in elementary p–p, p–p̄, and e^+–e^- collisions, the value $W_s \simeq 0.22$ is obtained. In order to estimate the possible influence of the annihilation process on p–p̄ reactions, two values of W_s are shown per energy, calculated with initial valence quarks and antiquarks (lower points) and without (upper points). For e^+–e^- interactions, the leading s quarks in $e^+e^- \to s\bar{s}$ have been subtracted [61]. Strangeness is thus relatively strongly suppressed. On the other hand, we also see that, in nuclear A–A′ collisions, W_s more than doubles compared with that in p–p interactions considered at the same energy.

To explain the two-fold increase in yield of strangeness a kinetic model of particle production requires a shift toward production of strangeness in all particle-formation processes. In other words, for modeling the enhanced yields of strangeness within a variety of approaches, see section 6.1, in each model a new reaction mechanism that favors production of strangeness must be introduced. Even at this relatively elementary level of counting abundances of hadrons, new physics is evident. In a model with the deconfined phase this new reaction mechanism is due to the presence of mobile gluons, which, as we have seen, are most effective at making pairs of strange quarks. Moreover, because the conditions created in the QGP become more extreme with increasing collision energy, e.g., the initial temperature exceeding substantially the mass of the strange quark, we expect an increase in W_s.

Several among the cascade models we listed in section 6.1 have been tuned to produce not only enough strangeness, but also the observed antinucleons and $\overline{\Lambda}$. However, once this has been done, these hadronic models predict wrong abundances of the rarely produced particles such as Ξ and $\overline{\Omega}$. We are not aware of any kinetic hadron model with or without 'new physics' that is capable of reproducing the pattern of production of rare hadrons, along with the enhancement in production of strangeness and hadron multiplicity. Moreover, if rapidity spectra are modeled, usually the transverse momentum spectra are incorrect, or vice-versa. It seems impossible, in a collision model based on confined hadron interactions, to find sufficiently many hadron–hadron collisions to occupy by hadrons the large phase space (high $p_\perp$, high y) filled by products of nuclear collision. If indeed a non-QGP reaction picture to explain heavy-ion-collision data exists, the current situation suggests that some essential reaction mechanism has been overlooked for 20 years. In short, a lot more effort needs to be expended on hadron models in order to reach satisfactory agreement with the experimental results, even regarding rather simple observables such as hadron multiplicities, transverse-energy production, and yield of strangeness.

Another way within the kinetic approach, and within the realm of quite conventional physics, to acquire an excess of strangeness in heavy-ion collisions compared with elementary reactions is to cook dense hadronic matter for a long time. In hadronic gas, strangeness can be produced in unusual (but not novel) reactions such as $\pi + \pi \rightarrow K + \overline{K}$. Even in the presence of an abundant yield of pions, the high-mass threshold, compared with the temperatures we are studying here, suggests that their reaction is 'slow'. Thus, an important piece in the puzzle is to know how long it takes for enhancement of strangeness to be created in a kinetic HG fireball. Otherwise, the fact that the final state in all reaction scenarios looks more or less the same will make the argument that the enhancement in production of strangeness is a signature of new physics difficult.

We therefore will consider, in the following section, the dynamics of production of strangeness in thermal processes in HG [164, 165], and we establish the time scales involved. Since the strange quarks are produced as constituents ('valence quarks') of the usual hadronic states, the direct mechanisms of production will not populate all hadronic states as the phase-space distribution would demand, and therefore, aside from the production, we will also encounter the (relatively rapid) process of stran- geness exchange (redistribution) in hadronic quark-exchange reactions. In general, the observed abundances of strange hadrons, and especially of the rarely produced ones, will be the result of multi-step processes.

18.2 Thermal hadronic strangeness production

In a thermally equilibrated central HG fireball the standard

$$N + N \rightarrow N + Y + K \tag{18.2}$$

nucleon-based reactions for production of strangeness are, surprisingly, not very important. The hyperon $Y = qqs$ may be either the iso-scalar $\Lambda = uds$ or the iso-triplet $\Sigma = (uus, uds, dds)$, and the kaon K may be either $K^+ = \bar{s}u$ or $K^0 = \bar{s}d$, which are found experimentally in one of the (almost) CP eigenstates K_S and K_L.

There are three reasons for the relative unimportance of the reaction Eq. (18.2):

1. The energy threshold, viz.,

$$\sqrt{s_{\text{th}}} = m_N + m_Y + m_K - 2m_N \sim 670 \, \text{MeV}, \tag{18.3}$$

is considerably higher than, e.g., the energy threshold for production of strangeness in reactions between a pion and a baryon:

$$\pi + N \rightarrow K + Y, \quad \sqrt{s_{\text{th}}} \sim 540 \, \text{MeV} . \tag{18.4}$$

2. Pions are the most abundant fireball particles, with the pion-to-baryon ratio of the central-rapidity region considerably exceeding unity in all central collisions at above 10 GeV per nucleon. It is for this reason that the reaction

$$\pi + \bar{\pi} \rightarrow K + \bar{K}, \quad \sqrt{s_{\text{th}}^{\pi\pi}} = 710 \, \text{MeV}, \tag{18.5}$$

between two pions, which has also a rather high threshold, is found to be important at temperature $T > 100$ MeV, as we shall discuss.

3. The final-state phase space of the two-particle reaction system above is more favorable than that of the three-body final state required in Eq. (18.2).

The common reaction feature of the hadronic production of strangeness is the $q\bar{q} \rightarrow s\bar{s}$ reaction, illustrated for the case of Eq. (18.4) in Fig. 18.2. Note that three of the five (light) quarks are spectators, a $q\bar{q}$ pair is annihilated, and an $s\bar{s}$ pair is formed. The experimental value of this

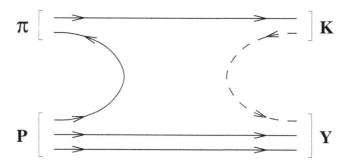

Fig. 18.2. The production of strangeness in reactions of the type $\pi + N \rightarrow K + Y$ in the HG phase. Solid lines indicate the flow of light quarks and the disappearance of one $\bar{q}q$ pair, the dashed line is for the added $\bar{s}s$ pair.

so-called Okubo–Zweig–Izuka-rule forbidden cross section is about 0.1–0.5 mb in the energy region of interest here (just above the threshold), thus of a magnitude comparable to the cross section for the QGP processes – but the threshold is considerably lower in the QGP processes, and most of the thermal collisions occur in this lower-energy region, as we have shown in Fig. 17.6.

Apart from production of strangeness, we also have 'strangeness-ex-change' reactions, as depicted in Fig. 18.3: we see that, in the process, the already existent strange quark can be moved from one particle 'carrier' to another, but the number of strange quarks remains unchanged. We show, in Fig. 18.3, the most relevant class of exchange reactions:

$$\bar{K} + N \rightarrow \pi + Y. \tag{18.6}$$

In an exchange reaction, new hadrons that are difficult to make in direct reactions can be produced. For example, the reaction

$$\Lambda + \bar{K} \rightarrow \pi + \Xi \tag{18.7}$$

produces the doubly strange particles $\Xi(ssq)$. More generally, exchange reactions distribute strangeness into all accessible particle 'carriers', and help establish the relative chemical equilibrium.

For every strangeness-reaction, there is the same type back-reaction, and one of the two has to be exothermic. Thus 50% of strangeness-exchange reactions are exothermic. Exothermic cross sections do not vanish at small collision energies, and the thermal average is actually nearly a constant $\mathcal{O}(1 \text{ mb})$, which is related to the geometric size and structure of hadrons. Consequently, thermally averaged reaction rates for exchange of strangeness are, especially at low temperatures, much greater than rates of production of strangeness. This assures that strange quarks

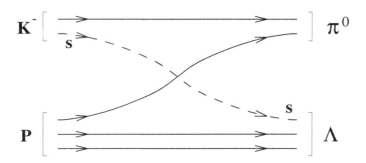

Fig. 18.3. An example of a strangeness-exchange reaction in the HG phase: $K^- + p \to \Lambda + \pi^0$. Solid lines, flow of u and d quarks; dashed line, exchange of an s quark between two hadrons.

produced in HG are rapidly distributed among many hadronic states, even if the absolute chemical equilibrium is not reached.

On the other hand, the cross section for production of strangeness in HG is relatively (to QGP) slow, being suppressed by smaller particle densities and the higher threshold of the production process. Some authors, pursuing explanation of the enhancement in production of strangeness within a HG scenario, try to overcome this in terms of partial restoration of chiral symmetry, which allows hadron masses to melt in part, lowering thresholds for production of strangeness; see [241] for the current status of the search for this effect. We will not pursue further in this book these interesting developments.

In the study of the evolution with time of the production of strange hadrons in the HG fireball, similarly to the case of QGP, only the total numbers of particles will be considered. We are assuming that the thermalization (kinetic equilibration) is a rapid process, in comparison with the relaxation time constant for strange hadrons, and also in comparison with the lifetime of the fireball. To quantitatively develop the kinetic evolution abundances of strange hadrons in the HG, we need to use a large number of hadronic cross sections.

Only limited experimental information on cross sections is available, since it is often impossible to study in the laboratory some of the processes which involve relatively short-lived particles (which of course are quite infinitely long-lived on the scale of hadronic collisions). There is no reliable theoretical framework allowing one to evaluate cross sections not accessible to direct measurement, since our understanding of the hadronic structure is incomplete. Consequently, we consider a method facilitating use of the accessible experimental information, in order to be able to estimate required reaction cross sections.

The cross section times the flux factor (velocity) for two-body reactions $1+2 \to n$, in which n particles are produced in the exit channel, takes the form Eq. (17.9). The generalization of the thermally averaged production cross section Eq. (17.16) to n final-state particles, can be cast into the form

$$R_n = \rho_1 \rho_2 \langle \sigma_{12} v_{12} \rangle = \frac{g_1 g_2}{(2\pi)^5} T \frac{|\mathcal{M}(1+2 \to n)|^2}{(2\pi)^{3n-3}}$$

$$\times \int_{s_{\mathrm{th}}}^{\infty} ds \sqrt{s} \, \mathrm{IMS}(s;2) \mathrm{IMS}(s;n) K_1(\sqrt{s}/T), \quad (18.8)$$

where the n-particle invariant phase space is a generalization of the two particle invariant phase space Eq. (17.14):

$$\mathrm{IMS}(s;n) = \int \prod_{i=1}^{n} d^4 p_i \, \delta_0(p_i^2 - m_i^2) \delta^4 \left(p_1 + p_2 - \sum_{i}^{n} p_i \right). \quad (18.9)$$

IMS is implicitly a function of the masses of the particles.

Many studies have shown that the hadronic-reaction matrix element $|\mathcal{M}(1 + 2 \to n)|^2$ for hadronic processes is insensitive to the relatively slow changes of $\sqrt{s}$ in the region of interest to us, and it can be assumed to be constant. For this reason, it is outside of the integral over s in Eq. (18.8). This is the crucial detail which permits us to relate many thermal cross sections to each other: Eq. (18.8) allows us to infer, from an example known experimentally, the thermal average of a family of cross sections. All we need do is adjust the applicable threshold and particle masses. Given the high range of temperatures we are considering, the isospin-breaking effects are relatively unimportant, and one does not distinguish between the u and d quarks. Consequently, one can study isospin-averaged cross sections. Appendix B of [164] gives a useful listing of relevant reactions.

We show now how the method works: the isospin-averaged, thermally averaged associate rate of production of strangeness, $\langle \sigma(\pi N \to K) v_{\pi N} \rangle$, can be determined from experimentally known cross sections. It allows us to infer the strength of other associate production processes,

$$\langle \sigma(\pi Y \to K\Xi) v_{\pi Y} \rangle = \langle \sigma(\pi N \to KY) v_{\pi N} \rangle P_1, \quad (18.10a)$$

$$P_1 = \frac{\langle \sigma(\pi Y \to K\Xi) v_{\pi Y} \rangle}{\langle \sigma(\pi N \to KY) v_{\pi N} \rangle}, \quad (18.10b)$$

where P_1 in Eq. (18.10b), with the assumption of constant $|\mathcal{M}|^2$, depends on T and μ_b only through its dependence on the particle phase space.

The thermal averages of cross sections falling under the categories 'strangeness exchange' and 'baryon annihilation' can also be dealt with

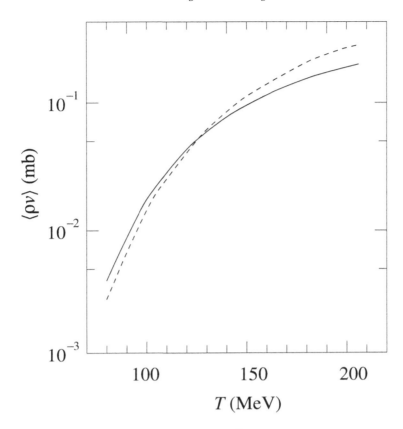

Fig. 18.4. The thermally averaged $\pi\pi \to K\bar{K}$ reaction cross section: the solid line is for a constant matrix element (see the text); the dashed line is for a constant value of the reaction cross section, $\sigma = 3$ mb [164].

in this way. The various strangeness-exchange cross sections are related to $\langle\sigma(\bar{K}N \to Y\pi)v_{\bar{K}N}\rangle$ and diverse baryon-annihilation cross sections to $\langle\sigma(p\bar{p} \to 5\pi)v_{p\bar{p}}\rangle$, which are known experimentally. Similarly, the matrix element $|\mathcal{M}|^2$ for reactions in which particles and antiparticles are interchanged is the same, and the average thermal cross sections of the reverse reactions are given by those of the forward reactions times appropriate phase-space factors.

This method leaves us with one important channel, Eq. (18.5), for production of strangeness in the thermally equilibrated hadronic gas, which can neither be measured nor be related to known reactions. Its strength has been derived from the reactions pp $\to$ pp$\pi\pi$, pp$K\bar{K}$ with the help of dispersion relations. Protopopescu *et al.* [210] have found that, above the threshold, pion-based production of strangeness has an approximately constant cross section of 3 ± 1 mb. The thermally averaged cross section

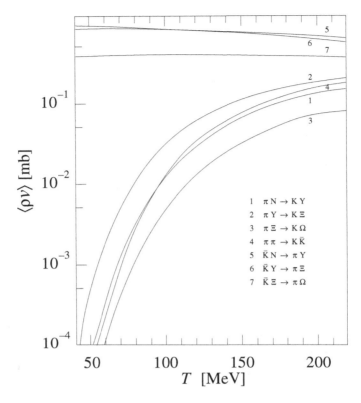

Fig. 18.5.　Thermally averaged cross sections for formation and exchange of strangeness in HG $\langle \sigma v \rangle /c$, based on the assumption of universal invariant matrix elements [164].

using as input a constant cross section of 3 mb is shown in Fig. 18.4 by the dashed line, as a function of temperature. In comparison, the solid line describes the result of an evaluation in which a constant matrix element $|\mathcal{M}|^2$ was used with the strength equal to the associate reaction for production of strangeness $\pi + \text{N} \rightarrow \text{K} + \text{Y}$.

In Fig. 18.5, we show the thermally averaged cross sections for formation and exchange of strangeness obtained by Koch *et al.* [164] within the framework outlined above. Note that the thermally averaged strangeness-exchange reactions which are capable of building up multiply strange hadrons are $\mathcal{O}(1 \text{ mb})$ at all temperatures. This is consistent with the intuition that this value should be at the level of a third of the geometric cross section, multiplied by the probability of exchanging one of the quarks, which is at the level of 10%, certainly not 100%, as is sometimes assumed by some authors. Even though (by coincidence) reactions 5 and 6 agree in strength, reaction 7, which is important for understanding the formation of Ω, is considerably smaller. This strongly breaks the $SU(3)$ flavor symmetry, and slows down the formation of the all-strange Ω and $\overline{\Omega}$ in HG, compared with the formation of other strange baryons and antibaryons.

We further see in Fig. 18.5 that strangeness-formation reactions in HG are dominated at high temperature by the $\pi\pi \to K\overline{K}$ reaction. Up to temperatures near $T = 150\,\mathrm{MeV}$, strangeness-exchange reactions 5 and 6 are more than an order of magnitude faster than is the production of strangeness. Thus, relative chemical equilibrium can be established 'instantaneously' in singly and doubly strange hadrons, but not for Ω and $\overline{\Omega}$, during the growth in abundance of strangeness in the fireball. For the same reason, it makes good sense to expect relative chemical equilibrium among strange hadrons to occur in an equilibrium model of hadronization of QGP, except for Ω and $\overline{\Omega}$.

18.3 The evolution of strangeness in hadronic phase

To address the issue of equilibration of strangeness in HG in a quantitative manner, we study the temporal evolution in the framework of a master equation. We consider all two-body reactions, which are predominant,

$$\frac{d\rho_i(t)}{dt} = \sum_{AB}\langle\sigma v_{AB\to i}\rangle_T\rho_A\rho_B - \sum_{C}\langle\sigma v_{iC\to X}\rangle_T\rho_i(t)\rho_C, \qquad (18.11)$$

where the collision of particles $A + B$ leads to production of the strange particle i, and i can be annihilated in collisions with (strange) particles C. Assuming the thermal momentum distribution of all particles Eq. (17.16) permits one to compute the evolution of particle populations using Eq. (18.11), once the thermal cross sections of the reaction are known; see section 18.2. Appendix A of [164] gives the explicit form of the master equation Eq. (18.11).

In HG antibaryons are difficult to produce, considering that the direct pair-production thermal cross section for $\pi + \pi \to N + \overline{N}$ is suppressed by a high threshold. Intermediate steps are necessary in order to build up heavy mesons, e.g., ρ and ω resonances, which are more capable of producing a baryon–antibaryon pair, and which also benefit from a larger elementary cross section. A recent kinetic-theory study of reactions involving fusion of three mesons into antibaryons suggests that this channel may be the dominant source of antibaryons [89].

In order to establish an upper limit for the abundance of multistrange baryons in HG, the results shown below were obtained assuming that the non-strange antibaryons are at the chemical HG equilibrium yield at the initial time, at which the evolution of the master equation is considered. In this approach, the build-up of strange antibaryons is a function of the developing abundance of strangeness, and the effectiveness of strangeness-exchange reactions, which as we have seen is particularly weak for the formation of the most enhanced $\overline{\Omega}$. Even allowing full occupancy of non-strange-antibaryon phase space, the relaxation time for yields of strange antibaryons in HG turns out to be much too long to be relevant in a dynamic heavy-ion-collision environment. In fact, the discovery of abundant yields of strange antimatter that sometimes exceed chemical equilibrium in heavy-ion collisions is proof that collective formation mechanisms akin to those operating in quark-soup hadronization have been operational.

Somewhat surprised by the high yield of (strange) antibaryons observed in heavy-ion collisions, some authors have proposed to use detailed balance considerations to infer an effective rate of five-hadron collisional production of a baryon–antibaryon pair, using as reference in detailed-balance analysis the reverse reaction $N + \overline{N} \rightarrow 5\pi$. In our opinion, use of detailed-balance arguments to infer a five-body collisional reaction rate is wrong physics. Certainly five-particle collisions occur more rarely than do four-particle collisions, which occur more rarely than do three-particle collisions, etc. One has first to establish the dominant forward *and* backward channels (if such exist, for equilibrium could be the result of multistep processes) when one is using the detailed-balance argument to estimate a rate. For the case of baryon–antibaryon formation and annihilation in HG, depending on temperature and density, two- and three-body reactions among mesons should dominate higher many-body processes.

We shall now review a few properties of the solutions of the master equations Eq. (18.11). To maximize the yield of strangeness that can be obtained, and to maximize the production of strange antibaryons, we assume that, after $\tau_0 \simeq 1$ fm, the densities both of pions and of antinucleons should have reached approximately their chemical-equilibrium values, and, at this point in time, the thermal processes for production of strangeness are turned on in HG. First, we turn our attention to the evolution of the total yield of strangeness in the HG phase assuming that the initial densities of strange particles are zero. Then, we consider the evolution of the yields of individual strange hadrons, including strange antibaryons.

We study the total strangeness in HG, to determine the time required for the total abundance of strangeness to reach the chemical equilibrium. In Fig. 18.6, the evolution of the total abundance of strangeness is shown, and we see that the production of strangeness in HG is roughly 100 times slower than the time for chemical relaxation in QGP, see Fig. 17.11, for the example of temperature $T = 160$ MeV. The considered range of the baryo-chemical potential $\mu_b \in [0, 450]$ MeV covers all regions of interest and, as we see, has a negligible impact on this result.

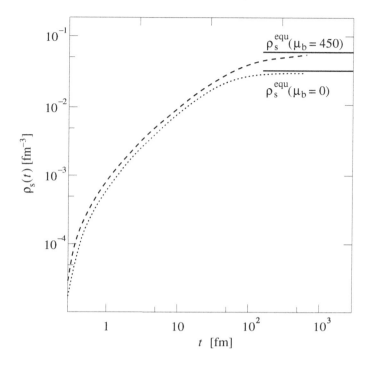

Fig. 18.6.   The thermal production of strangeness as a function of time in a confined hadron gas at $T = 160$ MeV. Results for two values of baryo-chemical potential ($\mu_b = 0$ and 450 MeV) are shown [164].

The yield of strangeness observed in experiments at the SPS and RHIC can not be produced by thermal HG processes. This becomes even clearer on considering individual particles with higher strangeness content such as (multi)strange antibaryons, shown in Fig. 18.7. For reasons we discussed above, strange antibaryons remain more distant from the equilibrium distribution than do strange mesons and strange baryons. In fact, to arrive within the time scale shown at a measurable yield, the result presented benefits from the rather optimistic assumptions of equilibrium abundances for the non-strange antibaryons. As expected, we see in Fig. 18.7 that kaons, $K = \bar{s}q$, antikaons, $\bar{K} = s\bar{q}$, and hyperons, $Y = qqs$, are the first to reach equilibrium abundance – the other $\bar{s}$ carriers are delayed by another factor of 3–5 in getting to their HG limits, with $\overline{\Omega} = \bar{s}\bar{s}\bar{s}$ trailing far behind. We expect that the HG-based production of multistrange antibaryons generates negligible background. This is one of the important reasons allowing multistrange antibaryons to signal the formation and hadronization of a deconfined quark–gluon phase in heavy-ion collisions.

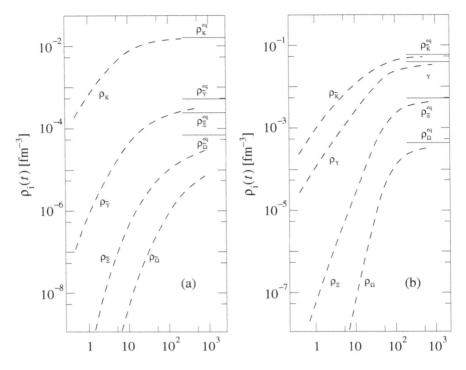

Fig. 18.7. The approach to chemical equilibrium by strange hadrons in hot hadronic matter at temperature $T = 160$, $\mu_b = 450\,\mathrm{MeV}$ – $\bar{\mathrm{s}}$ hadrons are shown in (a), whereas s hadrons are shown in (b) [164].

19 Hadron-freeze-out analysis

19.1 Chemical nonequilibrium in hadronization

In the final state, we invariably see many hadronic particles, and naturally we observe their spectra and yields only in restricted domains of phase space. An extra reaction step of 'hadronization' is required in order to connect the properties of the fireball of deconfined quark–gluon matter, and the experimental apparatus. In this process, the quark and gluon content of the fireball is transferred into ultimately free-flowing hadronic particles. In hadronization, gluons fragment into quarks, and quarks coalesce into hadrons.

Hadronization of course occurs in all reactions in which final-state hadrons are observed: for example, in high-energy $e^+e^- \rightarrow q\bar{q}$ reactions, we see jets of final-state hadrons carrying the energy and momentum of the two quarks produced. It is not yet clear whether there is a fundamental difference between the hadronization of a thermal fireball and that of a fast-moving quark. We refer to the process in which a thermal fireball of quarks and gluons turns into hadrons as statistical hadronization. This is an area of nonperturbative strong-interaction physics in rapid development, and this interesting topic could nearly fill a review of the size of this book.

The question we address is that of whether chemical-equilibrium hadron phase space (the so-called 'hadronic gas') can be used consistently to describe the physics of thermal hadronization. We first note that the production of entropy is small, or even null, in hadronization of thermal QGP. Color 'freezes', and the excess entropy of QGP has to find a way to get away, so any additional production is hindered. We consider the Gibbs–Duham relation for a unit volume, Eq. (10.26), and combine it with the instability condition of dynamic expansion, Eq. (3.31), cast in the form

$$0 = P|_h + (P|_h + \epsilon|_h)\frac{\kappa v_c^2}{1 - v_c^2}. \tag{19.1}$$

The result is

$$\frac{\epsilon}{\sigma}\bigg|_h = \left(T|_h + \frac{\mu_b \nu_b}{\sigma}\bigg|_h\right)\left(1 + \frac{\kappa v_c^2}{1 - v_c^2}\right). \tag{19.2}$$

Using the energy, E, the entropy S, and the baryon number b as variables, we obtain [222]

$$\frac{E}{S}\bigg|_h = (T|_h + \delta T|_h)\left(1 + \frac{\kappa v_c^2}{1 - v_c^2}\right), \quad \delta T = \mu_b\frac{\nu_b}{\sigma} = \frac{\mu_b}{S/b}. \tag{19.3}$$

For the RHIC, we have $\delta T|_h < 0.4$ MeV, considering that $\mu_b < 40$ MeV and $S/b > 100$, whereas at the top SPS energy we have $\mu_b \simeq 200$–250 MeV and $S/b \simeq 25$–45, and thus $\delta T|_h \simeq 5$–8 MeV. Eq. (19.3) shows that

$$\frac{E}{S}\bigg|_h > T|_h \tag{19.4}$$

when super-cooling occurs prior to hadronization. Since all three quantities, $E|_h$, $S|_h$, $T|_h$ can be obtained from particle abundances, this condition can be verified. More generally, we can transcribe the Gibbs–Duham relation, Eq. (10.26), to obtain the relation

$$\frac{E}{S} + \frac{PV}{S} = T + \delta T > T. \tag{19.5}$$

Solving for P/ϵ, we obtain

$$\boxed{\frac{P}{\epsilon} = \frac{T + \delta T - E/S}{E/S}.} \tag{19.6}$$

The study of hadron production, which will follow, allows an evaluation of both the freeze-out temperature and the energy per entropy E/S. The Gibbs–Duham relation in the form Eq. (19.6) determines a key property of the equation of state, the ratio of the pressure to the energy density. We expect to find P/ϵ negative and small in magnitude, in nonequilibrium-sudden hadronization. In an equilibrium transformation, the range $\frac{1}{3} < P/\epsilon < \frac{1}{7}$, see Fig. 11.3 on page 211, is expected, which spans the domain of highly relativistic matter and realistic hadronic gas.

Chemical nonequilibrium is naturally connected to the sudden (statistical) hadronization, which is the favored reaction mechanism in view of the results seen at the RHIC, and also has been applied successfully to explain the results obtained at the top energy at the SPS. The reader should keep in mind that, as the collision energy is reduced, the transverse expansion of the fireball of dense matter diminishes, and, at a sufficiently low collision energy, a more adiabatic hadronization has to occur. In this 'AGS limit', some of the signatures of the QGP phase we are discussing in this chapter may be erased.

The word sudden refers to the time hadrons have available, following their formation, to rescatter from other hadrons. The observed high abundances of short-lived hadronic resonances such as K*(892) [279], which has a natural half life of $\tau_{K^*} \simeq \ln 2/\Gamma_{K^*} = 2.8$ fm, implies that the decay occurs mostly outside of the hadronic environment. Had the decay products undergo rescattering, reconstruction of the K*(892) would in most cases be impossible. Conversely, reconstruction of abundant K*(892) implies that the two decay products (π and K) did not propagate through dense hadron matter [224, 260]. The HBT results, section 9.3, also place a very severe constraint on the size and life span of the pion source. The source is smaller than one would expect if the prolonged expanding HG phase were to exist. Finally, the hadron spectra are described in terms of a source breaking up at $T = 165 \pm 7$ MeV and expanding with $v_\perp = 0.52c$ [84], with chemical and thermal freeze-outs coinciding.

It has been proposed that a mechanical instability associated with super-cooling of QGP is at the origin of the sudden-breakup mechanism. In section 3.5, we have seen that the motion of the quark–gluon fluid adds to the pressure exercised against the vacuum. Equation (3.31) describes the balance condition when the dynamic expansion has run out of 'speed'. The normal stable case is that $v_{\rm c} = 0$. However, if the initial condition generated by the great collisional compression produces a fast expansion with a finite velocity $v_{\rm c} \neq 0$ when the condition Eq. (3.31) is satisfied, this means that the outward flowing QGP matter is at a pressure $P = P_{\rm p} - \mathcal{B} < 0$, a highly unstable situation, in the (locally at rest) frame of reference. The fireball matter reaching this super-cooled condition breaks up into smaller clusters. These drops of QGP separate and hadronize into free-streaming hadrons. A HG phase is never formed. However, production of particles occurs as dictated by the phase space available. Their yields are controlled by abundances of quarks in the hadronizing phase.

19.2 Phase space and parameters

Before proceeding with this section, the reader should refresh his memory about the role chemical potentials play in counting hadrons (sections 11.2 and 11.4). The approach we present is in its spirit a generalization of Fermi's statistical model of hadron production [121, 122], in that the yield of hadrons is solely dictated by the study of the magnitude of the phase space available.

The relative number of final-state hadronic particles freezing out from, e.g., a thermal quark–gluon source, is obtained by noting that the fugacity f_i of the ith emitted composite hadronic particle containing k components is derived from fugacities λ_k and phase-space occupancies γ_k:

$$N_i \propto e^{-E_i/T_f} f_i = e^{-E_i/T_f} \prod_{k \in i} \gamma_k \lambda_k. \tag{19.7}$$

In most cases, we study chemical properties of the light quarks u and d jointly, though, on occasion, we will introduce the isospin asymmetry, Eq. (11.11). As seen in Eq. (19.7), we study particle production in terms of five statistical parameters, $T, \lambda_q, \lambda_s, \gamma_q$, and γ_s. In addition, to describe the shape of spectra, one needs matter-flow-velocity parameters; these become irrelevant when only total abundances of particles are studied; these are obtained by integrating over all of phase space, or equivalently in the presence of strong longitudinal flow, when we are looking at a yield per unit rapidity.

Assuming a QGP source, several of the statistical parameters have natural values.

1. λ_s. The fugacity of the strange quark λ_s can be obtained from the requirement that strangeness balances, $\langle n_s \rangle - \langle n_{\bar{s}} \rangle = 0$, which, for a source in which all s and $\bar{s}$ quarks are unbound and have symmetric phase space, Eq. (11.13), implies $\lambda_s = 1$. However, the Coulomb distortion of the strange-quark phase space plays an important role in the understanding of this constraint for Pb–Pb collisions, leading to the Coulomb-deformed value $\lambda_s \simeq 1.1$; see section 11.3.

2. γ_s. The strange-quark phase-space occupancy γ_s can be computed, section 17.5, within the framework of kinetic theory and $\gamma_s \simeq 1$. Recall that the difference between the two different types of chemical parameters λ_i and γ_i is that the phase-space-occupancy factor γ_i regulates

the number of pairs of flavor 'i', and hence applies in the same manner to particles and antiparticles, whereas the fugacity λ_i applies only to particles, while λ_i^{-1} is the antiparticle fugacity.

3. λ_q. The light-quark fugacity λ_q, or, equivalently, the baryo-chemical potential, Eq. (11.2a), regulates the baryon density of the fireball and hadron freeze-out. This density can vary depending on the energy and size of colliding nuclei, and the value of λ_q is not easily predicted. However, it turns out that this is the most precisely measurable parameter, with everybody obtaining the same model-independent answer, for it directly enters all highly abundant hadrons. Since T differs depending on the strategy of analysis, the value of μ_b is not so well determined.

4. γ_q. The equilibrium phase-space occupancy of light quarks γ_q is expected to significantly exceed unity in order to accommodate the excess entropy content in the plasma phase. There is an upper limit, Eq. (7.20). We addressed this in section 7.5.

5. T_f. The freeze-out temperature T_f is expected to be within 10% of the Hagedorn temperature $T_H \simeq 160\,\mathrm{MeV}$, which characterized the production of particles in proton–proton reactions; see chapter 12.

6. v_c. The collective-expansion velocity v_c is expected to remain near to the relativistic velocity of sound,

$$v_c \leq 1/\sqrt{3},$$

the natural speed of flow of information in the QGP phase. There is a longitudinal velocity, which is needed in order to describe rapidity spectra, section 8.3, and there is a motion of the hadronization surface, aside from many further parameters one may wish to use to model the velocity profile of flowing matter.

The resulting yields of final-state hadronic particles are most conveniently characterized by taking the Laplace transform of the accessible phase space. This approach generates a function that, in terms of its mathematical properties, is identical to the partition function. For example, for the open-strangeness sector, we find (with no flow)

$$\mathcal{L}\left(e^{-E_i/T_f} \prod_{k \in i} \gamma_k \lambda_k \right) \propto \ln \mathcal{Z}_s^{\mathrm{HG}}, \qquad (19.8)$$

with $\ln \mathcal{Z}_s^{\mathrm{HG}}$ given in Eq. (11.19).

It is important to keep in mind that

- Eq. (19.8) does not require formation of a phase comprising a gas of hadrons, but is not inconsistent with such a step in the evolution of the matter; Eq. (19.8) describes not a partition function, but just a look-alike object arising from the Laplace transform of the accessible phase space;

- the final abundances of particles measured in an experiment are obtained after all unstable hadronic resonances 'j' have been allowed to disintegrate, contributing to the yields of stable hadrons; and

- in some experimental data, it is important to distinguish between the two light-quark flavors, for example experiments are sensitive only to Ξ^-, not to Ξ^0, and averaging over isospin does not occur.

The unnormalized particle multiplicities arising are obtained by differentiating Eq. (19.8) with respect to the particle fugacity. The relative particle yields are simply given by ratios of corresponding chemical factors, weighted by the size of the momentum phase space accepted by the experiment. The ratios of strange antibaryons to strange baryons *of the same type of particles* are, in our approach, simple functions of the quark fugacities; see Eqs. (11.21a)–(11.21e). When particles of unequal mass are considered, and are fed by decay of other hadrons, considerable numerical effort is required to evaluate yield ratios for particles, in particular, if these are available in a fraction of phase space only. To the best of our knowledge, the numerical results of various groups working with the statistical-hadronization method are consistent, though the physics content can vary widely, depending on assumptions introduced.

19.3 SPS hadron yields

We have argued that, in general, we must expect $\gamma_q \neq 1$, section 19.1, i.e., chemical nonequilibrium at hadron freeze-out is an expected ingredient in a precise interpretation of the experimental results on particle ratios obtained in the SPS energy range. For strangeness, it has been expected and was seen early on in experimental data [216]. Full chemical nonequilibrium was first noted in the study of the S–Au/W/Pb collisions at $200A$ GeV [176]. On fitting the yields of hadrons observed, it was noted that the statistical significance increased when chemical nonequilibrium was introduced. The statistical significance is derived from the total statistical error:

$$\chi^2 \equiv \frac{\sum_j (R_{\mathrm{th}}^j - R_{\mathrm{exp}}^j)^2}{(\Delta R_{\mathrm{exp}}^j)^2}. \tag{19.9}$$

It is common to normalize χ^2 by the difference between the number of data points and parameters used, the so-called 'dof' (degrees-of-freedom) quantity.

Table 19.1.   Statistical parameters obtained from fits of data for S–Au/W/Pb collisions at 200A GeV, without enforcing conservation of strangeness [176]

T_f [MeV]	λ_q	λ_s	γ_s	γ_q	$\chi^2/$dof
145 ± 3	1.52 ± 0.02	1^*	1^*	1^*	17
144 ± 2	1.52 ± 0.02	0.97 ± 0.02	1^*	1^*	18
147 ± 2	1.48 ± 0.02	1.01 ± 0.02	0.62 ± 0.02	1^*	2.4
144 ± 3	1.49 ± 0.02	1.00 ± 0.02	0.73 ± 0.02	1.22 ± 0.06	0.90

* denotes fixed (input) values

We show the resulting statistical parameters obtained in hadron-yield fits in table 19.1, for S–Au/W/Pb collisions at 200A GeV. Asterisks (*) mark fixed (input) values, thus the first column assumes not only chemical equilibrium, but also the symmetric value $\lambda_s = 1$ for the QGP phase space. Interestingly, little is gained by allowing λ_s to vary, and all different fitting strategies point to $\lambda_s = 1$. However, allowing for strange $\gamma_s \neq 1$ and then light-quark $\gamma_q \neq 1$, nonequilibrium brings the result of the fit progressively to statistical significance. For systems we study, with a few degrees of freedom (typically 5–15), a statistically significant fit requires that $\chi^2/$dof < 1. For just a few 'dof', the error should be as small as $\chi^2/$dof < 0.5. The usual requirement $\chi^2 \to 1$ applies only for infinitely large 'dof'. We learn from these results that the chemical nonequilibrium factor γ_i for both strange and light quarks is a required ingredient in the statistical hadronization model.

Turning now to the Pb–Pb system at collision energy 158A GeV, we consider the particles listed in the top section of table 19.2 from the experiment WA97, for $p_\perp > 0.7\,$GeV, within a narrow central-rapidity window $\Delta y = 0.5$. Further below are shown results from the large-acceptance experiment NA49, extrapolated by the collaboration to full 4π coverage of phase space. The total error χ^2 for the two columns of results is shown at the bottom of this table along with the number of data points 'N' and parameters 'p' used, and the number of (algebraic) redundancies 'r' connecting the experimental results. For $r \neq 0$, it is more appropriate to quote the total χ^2, since the condition for statistical relevance is more difficult to establish given the constraints, but since $\chi^2/(N-p-r) < 0.5$, we are certain to have a valid description of hadron multiplicities. We will return to discuss the yields of Ω and $\overline{\Omega}$ at the end of this section.

In table 19.2 second from last column, the superscript 's' means that λ_s is fixed by strangeness balance. The superscript 'γ_q', in the two last columns means that $\gamma_q = \gamma_q^c = e^{m_\pi/(2T_f)}$ is fixed in such a way as to maximize the entropy content in the hadronic phase space. The fits presented were obtained with the updated NA49 experimental results, i.e., they include the updated h$^-/b$, newly published yield of ϕ [43], and pre-

Table 19.2.   *WA97 (top) and NA49 (bottom) Pb–Pb 158A-GeV-collision hadron ratios compared with phase-space fits*

Ratios	Reference	Experimental data	$Pb|^{s,\gamma_q}$	$Pb|^{\gamma_q}$
Ξ/Λ	[171]	0.099 ± 0.008	0.096	0.095
$\overline{\Xi}/\overline{\Lambda}$	[171]	0.203 ± 0.024	0.197	0.199
$\overline{\Lambda}/\Lambda$	[171]	0.124 ± 0.013	0.123	0.122
$\overline{\Xi}/\Xi$	[171]	0.255 ± 0.025	0.251	0.255
K^+/K^-	[79]	1.800 ± 0.100	1.746	1.771
K^-/π^-	[248]	0.082 ± 0.012	0.082	0.080
K_s^0/b	[152]	0.183 ± 0.027	0.192	0.195
h^-/b	[43]	1.970 ± 0.100	1.786	1.818
ϕ/K^-	[21]	0.145 ± 0.024	0.164	0.163
$\overline{\Lambda}/\overline{p}$	$y=0$		0.565	0.568
$\overline{p}/\pi^-$	All y		0.017	0.016
χ^2			1.6	1.15
$N;\ p;\ r$			9; 4; 1	9; 5; 1

dict the $\overline{\Lambda}/\overline{p}$ ratio. b is here the number of baryon participants, and $h^- = \pi^- + K^- + \overline{p}$ is the yield of stable negative hadrons, which includes pions, kaons, and antiprotons. We see, on comparing the two columns, that conservation of strangeness (which is enforced in the second from last column) is consistent with the experimental data shown; enforcing it does not change much the results for particle multiplicities.

The six parameters ($T, v_c, \lambda_q, \lambda_s, \gamma_q$ and γ_s) describing the abundances of particles are shown in the top section of table 19.3. Since the results of the WA97 experiment do not cover the full phase space, one finds a reasonably precise value for one velocity parameter, taken to be the spherical surface-flow velocity v_c of the fireball hadron source.

As in S-induced reactions in which $\lambda_s = 1$ [176], so also in Pb-induced reactions, a value $\lambda_s^{Pb} \simeq 1.1$ characteristic of a source of freely movable strange quarks with balancing strangeness, i.e., with $\tilde{\lambda}_s = 1$, is obtained; see Eq. (11.17). Since all chemical-nonequilibrium studies of the Pb–Pb system converge to the case of maximum entropy, see Fig. 7.7 on page 125, we have presented the results with fixed $\gamma_q = \gamma_q^c = e^{m_\pi/(2T_f)}$. The large values of $\gamma_q > 1$ seen in table 19.3 confirm the need to hadronize the excess entropy of the QGP that possibly is formed. This value is derived both from the specific abundance of negative hadrons h^-/b and from the relative yields of strange hadrons.

The results shown in table 19.3 allow us to evaluate P/ϵ, section 19.1. Using Eq. (19.6) with $\delta T = 8$ MeV (see Eq. (19.3)), $T = 148$ MeV, and

Table 19.3.   Upper section; the statistical model parameters which best describe the experimental results for Pb–Pb data seen in table 19.2; bottom section, energy per entropy, antistrangeness, and net strangeness of the full hadron phase space characterized by these statistical parameters. In column two, we fix λ_s by the requirement of conservation of strangeness

	$\mathrm{Pb}\big|_v^{\mathrm{s},\gamma_q}$	$\mathrm{Pb}\big|_v^{\gamma_q}$
T [MeV]	151 ± 3	147.7 ± 5.6
v_c	0.55 ± 0.05	0.52 ± 0.29
λ_q	1.617 ± 0.028	1.624 ± 0.029
λ_s	1.10^*	1.094 ± 0.02
γ_q	$\gamma_q^{c*} = e^{m_\pi/(2T_{\mathrm{f}})} = 1.6$	$\gamma_q^{c*} = e^{m_\pi/(2T_{\mathrm{f}})} = 1.6$
γ_s/γ_q	1.00 ± 0.06	1.00 ± 0.06
E/b [GeV]	4.0	4.1
s/b	0.70 ± 0.05	0.71 ± 0.05
E/S [MeV]	163 ± 1	160 ± 1
$(\bar{s} - s)/b$	0^*	0.04 ± 0.05

* indicates values resulting from constraints.

$E/S = 160$ MeV, we obtain $P/\epsilon \sim -1/40$. This negative and small value is consistent with the super-cooling hypothesis. One other interesting quantitative results of this analysis is shown in the bottom section of table 19.3: the yield of strangeness per baryon, $s/b \simeq 0.7$. We see, in the lower portion of table 19.3, that near strangeness balance is obtained without constraint.

The most rarely produced hadrons are the triply strange $\Omega(sss)$ and $\overline{\Omega}(\bar{s}\bar{s}\bar{s})$, which are the heaviest stable hadrons, $M_\Omega = 1672$ MeV. The phase space for $\overline{\Omega}$ is ten times smaller than that for $\overline{\Xi}$ under the conditions of chemical freeze-out we have obtained, and any contribution from non-statistical hadronization would be visible first in the pattern of production of Ω and $\overline{\Omega}$. For the parameters in table 19.3, the yields of $\overline{\Omega}$ reported for the experiment WA97 are underpredicted by nearly a factor of two. This excess yield originates at the lowest $m_\perp$, as can be seen in Fig. 8.11 on page 155. The 'failure' of a statistical-hadronization model to describe yields of $\overline{\Omega}$ (and, by 30%, Ω) has several possible explanations.

One is the possibility that an enhancement in production of Ω and $\overline{\Omega}$ is caused by pre-clustering of strangeness in the deconfined phase [215]. This would enhance the production of all multistrange hadrons, but most prominently the phase-space-suppressed yields of Ω and $\overline{\Omega}$. This mechanism would work only if pairing of strange quarks near to the phase transition were significant. Current models of 'color superconductivity' support such a clustering mechanism [30, 185, 206, 237]. We have men-

tioned, at the end of section 11.5, the possibility that distillation of strangeness followed by breakup of strangelets could contribute to production
of Ω and $\overline{\Omega}$. The decay of disoriented chiral condensates has also been
proposed as a source of soft Ω and $\overline{\Omega}$ [158].

A more conventional explanation for the excess production of Ω and $\overline{\Omega}$
is obtained by noting that, due to the low reaction cross section, Ω and $\overline{\Omega}$
could decouple from the HG background somewhat sooner than do all the
other hadrons [55, 263]. To augment the yields by factor k, it is sufficient
to take an incrementally δT_Ω higher freeze-out temperature, determined
from studying the Ω phase space:

$$\delta T_\Omega \simeq T \frac{\ln k}{M_\Omega / T}. \tag{19.10}$$

In order to increase the yield by a factor of two, the freeze-out of Ω
would need to occur at $T_\Omega = 160\,\text{MeV}$ rather than at $T = 150\,\text{MeV}$.
Since the temperature drops as the expansion of the fireball develops, a
higher freeze-out temperature means freeze-out occurring slightly earlier
in time. Even though the required staging in time of hadron production
is apparently small, a consistent picture requires fine-tuning and it seems
unnatural, considering that yields of all the other particles are perfectly
consistent with just one sudden chemical-freeze-out condition.

In view of these pre- and post-dictions of the anomalous yield of Ω and
$\overline{\Omega}$, one should abstain from introducing these particles into statistical-
hadronization-model fits. We note that the early statistical descriptions
of yields of Ω and $\overline{\Omega}$ have not been sensitive to the problems we described
[61, 180]. In fact, as long as the parameter γ_q is not considered, it is not
possible to describe the experimental data at the level of precision that
would allow recognition of the excess yield of Ω and $\overline{\Omega}$ within statistical hadronization model. For example, a chemical-equilibrium fit, which
includes the yield of Ω and $\overline{\Omega}$ has for 18 fitted data points with two parameters a $\chi^2/\text{dof} = 37.8/16$ [80]. Such a fit is quite unlikely to contain
all the physics even if its appearance to the untrained eye suggests that
a very good description of experimental data as been achieved.

19.4 Strangeness as a signature of deconfinement

We have found that the rate of production of strangeness in QGP is
sensitive to the temperature achieved at the time gluons reach chemical and thermal equilibrium. There is considerable uncertainty regarding
how short the time required to relax the strangeness flavor in the thermal
gluon medium is, with the upper limit being the hatched area in Fig. 17.11.
Consideration of the small mass for strangeness found in lattice studies
of strange hadrons has yielded much smaller τ_s (the thin-dotted line in

Fig. 17.11). There is also the probable further reduction due to the next-to-leading-order effects (the K-factor). In view of this, we now establish a benchmark yield of strangeness, assuming that the equilibration process leads to near-chemical-equilibrium conditions for hadronizing QGP. Specifically, the abundance of light quarks in the QGP phase may be considered near to the equilibrium yield, $\gamma_q^{QGP} \to 1$, whereas the yield of strange quarks characterized by the QGP phase-space occupancy before hadronization, γ_s^{QGP}, may differ appreciably from equilibrium.

We consider the ratio of the equilibrium density of strangeness, arising in the Boltzmann-gas limit, Eq. (10.51), to the baryon density in a fireball of QGP given in Eq. (10.75):

$$\boxed{\frac{\rho_s}{\rho_b} = \frac{s}{b} = \frac{s}{q/3} = \frac{\gamma_s^{QGP}}{\gamma_q^{QGP}} \frac{(3/\pi^2)T^3 W(m_s/T)}{\frac{2}{3}\left(\mu_q T^2 + \mu_q^3/\pi^2\right)}.}$$

(19.11)

W is as defined in Eq. (10.50a) and shown in Fig. 10.1. The equilibrium strange-quark density Eq. (10.51), with $g_s = 6$, is the first term in the expansion Eq. (10.63). Higher-order quantum-statistics correction terms are negligible, given $m_s/T = \mathcal{O}(1)$. To a first approximation, perturbative thermal QCD corrections, see Eq. (16.24), cancel out in the ratio. For $m_s = 200$ MeV and $T = 150$ MeV, we have

$$\frac{s}{b} \simeq \frac{\gamma_s^{QGP}}{\gamma_q^{QGP}} \frac{0.7}{\ln \lambda_q + (\ln \lambda_q)^3/\pi^2}.$$

(19.12)

The relative yield s/b is mainly dependent on the value of λ_q. In the approximation considered, it is nearly temperature-independent. This light-quark fugacity pertinent to the final-state hadrons is well determined and does not vary depending on the strategy used for analysis of data.

We show in Fig. 19.1, as a function of $\lambda_q - 1$ (the variable chosen to enlarge the interesting region $\lambda_q \to 1$), the expected relative yield per baryon originating from the QGP, defined in Eq. (19.12) with $\gamma_s^{QGP} = \gamma_q^{QGP} = 1$. At the top SPS energy, we see that the equilibrium yield is 1.5 pairs of strange quarks per participating baryon (for $\lambda_q \simeq 1.5$–1.6). Considering the experimental yield in table 19.3, $\gamma_s^{QGP} \sim 0.5$. The explanation of this is that, if formed, the QGP system did not become hot enough for long enough. In p–p collisions at the corresponding energy, the yield is below 0.3 pairs of strange quarks per participant [277], which is 40% of the Pb–Pb yield.

For the RHIC 130-GeV run, the value $\lambda_q = 1.09$ allows one to understand many particle yields at central rapidity. We see, in Fig. 19.1, that the specific yield of strangeness in a fireball of QGP at equilibrium is an order of magnitude greater than that currently observed at the SPS top

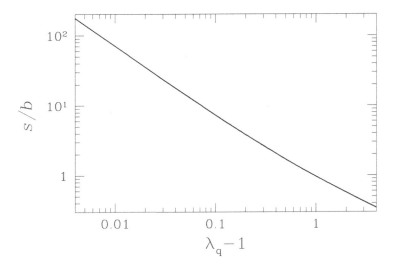

Fig. 19.1.   The yield of strangeness per baryon as a function of λ_q in equilibrated QGP.

energy. This remarkable increase is due to the expected chemical equilibration ($\gamma_s \rightarrow 1$) at the RHIC, see section 17.5, as well as to a substantial reduction of baryon density at central rapidity. In comparison with the general hadron multiplicity, only a modest enhancement of production of strangeness at most can be expected at the RHIC, compared to SPS; the remarkable feature of the RHIC situation is that this enhancement is found in the abundance of (multistrange) baryons. Given the large strangeness-per-baryon ratio, baryons and antibaryons produced at the RHIC are mostly strange [221]. We are not aware of any reaction model other than formation of QGP and its hadronization that could produce this type of yield anomaly.

While the specific yield of strangeness s/b is a clear indicator for the extreme conditions reached in heavy-ion collisions, a more directly accessible observable is the occupancy of the hadron-strangeness phase space, γ_s^{HG}. Due to the need to hadronize into a strangeness-poor phase, γ_s^{HG} can be appreciably greater than unity. To understand this, we compare the phase space of strangeness in QGP with that of the resulting HG. The absolute yields must be the same in both phases. This hadronization condition allows us to relate the two phase-space occupancies in HG and QGP, by equating the strangeness content in the two phases. On canceling out the common normalization factor $T^3/(2\pi^2)$, Eq. (11.26), we obtain

$$\gamma_s^{\text{QGP}} V^{\text{QGP}} g_s W\left(\frac{m_s}{T^{\text{QGP}}}\right) \simeq \gamma_s^{\text{HG}} V^{\text{HG}} \left(\frac{\gamma_q \lambda_q}{\lambda_s} F_K + \frac{\gamma_q^2}{\lambda_q^2 \lambda_s} F_Y\right). \quad (19.13)$$

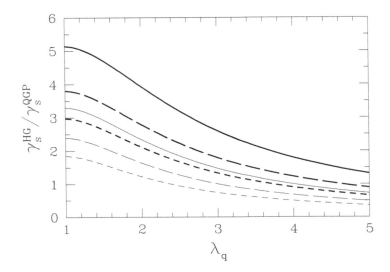

Fig. 19.2. The HG/QGP strangeness-occupancy γ_s ratio in sudden hadroniza-
tion as a function of λ_q. Solid lines, $\gamma_q^{HG} = 1$; long-dashed lines, $\gamma_q^{HG} = 1.3$; and
short-dashed lines, $\gamma_q^{HG} = 1.6$. Thin lines are for $T = 170$ and thick lines for
$T = 150$ MeV, for both phases.

Here we have, without loss of generality, followed the $\bar{s}$-carrying hadrons in
the HG phase space, and we have omitted the contribution of multistrange
antibaryons for simplicity. We now use the condition that strangeness is
conserved, Eq. (11.28), to eliminate λ_s from Eq. (19.13), and obtain

$$\frac{\gamma_s^{HG}}{\gamma_s^{QGP}}\frac{V^{HG}}{V^{QGP}} = \frac{g_s W(m_s/T^{QGP})}{\sqrt{(\gamma_q F_K + \gamma_q^2 \lambda_q^{-3} F_Y)(\gamma_q F_K + \gamma_q^2 \lambda_q^3 F_Y)}}. \tag{19.14}$$

In sudden hadronization, $V^{HG}/V^{QGP} \simeq 1$, the growth of volume is negligible, $T^{QGP} \simeq T^{HG}$, the temperature is maintained across the hadronization front, and the chemical occupancy factors in both states of matter
accommodate the different magnitude of the particle phase space. In this
case, the 'squeezing' of the strangeness of the QGP into the smaller HG
phase space results in $\gamma_s^{HG}/\gamma_s^{QGP} > 1$. We show, in Fig. 19.2, the enhancement of phase-space occupancy expected in sudden hadronization of the
QGP. The temperature ranges $T = 150$ MeV (thick lines) and $T = 170$
MeV (thin lines) span the ranges being considered today at the SPS and
RHIC. The value of γ_q is in the range of the chemical equilibrium in HG,
$\gamma_q = 1$ (solid lines), to the expected excess in sudden hadronization, see
section 7.5, $\gamma_q = 1.6$ (short-dashed lines), with the intermediate value
$\gamma_q = 1.3$ (long-dashed lines).

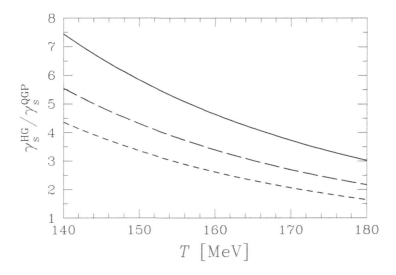

Fig. 19.3. The HG/QGP strangeness-occupancy γ_s ratio in sudden hadroniza-tion, as a function of T for $\lambda_q = 1$. Solid line, $\gamma_q^{HG} = 1$; long-dashed line, $\gamma_q^{HG} = 1.3$; and short-dashed line, $\gamma_q^{HG} = 1.6$.

We note that, for the top SPS energy range, for which $\lambda_q = 1.5$–1.6, sudden-hadronization analysis of data implies $T \simeq 150\,\mathrm{MeV}$ and $\gamma_q \simeq 1.6$, and the value of γ_s increases across hadronization by factor 2.7. Since the yield of strangeness seen at the SPS implies that $\gamma_s^{QGP} \simeq 0.6$, this in turn implies that $\gamma_s^{HG} \simeq 1.6 \simeq \gamma_q^{HG}$, as is indeed found in hadron-production analysis in the sudden-hadronization picture; see table 19.3.

Because $\gamma_s^{HG}/\gamma_q^{HG} \simeq 1$, one can also model the hadronization at the SPS energy in terms of an equilibrium-hadronization model. The en-hancement in production of pions associated with the high-entropy phase can be accommodated by use of two temperatures, one for the determi-nation of absolute yields of particles, and another for determination of the spectral shape. Such an approach has a similar number of param-eters, and comparable predictive power. However, the SPS condition, $\gamma_s^{HG}/\gamma_q^{HG} \simeq 1$, does not hold for the RHIC energy range. We therefore expect that a much clearer picture about the dynamics of hadronization of QGP should emerge from the study of yields of strange particles at the RHIC.

As can be seen in Fig. 19.2, near $\lambda_q \to 1$ (the condition at the RHIC) there is practically no variation in the ratio $\gamma_s^{HG}/\gamma_q^{HG}$. Therefore, we consider, in Fig. 19.3 for $\lambda_q = 1$, the dependence on the hadronization temperature of $\gamma_s^{HG}/\gamma_q^{HG}$, with the three cases shown: solid line, $\gamma_q = 1$; long-dashed line, $\gamma_q = 1.3$; and short-dashed line, $\gamma_q = 1.6$. We now compare the hadronization conditions for the RHIC in the range between

$T = 180$ MeV and $\gamma_q = 1$ and $T = 155$ MeV and $\gamma_q = 1.6$. Across this domain, we see that the phase space of strangeness in the HG is three times smaller than that in QGP, or, as shown in Fig. 19.3, there is this large increase in occupancy of the strange-quark phase space.

In the likely event that the QGP formed at the RHIC evolves toward the chemical-equilibrium abundance of strangeness, or possibly even exceeds it, section 17.5, we should expect a noticeable over-occupancy of strangeness to be measured in terms of the chemical-equilibrium abundance of final-state hadrons. This high phase-space occupancy is one of the requirements for the enhancement of production of multistrange (anti)baryons, which is an important hadronic signal of QGP phenomena [213–215, 226]. In particular, we hope that hadrons produced in phase space with a small probability, such as Ω and $\overline{\Omega}$, will be observed with yields above expectations, continuing the trend seen at the SPS. Because much of the strangeness is in the baryonic degrees of freedom, the kaon-to-pion ratio appears suppressed, in Fig. 1.5 on page 17, relative to SPS results, however we will show below that this is not the case.

Many results from the RHIC run at $\sqrt{s_{NN}} = 130$ GeV are still preliminary and the following quantitative discussion is probably not the final word on this matter. However, the results we find are very interesting, and in qualitative agreement with the sudden-breakup reaction picture of QGP predominantly presented in this book. The data is mainly obtained in the central-rapidity region, where, due to approximate longitudinal scaling, the effects of flow cancel out and we can evaluate the full phase-space yields in order to obtain particle ratios. We do not explore trivial results such as $\pi^+/\pi^- = 1$, since the large hadron yield combined with the flow of baryon isospin asymmetry toward the fragmentation rapidity region assures us that this result will occur to within a great precision. We also do not use the results for K^* and $\overline{K}^*$ since these yields depend on the degree of rescattering of resonance decay products [224, 259]. The data we use has been reported in conference reports of the STAR collaboration of summer 2001, which has been combined with data of PHENIX, BRAHMS, and PHOBOS; for more discussion of the origin of data, see [81]. We assume, in our fit in table 19.4, that the multistrange weak-interaction cascading $\Xi \to \Lambda$, in the STAR result we consider, is cut by vertex discrimination and thus we use these uncorrected yields.

We first look at the last column in table 19.4, the chemical-equilibrium fit. Its large χ^2 originates from the inability to account for production of multistrange $\overline{\Xi}$ and Ξ. Similar results are presented in [81], in an equilibrium fit that does not include multistrange hadrons. The equilibrium fit yields $E/S = 159$ MeV $< T = 183$ MeV. With a negligible contribution from δT, Eq. (19.6) implies that $P/\epsilon \simeq 1/6.6$ as is expected for the high-freeze-out temperature.

Table 19.4. Fits of central-rapidity hadron ratios for the RHIC run at $\sqrt{s_{\mathrm{NN}}} =$ 130 GeV. Top section: experimental results, followed by chemical parameters, physical properties of the phase space, and the fitting error. Columns: data, the full nonequilibrium fit, the nonequilibrium fit constrained by conservation of strangeness and supersaturation of pion phase space, and, in the last column, the equilibrium fit constrained by conservation of strangeness, superscript * indicates quantities fixed by these considerations

	Data	Fit	Fit $s - \bar{s} = 0$	Fit^{eq} $s - \bar{s} = 0$
$\bar{p}/p$	0.64 ± 0.07	0.637	0.640	0.587
$\bar{p}/h^-$		0.068	0.068	0.075
$\bar{\Lambda}/\Lambda$	0.77 ± 0.07	0.719	0.718	0.679
Λ/h^-	0.059 ± 0.001	0.059	0.059	0.059
$\bar{\Lambda}/h^-$	0.042 ± 0.001	0.042	0.042	0.040
$\bar{\Xi}/\Xi$	0.83 ± 0.08	0.817	0.813	0.790
Ξ^-/Λ	0.195 ± 0.015	0.176	0.176	0.130
$\bar{\Xi}^-/\bar{\Lambda}$	0.210 ± 0.015	0.200	0.200	0.152
K^-/K^+	0.88 ± 0.05	0.896	0.900	0.891
K^-/π^-	0.149 ± 0.020	0.152	0.152	0.145
K_S/h^-	0.130 ± 0.001	0.130	0.130	0.124
Ω/Ξ^-		0.222	0.223	0.208
$\bar{\Omega}/\bar{\Xi}^-$		0.257	0.256	0.247
$\bar{\Omega}/\Omega$		0.943	0.934	0.935
T		158 ± 1	158 ± 1	183 ± 1
γ_{q}		1.55 ± 0.01	1.58 ± 0.08	1^*
λ_{q}		1.082 ± 0.010	1.081 ± 0.006	1.097 ± 0.006
γ_{s}		2.09 ± 0.03	2.1 ± 0.1	1^*
λ_{s}		1.0097 ± 0.015	1.0114^*	1.011^*
E/b [GeV]		24.6	24.7	21
s/b		6.1	6.2	4.2
S/b		151	152	131
E/S [MeV]		163	163	159
χ^2/dof		$2.95/(10-5)$	$2.96/(10-4)$	$73/(10-2)$

The chemical-nonequilibrium fit appears more internally consistent. The value of the hadronization temperature $T = 158$ MeV is below the central expected equilibrium phase-transition temperature for the case of $2 + 1$ flavors, section 15.5. This is also near to the $P = 0$ condition for the quark–gluon-liquid model developed in chapter 16, see Fig. 4.2 on page 70, and, as is appropriate, a little above the temperatures seen for the SPS baryon-rich freeze-out; see table 19.3. We find, Eq. (19.6) $P/\epsilon \simeq -1/33$, a value consistent with super-cooling and sudden QGP hadronization. This reaction picture is in agreement with the relatively large γ_{s} and $\gamma_{\mathrm{q}} > 1$

obtained. Comparing the two types of results, we conclude that it is the inclusion of the yields of the multistrange antibaryons in the RHIC data analysis, along with the hypothesis of chemical nonequilibrium, which allows us to discriminate between the different scenarios of reaction.

We look next at the strangeness content, $s/b = 6$, in table 19.4: the full QGP phase space would have yielded 8.6 pairs of strange quarks per baryon at $\lambda_q = 1.085$, as is seen in Fig. 19.1, and $\gamma_s^{QGP} = 6/8.6 = 0.7$. With this value, and using the fitted value $\gamma_s^{HG} = 2.1$, we compute $\gamma_s^{HG}/\gamma_s^{QGP} = 2.1/0.7 = 3$ and, as we see in Fig. 19.3, this is, for $T = 158$ MeV and $\gamma_q = 1.55$, the expected condition for hadronization of QGP.

The fact that the inferred strangeness phase space in QGP is not fully saturated is, on a second careful look, in qualitative agreement with predictions of kinetic theory, Fig. 17.13. Namely, particle multiplicities observed and the shape of particle spectra suggest that the initial conditions are near $T = 300$ MeV in the RHIC run at $\sqrt{s_{NN}} = 130$ GeV, which pushes the value of γ_s^{QGP}, shown in Fig. 17.13, toward 0.9.

A further reduction in this prediction, more applicable to the highest RHIC energy range, arises since the initial volume of the fireball assumed in order to obtain the result in Fig. 17.13 is up to twice as large as that implied by the RHIC 130A-GeV results. This is understood by considering the total experimental hadron multiplicity entropy, which we can derive from the fitted entropy per baryon, $S/b \simeq 150$. We cannot multiply this by the total number of participants, since many of the 350 participating nucleons are found in the fragmentation regions. If we assume that the central-rapidity fraction is 100 baryons, the central entropy content is 15 000, while the calculations for Fig. 17.13 are based on 38 000 entropy units. Thus, the initial volume of the fireball populating the particle yields in the central-rapidity region is about half that used to obtain the results in Fig. 17.13. A smaller system lives for a shorter time, and, since we are not yet in the regime of full equilibration of strangeness, a smaller value of γ_s^{QGP} than that seen in Fig. 17.13 is expected. Also, the smaller volume of the system is more consistent with the HBT results.

The value of the thermal energy content, $E/b = 25$ GeV, seen in the bottom portion of table 19.4 is also in very good agreement with expectations once we allow for the kinetic-energy content associated with longitudinal and transverse motion. The energy of each particle is 'boosted' by the factor $\gamma_\perp^v \cosh y_\parallel$, see, e.g., Eq. (8.39). For $v_\perp = c/\sqrt{3}$, we have $\gamma_\perp^v = 1.22$. The range of longitudinal flow is about ± 2.3 rapidity units; see Fig. 9.19. To obtain the the increase in energy due to longitudinal flow, we have to multiply by the average, $\int dy_\parallel \cosh y_\parallel / y_\parallel \to \sinh(2.3)/2.3 = 2.15$, for a total average increase in energy by factor 2.62, which takes the full energy content to $E^v/b \simeq 65$ GeV as expected.

We now consider what these results imply about the total yield of strangeness in the RHIC fireball. First, we sum up the yield of strange quarks contained in hyperons. Recalling Fig. 2.6 on page 32, we have in singly strange hyperons 1.5 times the yield observed in Λ. Also, accounting for the doubly strange Ξ^-, which are half of all the Ξ, we have

$$\frac{\langle s \rangle_Y}{h^-} = 1.5 \times 0.059 + 2 \times 2 \times 0.195 \times 0.059 = 0.133.$$

Allowing for the unobserved Ω at the theoretical rate, this number increases to $\langle s \rangle_Y / h^- = 0.14$. Repeating the same argument for antihyperons, the result is 0.10. The s and $\bar{s}$ content in kaons is four times that in K_S and thus

$$\frac{\langle s + \bar{s} \rangle}{h^-} = 0.76,$$

with 32% of this yield contained in hyperons and antihyperons. Up to this point, the analysis is practically solely based on direct measurements and established yields of particles.

We now estimate the increase in the 'strangeness-suppression' factor W_s, Eq. (18.1). Correcting for the presence of K^- among negatively charged hadrons, and assuming that all three pions are equally abundant, we find

$$\frac{\langle s + \bar{s} \rangle}{\pi^- + \pi^+ + \pi^0} \simeq 0.30.$$

The total number of pions produced comprises pions arising from resonance decays and from fragments of the projectile and target. Thus, as few as half of the pions originate from the newly made $q\bar{q}$ pairs. In the RHIC run at $\sqrt{s_{NN}} = 130$ GeV, we estimate $W_s \simeq 0.6$. The increase compared with the SPS energy is largely due to the strangeness content in hyperons. Considering that $\gamma_s^{QGP} \simeq 0.7$ at $\sqrt{s_{NN}} = 130$ GeV, there is still space for a further rise in yield of strangeness at the highest RHIC energy, and we hope and expect that $W_s \to 1$ when the initial temperature rises to well above the mass of the strange quark for a sufficient length of time.

We have learned to appreciate, in this last part, how the deconfined thermal phase, through its gluon content, manifests itself as a strangeness signature of QGP. The presence of gluons is essential for rapid thermal production of strangeness. The SPS strangeness results decisively show interesting new physics, with a significant excess of strangeness and strange antibaryons. We see, at the SPS and at the RHIC, considerable convergence of the understanding of the production of hadrons around properties of suddenly hadronizing entropy and strangeness-rich QGP. We see hadronization into pions, at $\gamma_q \to \gamma_q^c = e^{m_\pi/(2T_f)} \simeq 1.6$, which is an effective way to convert the excess of entropy in the deconfined state into hadrons. We have seen that strange-particle signatures of hadronization of QGP become more extreme and clearer at the RHIC.

It is important to remember that it is not only at the RHIC, and in the near future at the LHC, that QGP can be studied. An alternative energy domain for investigating QGP is the phase-transition region at high baryon density. It is very probable that the onset of deconfinement occurs at modest collision energies, perhaps in collisions of 20–40A-GeV Pb projectiles with a laboratory target. The formation of the QGP phase is an endothermic process. In experiments near to the condition for phase transformation, energy balance, lack of explosive flow, and the onset of abundant production of entropy and strangeness as a function of the energy and reaction volume should provide good global signatures of new physics. Anomalous production of multistrange baryons and antibaryons should help to pinpoint the deconfined phase.

We hear sometimes the following question: "Where is the 'smoking gun' signature of QGP?". The disappearance of the suppression of the production of strangeness is surely one such observation. However, we must remember that the discovery of QGP, unlike the discovery of a new particle, requires a change in our understanding of the fundamental hadronic degrees of freedom. This is a deductive process, and requires a global cross check of consistency at each step in its development. It is unlikely that our detailed arguments will persuade the skeptic, however, we hope to show the new way to the uninitiated.

QGP discovery is similar to the slow and painful path to the understanding of electricity. The reader is invited to think through how he/she would proceed without the plug in the wall, the instrument in the laboratory, and the battery in the drawer, and without comprehension of the principles of conductivity, to introduce in a lecture the discovery of the existence of electricity. The challenge of understanding the 'ionized' quark–gluon form of matter is certainly more complex than the unraveling of electricity.

References

[1] S. Abatzis *et al.*, WA85 collaboration, 1990. Λ and $\overline{\Lambda}$ production in sulphur–tungsten interactions at 200 GeV/c per nucleon. *Phys. Lett. B,* **244**, 130.

[2] S. Abatzis *et al.*, WA85 collaboration, 1991. Production of multistrange baryons and anti-baryons in sulphur–tungsten interactions at 200 GeV/c per nucleon. *Phys. Lett. B,* **259**, 508.

[3] S. Abatzis *et al.*, WA85 collaboration, 1991. Ξ^-, $\overline{\Xi^-}$, Λ and $\overline{\Lambda}$ production in sulphur–tungsten interactions at 200 GeV/c per nucleon. *Phys. Lett. B,* **270**, 123.

[4] S. Abatzis *et al.*, WA85 collaboration, 1993. Observation of omega and anti-omega in sulphur–tungsten interactions at 200 GeV/c per nucleon. *Phys. Lett. B,* **316**, 615.

[5] S. Abatzis *et al.*, WA85 collaboration, 1995. Measurement of the Ω/Ξ production ratio in central S–W interactions at 200A GeV. *Phys. Lett. B,* **347**, 158.

[6] S. Abatzis *et al.*, WA85 collaboration, 1996. Strangeness production in p + W and S + W interactions at 200A GeV. *APH N. S., Heavy Ion Physics,* **4**, 79.

[7] S. Abatzis *et al.*, WA85 collaboration, 1996. Study of K^0_S, Λ and $\overline{\Lambda}$ production in S–W collisions at 200 GeV/c per nucleon. *Phys. Lett. B,* **376**, 251.

[8] S. Abatzis *et al.*, WA94 collaboration, 1995. A study of a cascade and strange baryon production in sulphur–sulphur interactions at 200 GeV/c per nucleon. *Phys. Lett. B,* **354**, 178.

[9] F. Abe *et al.*, CDF collaboration, 1998. Observation of B_c mesons in p–$\bar{p}$ collisions at $\sqrt{s} = 1.8$ TeV. *Phys. Rev. D,* **58**, 112004.

[10] F. Abe *et al.*, CDF collaboration, 1998. Observation of the B_c meson in p–$\bar{p}$ collisions at $\sqrt{s} = 1.8$ TeV. *Phys. Rev. Lett.,* **81**, 2432.

[11] M. C. Abreu *et al.*, NA38 collaboration, 1998. Transverse momentum of J/Ψ, Ψ' and mass continuum muon pairs produced in ^{32}S–U collisions at 200 GeV/c per nucleon. *Phys. Lett. B*, **423**, 207.

[12] M. C. Abreu *et al.*, NA50 collaboration, 2000. Evidence for deconfinement of quarks and gluons from the J/Ψ suppression pattern measured in Pb + Pb collisions at the CERN SPS. *Phys. Lett. B*, **477**, 28.

[13] M. C. Abreu *et al.*, NA50 collaboration, 2001. Results on open charm from NA50. *J. Phys. G*, **27**, 677.

[14] M. C. Abreu *et al.*, NA50 collaboration, 2001. Transverse momentum distributions of J/Ψ, Ψ', Drell–Yan and continuum dimuons produced in Pb + Pb interactions at the SPS. *Phys. Lett. B*, **499**, 85.

[15] K. H. Ackermann *et al.*, STAR collaboration, 2001. Elliptic flow in Au + Au collisions at $\sqrt{s_{NN}} = 130$ GeV. *Phys. Rev. Lett.*, **86**, 402.

[16] K. Adcox *et al.*, PHENIX collaboration, 2001. Measurement of the mid-rapidity transverse energy distribution from $\sqrt{s_{NN}} = 130$ GeV Au + Au collisions at RHIC. *Phys. Rev. Lett.*, **87**, 52 301.

[17] K. Adcox *et al.*, PHENIX collaboration, 2002. Suppression of hadrons with large transverse momentum in central Au + Au collisions at $\sqrt{s_{NN}} = 130$ GeV. *Phys. Rev. Lett.*, **88**, 22 301.

[18] C. Adler *et al.*, STAR collaboration, 2001. Pion interferometry of $\sqrt{s_{NN}} = 130$ GeV Au + Au collisions at RHIC. *Phys. Rev. Lett.*, **87**, 82 301.

[19] C. Adler, STAR collaboration, 2001. Mid-rapidity anti-proton to proton ratio from Au + Au collisions at $\sqrt{s_{NN}} = 130$ GeV. *Phys. Rev. Lett.*, **86**, 4778.

[20] A. T. M. Aerts and J. Rafelski, 1984. QCD, bags and hadron masses. *Phys. Lett. B*, **148**, 337.

[21] S. V. Afanasev *et al.*, NA49 collaboration, 2000. Production of φ mesons in p + p, p + Pb and central Pb + Pb collisions at $E_{beam} = 158A$ GeV. *Phys. Lett. B*, **491**, 59.

[22] G. Agakishiev *et al.*, CERES collaboration, 1998. Low mass e$^+$ e$^-$ pair production in 158A GeV Pb–Au collisions at the CERN SPS, its dependence on multiplicity and transverse momentum. *Phys. Lett. B*, **422**, 405.

[23] J.-E. Alam, S. Raha, and B. Sinha, 1994. Successive equilibration in quark–gluon plasma. *Phys. Rev. Lett.*, **73**, 1895.

[24] T. Alber *et al.*, NA35 collaboration, 1994. Strange particle production in nuclear collisions at 200 GeV per nucleon. *Z. Phys. C*, **64**, 195.

[25] T. Alber *et al.*, NA35 collaboration, 1995. Two-pion Bose–Einstein correlations in nuclear collisions at 200 GeV per nucleon. *Z. Phys. C*, **66**, 77.

[26] T. Alber *et al.*, NA35 collaboration, 1998. Charged particle production in proton, deuteron, oxygen and sulphur–nucleus collisions at 200 GeV per nucleon. *Eur. Phys. J. C*, **2**, 643.

[27] T. Alber *et al.*, NA49 collaboration, 1995. Transverse energy production in Pb^{208} + Pb collisions at 158 GeV per nucleon. *Phys. Rev. Lett.*, **75**, 3814.

[28] R. Albrecht *et al.*, WA80 collaboration, 1989. Global and local fluctuations in multiplicity and transverse energy for central ultrarelativistic heavy ion interactions. *Z. Phys.* C, **45**, 31.

[29] R. Albrecht *et al.*, WA80 collaboration, 1995. Production of η mesons in 200A GeV S + S and S + Au. *Phys. Lett.* B, **361**, 14.

[30] M. Alford, K. Rajagopal, and F. Wilczek, 1998. Color superconductivity and signs of its formation. *Nucl. Phys.* A, **638**, 515c.

[31] M. Alford, K. Rajagopal, and F. Wilczek, 1999. Color flavor locking and chiral symmetry breaking in high density QCD. *Nucl. Phys.* B, **537**, 443.

[32] A. Ali Khan *et al.*, CP-PACS collaboration, 2001. Equations of state of finite-temperature QCD with two flavors of improved Wilson quarks. *Phys. Rev.* D, **64**, 74510.

[33] A. Ali Khan *et al.*, CP-PACS collaboration, 2001. Light hadron spectroscopy with two flavors of dynamical quarks on the lattice. *Hep-lat 0105015.*

[34] T. Altherr and D. Seibert, 1993. Thermal quark productions in pure glue and quark–gluon plasmas. *Phys. Lett.* B, **313**, 149.

[35] J. Ambjorn and P. Olesen, 1980. A color magnetic vortex condensate in QCD. *Nucl. Phys.* B, **170**, 60.

[36] J. O. Andersen, E. Braaten, and M. Strickland, 2000. Hard-thermal-loop resummation of the free energy of a quark–gluon plasma. *Phys. Rev.* D, **61**, 74016.

[37] E. Andersen *et al.*, 1989. A measurement of cross-sections for S^{32} interactions with Al, Fe, Cu, Ag and Pb at 200 GeV/c per nucleon. *Phys. Lett.* B, **220**, 328.

[38] E. Andersen *et al.*, WA97 collaboration, 1999. Strangeness enhancement at mid-rapidity in Pb–Pb collisions at 158A GeV. *Phys. Lett.* B, **449**, 401.

[39] B. Andersson, G. Gustafson, G. Ingelman, and T. Sjostrand, 1983. Parton fragmentation and string dynamics. *Phys. Rep.*, **97**, 31.

[40] B. Andersson, G. Gustafson, and H. Pi, 1993. The FRITIOF model for very high-energy hadronic collisions. *Z. Phys.* C, **57**, 485.

[41] F. Antinori *et al.*, WA85 collaboration, 1999. Enhancement of strange and multi-strange baryons and anti-baryons in S–W interactions at 200 GeV/c. *Phys. Lett.* B, **447**, 178.

[42] F. Antinori *et al.*, WA97 collaboration, 2000. Transverse mass spectra of strange and multi-strange particles in Pb–Pb collisions at 158A GeV. *Eur. Phys. J.* C, **14**, 633.

[43] H. Appelshäuser *et al.*, NA49 collaboration, 1999. Baryon stopping and charged particle distributions in central Pb + Pb collisions at 158 GeV per nucleon. *Phys. Rev. Lett.*, **82**, 2471.

[44] A. T. Armstrong *et al.*, E864 collaboration, 2001. Search for strange quark matter produced in relativistic heavy ion collisions. *Phys. Rev. C*, **63**, 54 903.

[45] R. Arsenescu *et al.*, NA52 collaboration, 2001. The NA52 strangelet search. *J. Phys. G*, **27**, 487.

[46] J. Bächler *et al.*, NA35 collaboration, 1991. Study of the energy flow in sulphur and oxygen–nucleus collisions at 60 and 200 GeV/nucleon. *Z. Phys. C*, **52**, 239.

[47] J. Bächler *et al.*, NA35 collaboration, 1992. Production of charged kaons in proton–nucleus and nucleus–nucleus collisions at 60 and 200 GeV/nucleon. *Nucl. Phys. A*, **544**, 609.

[48] B. B. Back *et al.*, PHOBOS collaboration, 2001. Charged-particle pseudorapidity density distributions from Au + Au collisions at $\sqrt{s_{NN}} = 130$ GeV. *Phys. Rev. Lett.*, **87**, 102 303.

[49] B. B. Back, PHOBOS collaboration, 2000. Charged particle multiplicity near midrapidity in central Au + Au collisions at $\sqrt{s} = 56$ and $130A$ GeV. *Phys. Rev. Lett.*, **85**, 3100.

[50] B. B. Back, PHOBOS collaboration, 2002. Energy dependence of particle multiplicity at central Au + Au collisions. *Phys. Rev. Lett.*, **88**, 22302.

[51] R. Balian and C. Bloch, 1970. Distribution of eigenfrequencies for the wave equation in a finite domain. I. Three-dimentional problem with smooth boundary surface. *Ann. Phys.*, **60**, 401.

[52] R. Balian and C. Bloch, 1971. Distribution of eigenfrequencies for the wave equation in a finite domain. II. Electromagnetic field. Riemannian spaces. *Ann. Phys.*, **64**, 271.

[53] B. C. Barrois, 1977. Superconducting quark matter. *Nucl. Phys. C*, **129**, 390. Ph. D. Thesis, Caltech 1979.

[54] J. Bartke *et al.*, NA35 collaboration, 1990. Neutral strange particle production in sulphur–sulphur and proton–sulphur collisions at 200 GeV/nucleon. *Z. Phys. C*, **48**, 191.

[55] S. Bass *et al.*, 1999. Group report: Last call for RHIC predictions. *Nucl. Phys A*, **661**, 205.

[56] I. A. Batalin, S. G. Matinyan, and G. K. Savvidy, 1977. Vacuum polarization by a source-free gauge field. *Sov. J. Nucl. Phys.*, **26**, 214.

[57] G. Baym and P. Braun-Munzinger, 1996. Physics of Coulomb corrections in Hanbury-Brown Twiss interferometry in ultrarelativistic heavy ion collisions. *Nucl. Phys. A*, **160**, 286c.

[58] G. Baym, B. L. Friman, J.-P. Blaizot, M. Soyeur, and W. Czyż, 1983. Hydrodynamics of ultrarelativistic heavy ion collisions. *Nucl. Phys.* A, **407**, 541.

[59] I. G. Bearden *et al.*, NA44 collaboration, 1997. Collective expansion in high-energy heavy ion collisions. *Phys. Rev. Lett.*, **78**, 2080.

[60] F. Becattini. Universality of thermal hadron production in pp, p̄p and e^+e^- collisions. In L. Cifarelli, A. Kaidalov, and V. A. Khoze., editors, *Universality Features in Multihadron Production and the Leading Effect.* World Scientific, Singapore, 1998.

[61] F. Becattini, M. Gaździcki, and J. Sollfrank, 1998. On chemical equilibrium in nuclear collisions. *Eur. Phys. J.* C, **5**, 143.

[62] F. Becattini and G. Pettini, 2001. Strangeness production in a statistical effective model of hadronization. In *Proceedings of QCD@Work: International Conference on QCD: Theory and Experiment, Martina Franca, 2001. Hep-ph/0108212* .

[63] S. Z. Belenkij, 1956. Connection between scattering and multiple production of particles. *Nucl. Phys.*, **2**, 259.

[64] E. Beth and G. E. Uhlenbeck, 1937. The quantum theory of non-ideal gas, II. Behavior at low temperatures. *Physica*, **4**, 915.

[65] A. Bialas, 1999. Fluctuations of the string tension and transverse mass distribution. *Phys. Lett.* B, **466**, 301.

[66] N. Bilić, J. Cleymans, I. Dadić, and D. Hislop, 1995. Gluon decay as a mechanism for strangeness production in a quark–gluon plasma. *Phys. Rev.* C, **52**, 401.

[67] T. Biró and J. Zimányi, 1982. Quarkochemistry in relativistic heavy ion collisions. *Phys. Lett.* B, **113**, 6.

[68] T. Biró and J. Zimányi, 1983. Quark–gluon plasma formation in heavy ion collisions and quarkochemistry. *Nucl. Phys.* A, **395**, 525.

[69] T. S. Biró, 2000. Quark coalescence and hadronic equilibrium. *Hep-ph/0005067*.

[70] T. S. Biró, P. Lévai, and B. Müller, 1990. Strangeness production with massive gluons. *Phys. Rev.* D, **42**, 3078.

[71] T. S. Biró, E. van Doorn, B. Müller, M. H. Thoma, and X.-N. Wang, 1993. Parton equilibrium in relativistic heavy ion collisions. *Phys. Rev.* D, **48**, 1275.

[72] J. D. Bjørken, 1982. Energy loss of energetic partons in quark–gluon plasma: Possible extinction of high $P_\perp$ jets in hadron–hadron collisions. *Fermilab-Pub-82/59-THY*.

[73] J. D. Bjørken, 1983. Highly relativistic nucleus–nucleus collisions: The central rapidity region. *Phys. Rev.* D, **27**, 140.

[74] J. D. Bjørken and S. D. Drell. *Relativistic Quantum Mechanics.* McGraw-Hill Book Co., New York, 1964.

[75] D. Boal, C. K. Gelbke, and B. Jennings, 1990. Intensity interferometry in subatomic physics. *Rev. Mod. Phys.*, **62**, 553.

[76] A. Bode *et al.*, ALPHA collaboration, 2001. First results on the running coupling in QCD with two massless flavors. *Phys. Lett.* B, **515**, 49.

[77] H. Boggild *et al.*, NA44 collaboration, 1996. Coulomb effect in single particle distributions. *Phys. Lett.* B, **372**, 339.

[78] P. N. Bogolioubov, 1967. Sur un modèle à quarks quasi-indépendants. *Ann. Inst. Henri Poincaré*, **8**, 163.

[79] C. Bormann, *et al.*, NA49 collaboration, 1997. Kaon, lambda and anti-lambda production in Pb + Pb collisions at 158 GeV per nucleon. *J. Phys.* G, **23**, 1817.

[80] P. Braun-Munzinger, I. Heppe, and J. Stachel, 1999. Chemical equilibration in Pb+Pb collisions at the SPS. *Phys. Lett.* B, **465**, 15.

[81] P. Braun-Munzinger, D. Magestro, K. Redlich, and J. Stachel, 2001. Hadron production in Au–Au collisions at RHIC. *Phys. Lett.* B, **518**, 41.

[82] P. Braun-Munzinger and J. Stachel, 1996. Probing the phase boundary between hadronic matter and the quark–gluon plasma in relativistic heavy ion collisions. *Nucl. Phys.* A, **606**, 320.

[83] L. Bravina, L. P. Csernai, P. Levai, and D. Strottman, 1994. Collective global dynamics in Au + Au collisions at the BNL AGS. *Phys. Rev.* C, **50**, 2161.

[84] W. Broniowski and W. Florkowski, 2001. Description of the RHIC p_T-spectra in a thermal model with expansion. *Phys. Rev. Lett.*, **87**, 272302.

[85] W. Busza, R. L. Jaffe, J. Sandweiss, and F. Wilczek, 2000. Review of speculative 'disaster scenarios' at RHIC. *Rev. Mod. Phys.*, **72**, 1125.

[86] E. Byckling and K. Kajantie. *Particle Kinematics.* J. Wiley, New York, 1973.

[87] N. Carrer, NA57 collaboration, 2001. First results on strange baryon production from the NA47 experiment. In [146].

[88] A. Casher, H. Neuberger, and S. Nussinov, 1979. Chromoelectric-flux-tube model of particle production. *Phys. Rev.* D, **20**, 179.

[89] W. Cassing, 2001. Antibaryon production in hot and dense nuclear matter. *Nucl. Phys.* A, *Nucl-th/0105069*.

[90] P. Chen, N. Christ, G. Fleming, A. Kaehler, C. Malureanu, R. Mawhinney, G. Siegert, C. Sui, L. Wu, Y. Zhestkov, and P. Vranas, 2001. The finite temperature QCD phase transition with domain wall fermions. *Phys. Rev.* D, **64**, 14503.

[91] S. A. Chin, 1978. Transition to hot quark matter in relativistic heavy-ion collision. *Phys. Lett.* B, **78**, 552.

[92] A. Chodos, R. L. Jaffe, K. Johnson, and C. B. Thorn, 1974. Baryon structure in the bag theory. *Phys. Rev.* D, **10**, 2599.

[93] A. Chodos, R. L. Jaffe, K. Johnson, C. B. Thorn, and V. F. Weisskopf, 1974. New extended model of hadrons. *Phys. Rev.* D, **9**, 3471.

[94] A. Chodos and C. B. Thorn, 1975. Chiral invariance in a bag theory. *Phys. Rev.* D, **12**, 2733.

[95] B. Combridge, 1979. Associated production of heavy flavour states in pp and $\bar{\text{p}}$p interactions: Some QCD estimates. *Nucl. Phys.* B, **151**, 429.

[96] F. Cooper and G. Frye, 1974. Single-particle distribution in the hydrodynamic and statistical thermodynamic models of multiparticle production. *Phys. Rev.* D, **10**, 186.

[97] M. Creutz. *Quarks, Gluons and Lattices*. Cambridge University Press, Cambridge, 1983.

[98] L. P. Csernai. *Introduction to Relativistic Heavy Ion Collisions*. J. Wiley and Sons, New York, 1994.

[99] T. DeGrand, R. L. Jaffe, K. Johnson, and J. Kiskis, 1975. Masses and other parameters of the light hadrons. *Phys. Rev.* D, **12**, 2060.

[100] M. D'Elia, A. Di Giacomo, and E. Meggiolaro, 1997. Field strength correlators in full QCD. *Phys. Lett.* B, **408**, 315.

[101] M. D'Elia, A. Di Giacomo, and E. Meggiolaro, 1999. Gauge invariant quark–anti-quark nonlocal condensates in lattice QCD. *Phys. Rev.* D, **59**, 54503.

[102] C. Derreth, W. Greiner, H.-Th. Elze, and J. Rafelski, 1985. Strangeness abundances in $\bar{\text{p}}$–nucleus annihilations. *Phys. Rev.* C, **31**, 1360.

[103] C. DeTar. Quark gluon plasma in numerical simulations of lattice QCD. In R. C. Hwa, editor, *Quark Gluon Plasma*, volume II, page 1. World Scientific, Singapore, 1995.

[104] D. Di Bari *et al.*, WA85 collaboration, 1995. Results on the production of baryons with $|s| = 1, 2, 3$ and strange mesons in S–W collisions at 200 GeV/c per nucleon. *Nucl. Phys.* A, **590**, 307c.

[105] H. G. Dosch and S. Narison, 1998. Direct extraction of the chiral quark condensate and bounds on the light quark masses. *Phys. Lett.* B, **417**, 173.

[106] M. S. Dubovikov and A. V. Smilga, 1981. Analytical properties of the quark polarization operator in an external selfdual field. *Nucl. Phys.* B, **185**, 109.

[107] G. V. Efimov and S. N. Nedelko, 1998. (Anti-)selfdual homogeneous vacuum gluon field as an origin of confinement and $SU_\text{L}(N_\text{F}) \times SU_\text{R}(N_\text{F})$ symmetry breaking in QCD. *Eur. Phys. J.* C, **1**, 343.

[108] D. Elia *et al.*, NA57 collaboration. Results on cascade production in Pb–Pb interactions from the NA57 experiment. In *36th Rencontres de Moriond on QCD and Hadronic Interactions, Les Arcs, France*, 2001. Hep-ex/0105049.

[109] E. Elizalde and J. Soto, 1986. A field configuration closer to the true QCD vacuum. *Z. Phys.* C, **31**, 237.

[110] R. K. Ellis, W. J. Stirling, and B. R. Webber. *QCD and Collider Physics*. Cambridge University Press, New York, 1996.

[111] H.-Th. Elze, W. Greiner, and J. Rafelski, 1983. On the color-singlet quark–gluon plasma. *Phys. Lett.* B, **124**, 515.

[112] H.-Th. Elze, W. Greiner, and J. Rafelski, 1984. Color degrees of freedom in a quark–gluon plasma at finite baryon density. *Z. Phys.* C, **24**, 361.

[113] H.-Th. Elze, J. Rafelski, and W. Greiner, 1980. The relativistic ideal Fermi gas revisited. *J. Phys.* G, **6**, L149.

[114] H.-Th. Elze, J. Rafelski, and L. Turko, 2001. Entropy production in relativistic hydrodynamics collisions. *Phys. Lett.* B, **506**, 123.

[115] J. Eshke, 1996. Strangeness enhancement in sulphur–nucleus collisions at 200 GeV/N. *Heavy Ion Phys.*, **4**, 105.

[116] D. Evans *et al.*, WA85 collaboration, 1994. New results from WA85 on multistrange hyperon production in 200*A* GeV S–W interactions. *Nucl. Phys.* A, **566**, 225c.

[117] D. Evans *et al.*, WA85 collaboration, 1995. Review of strange particle production from the WA85 collaboration. In [217], page 234.

[118] D. Evans *et al.*, WA85 collaboration, 1996. Strangeness production in p–W and S–W interactions at 200*A* GeV. *Heavy Ion Phys.*, **4**, 79.

[119] Y. Hama, F. Grassi, and T. Kodama, 1996. Particle emission in the hydrodynamical description of relativistic nuclear collisions. *Z. Phys.* C, **73**, 153.

[120] E. L. Feinberg, 1976. Direct production of photons and dileptons in thermodynamical models of multiple hadron production. *Nuovo Cimento* A, **34**, 39.

[121] E. Fermi, 1950. High-energy nuclear events. *Prog. Theor. Phys.*, **5**, 570.

[122] E. Fermi, 1953. Multiple production of pions in nucleon–nucleon collisions at cosmotron energies. *Phys. Rev.*, **92**, 452.

[123] H. Fritzsch, M. Gell-Mann, and H. Leutwyler, 1973. Advantages of the color octet gluon picture. *Phys. Lett.* B, **47**, 365.

[124] S. Frixione, M. L. Mangano, P. Nason, and G. Ridolfi, 1998. Heavy quark production. *Adv. Ser. Direct. High Energy Phys.*, **15**, 609.

[125] J. Gasser and H. Leutwyler, 1982. Quark masses. *Phys. Rep.*, **87**, 77.

[126] M. Gaździcki, 1995. Entropy in nuclear collisions. *Z. Phys.* C, **66**, 659.

[127] M. Gaździcki and D. Röhrich, 1995. Pion multiplicity in nuclear collisions. *Z. Phys.* C, **65**, 215.

[128] M. Gaździcki and D. Röhrich, 1996. Strangeness in nuclear collisions. *Z. Phys.* C, **71**, 55.

[129] K. Geiger, 1992. Thermalization in ultrarelativistic nuclear collisions. II. entropy production and energy densities at BNL relativistic Heavy Ion Collider and the CERN Large Hadron Collider. *Phys. Rev.* D, **46**, 4986.

[130] K. Geiger, 1995. Space–time description of ultrarelativistic nuclear collisions in the QCD parton picture. *Phys. Rep.*, **258**, 238.

[131] K. Geiger, 1997. VNI 3.1: MC simulation program to study high-energy particle collisions in QCD by space–time evolution of parton cascades and parton–hadron conversion. *Comput. Phys. Commun.*, **104**, 70.

[132] N. K. Glendenning and T. Matsui, 1983. Creation of $q\bar{q}$ pairs in a chromoelectric flux tube. *Phys. Rev.* D, **28**, 2890.

[133] M. I. Gorenstein, A. P. Kostyuk, H. Stöcker, and W. Greiner, 2001. Statistical coalescence model with exact charm conservation. *Phys. Lett.* B, **509**, 277.

[134] C. Greiner, P. Koch, and H. Stöcker, 1987. Separation of strangeness from antistrangeness in the phase transition from quark to hadron matter: Possible formation of strange quark matter in heavy-ion collisions. *Phys. Rev. Lett.*, **58**, 1825.

[135] C. Greiner and H. Stöcker, 1991. Distillation and survival of strange quark matter droplets in ultrarelativistic heavy ion collisions. *Phys. Rev.* D, **44**, 3517.

[136] D. E. Groom *et al.* (Particle Data Group), 2000. Review of particle properties. *Eur. Phys. J.* C, **15**, 1.

[137] H. Grote, R. Hagedorn, and J. Ranft. *Particle Spectra*. CERN black report, 1970.

[138] J. Günther *et al.*, NA35 collaboration, 1995. Anti-baryon production in S^{32} + nucleus collisions at 200 GeV/nucleon. *Nucl. Phys.* A, **590**, 487c.

[139] H. H. Gutbrod and J. Rafelski, editors. *Particle Production in Highly Excited Matter, Proceedings of Il Ciocco NATO Summer School*, volume 303. NATO Physics series B, Plenum Press, New York, 1993.

[140] R. Hagedorn, 1965. Statistical thermodynamics of strong interactions at high energies. I. *Suppl. Nuovo Cimento*, **3**, 147.

[141] R. Hagedorn. *Lectures on Thermodynamics of Strong Interactions*. CERN yellow report 71-12, 1971.

[142] R. Hagedorn, 1983. The pressure ensemble as a tool for describing the hadron–quark phase transition. *Z. Phys.* C, **17**, 265.

[143] R. Hagedorn, I. Montvay, and J. Rafelski. Thermodynamics of nuclear matter from the statistical bootstrap model. In N. Cabibbo and L. Sertorio,

editors, *Hadronic Matter at Extreme Energy Density*. Plenum Press, New York, 1980.

[144] R. Hagedorn and J. Rafelski, 1980. Hot hadronic matter and nuclear collisions. *Phys. Lett.* B, **97**, 136.

[145] R. Hagedorn and J. Ranft, 1968. Statistical thermodynamics of strong interactions at high energies. II Momentum spectra of particles produced in pp-collisions. *Suppl. Nuovo Cimento*, **6**, 169.

[146] T. J. Hallman *et al.*, editor. *Quark Matter'01, Brookhaven*. North Holland, Amsterdam, 2001.

[147] S. Hamieh, J. Letessier, and J. Rafelski, 2000. Quark–gluon plasma fireball. *Phys. Rev.* C, **62**, 64901.

[148] S. Hamieh, K. Redlich, and A. Tounsi, 2000. Canonical description of strangeness enhancement from p–A to Pb–Pb collisions. *Phys. Lett* B, **486**, 61.

[149] J. Hughes, 1980. Some comments on asymptotic freedom. *Phys. Lett.* B, **97**, 246.

[150] M. Jamin and M. Münz, 1995. The strange quark mass from QCD sum rules. *Z. Phys.* C, **66**, 633.

[151] K. Johnson, 1975. The M.I.T. bag model. *Acta Phys. Pol.* B, **6**, 865.

[152] P. G. Jones *et al.*, NA49 collaboration, 1996. Hadron yields and hadron spectra from the NA49 experiment. *Nucl. Phys.* A, **610**, 188c.

[153] S. Kabana *et al.*, NA52 collaboration, 1999. Centrality dependence of $\pi^\pm$, $K^\pm$, baryon and antibaryon production in Pb + Pb collisions at $158A$ GeV. *J. Phys.* G, **25**, 217.

[154] K. Kajantie, J. Kapusta, M. Kataja, L. McLerran, and A. Mekjian, 1986. Dilepton emission and the QCD phase transition in ultrarelativistic nuclear collisions. *Phys. Rev.* D, **34**, 2746.

[155] D. B. Kaplan, 1992. A method for simulating chiral fermions on the lattice. *Phys. Lett.* B, **288**, 342.

[156] J. I. Kapusta, 1979. Quantum chromodynamics at high temperature. *Nucl. Phys.* B, **148**, 461.

[157] J. I. Kapusta. *Finite-Temperature Field Theory*. Cambridge University Press, Cambridge, 1989.

[158] J. I. Kapusta and S. M. H. Wong, 2001. Is anomalous production of Ω and $\overline{\Omega}$ evidence for disoriented chiral condensates? *Phys. Rev. Lett.*, **86**, 4251.

[159] F. Karsch. Lattice QCD at high temperature and density. In *Lectures given at 40th Internationale Universitätswochen für Theoretische Physik: Dense Matter (IUKT 40), Schladming*, 2001. Hep-lat 106019.

[160] F. Karsch, E. Laermann, and A. Peikert, 2000. The pressure in 2, 2 + 1 and 3 flavour QCD. *Phys. Lett.* B, **478**, 447.

[161] F. Karsch, E. Laermann, and A. Peikert, 2001. Quark mass and flavor dependence of the QCD phase transition. *Nucl. Phys.* B, **605**, 579.

[162] F. Karsch, E. Laermann, A. Peikert, and B. Sturm, 1999. The three flavour chiral phase transition with an improved quark and gluon action in lattice QCD. *Nucl. Phys. Proc. Suppl.*, **73**, 468.

[163] M. Kataja, J. Letessier, P. V. Ruuskanen, and A. Tounsi, 1992. Equation of state and transverse expansion effects in heavy ion collisions. *Z. Phys.* C, **55**, 153.

[164] P. Koch, B. Müller, and J. Rafelski, 1986. Strange quarks in relativistic nuclear collisions. *Phys. Rep.*, **142**, 167.

[165] P. Koch and J. Rafelski, 1985. Time evolution of strange-particle densities in hot hadronic matter. *Nucl. Phys.* A, **444**, 678.

[166] P. Koch and J. Rafelski, 1986. Why the hadronic gas description of hadronic reactions works: The example of strange hadrons. *S. Afr. J. Phys.*, **9**, 8.

[167] P. Koch, J. Rafelski, and W. Greiner, 1983. Strange hadrons in hot nuclear matter. *Phys. Lett.* B, **123**, 151.

[168] J. Kogut and L. Susskind, 1975. Hamiltonian formulation of Wilson's lattice gauge theories. *Phys. Rev.* D, **11**, 395.

[169] K. Kolodziej and R. Ruckl, 1998. On the energy dependence of hadronic B_c production. *Nucl. Instrum. Methods* A, **408**, 33.

[170] A. Kostyuk, M. I. Gorenstein, H. Stöcker, and W. Greiner, 2001. Second cluster integral and excluded volume effects for the pion gas. *Phys. Rev.* C, **63**, 44 901.

[171] I. Králik, *et al.*, WA97 collaboration, 1998. Λ, Ξ and Ω production in Pb–Pb collisions at $158A$ GeV. *Nucl. Phys.* A, **638**, 115.

[172] L. D. Landau, 1953. On the multiparticle production in high-energy collisions. *Izv. Akad. Nauk SSSR, Ser. Fiz.*, **17**, 51. Reprinted in English translation in [173].

[173] L. D. Landau. *Collected Papers of L. D. Landau*, D. Ter Haar, editor. Pergamon, Oxford, 1965.

[174] L. D. Landau and S. Z. Belenkij, 1956. Hydrodynamic theory of multiple production of particles. *Usp. Phys. Nauk*, **56**, 309. Reprinted in English translation in [173].

[175] L. D. Landau and E. M. Lifshitz. *Statistical Physics*. Pergamon, Oxford, 1985.

[176] J. Letessier and J. Rafelski, 1999. Chemical non-equilibrium and deconfinement in $200A$ GeV sulphur induced reactions. *Phys. Rev.* C, **59**, 947.

[177] J. Letessier and J. Rafelski, 1999. Chemical non-equilibrium in high energy nuclear collisions. *J. Phys.* G, **25**, 295.

[178] J. Letessier and J. Rafelski, 2002. Rapidity particle spectra in sudden hadronization of QGP. *J. Phys.* G, **28**, 183.

[179] J. Letessier, J. Rafelski, and A. Tounsi, 1994. Strange particle abundance in QGP formed in $200A$ GeV nuclear collisions. *Phys. Lett.* B, **323**, 393.

[180] J. Letessier, J. Rafelski, and A. Tounsi, 1997. Strangeness in Pb–Pb collisions at $158A$ GeV. *Phys. Lett.* B, **410**, 315.

[181] J. Letessier, J. Rafelski, and A. Tounsi, 2000. Low-$m_\perp$ π^+–π^- asymmetry enhancement from hadronization of QGP. *Phys. Lett.* B, **475**, 213.

[182] J. Letessier, A. Tounsi, U. Heinz, J. Sollfrank, and J. Rafelski, 1993. Evidence for a high entropy phase in nuclear collisions. *Phys. Rev. Lett.*, **70**, 3530.

[183] J. Letessier, A. Tounsi, U. Heinz, J. Sollfrank, and J. Rafelski, 1995. Strange antibaryons and high entropy phase. *Phys. Rev.* D, **51**, 3408.

[184] S. Y. Lo, editor. *Geometrical Pictures in Hadronic Collisions*. World Scientific, Singapore, 1987.

[185] J. Madsen, 2001. Color-flavor locked strangelets. *Phys. Rev. Lett.*, **87**, 172 003.

[186] N. M. Mar *et al.*, 1996. Improved search for elementary particles with fractional electric charge. *Phys. Rev.* D, **53**, 1.

[187] S. G. Matinian and G. K. Savvidy, 1978. On the radiative corrections to classical lagrangian and dynamical symmetry breaking. *Nucl. Phys.* B, **134**, 539.

[188] T. Matsui and H. Satz, 1986. J/Ψ suppression by quark–gluon plasma formation. *Phys. Lett.* B, **178**, 416.

[189] T. Matsui, B. Svetitsky, and L. D. McLerran, 1986. Strangeness production in ultrarelativistic heavy-ion collisions. *Phys. Rev.* D, **34**, 783.

[190] H. E. Miettinen and P. M. Stevenson, 1987. Hadron–nucleus scattering as a function of nuclear size. *Phys. Lett.* B, **199**, 591.

[191] D. P. Morrison and S. Sorenson *et al.*, WA98 collaboration. Private communication. 1996.

[192] B. Müller. The physics of the quark–gluon plasma. In *Lecture Notes in Physics*, volume 225. Springer-Verlag, Berlin, 1984.

[193] B. Müller and J. Rafelski, 1975. Stabilization of the charged vacuum created by very strong electrical fields in nuclear matter. *Phys. Rev. Lett.*, **34**, 349.

[194] T. Muta. *Foundations of Quantum-Chromodynamics*. World Scientific, Singapore, 1987.

[195] Y. Nambu and G. Jona-Lasinio, 1961. Dynamical model of elementary particles based on an analogy with superconductivity. I. *Phys. Rev.*, **122**, 345.

[196] Y. Nara, 1998. A parton–hadron cascade approach in high-energy nuclear collisions. *Nucl. Phys.* A, **638**, 555c.

[197] S. Narison, 1995. Model independent determination of $m(s)$ from τ-like inclusive decays in e^+e^- and implications for the χSB parameters. *Phys. Lett.* B, **358**, 113.

[198] S. Narison, 1996. Heavy quarkonia mass splittings in QCD: Gluon condensate, α_s and $1/m$ expansion. *Phys. Lett.* B, **387**, 162.

[199] H. B. Nielsen and P. Olesen, 1979. A quantum liquid model for the QCD vacuum: Gauge and rotational invariance of domained and quantized homogeneous color fields. *Nucl. Phys.* B, **160**, 380.

[200] N. K. Nielsen and P. Olesen, 1978. An unstable Yang–Mills field mode. *Nucl. Phys.* B, **144**, 376.

[201] I. Otterlund. Physics of relativistic nuclear collisions. In [139], page 57, 1993.

[202] Y. Pang, T. J. Schlagel, and S. H. Kahana, 1992. ARC: A relativistic cascade. *Nucl. Phys.* A, **544**, 435c.

[203] Y. Pang, T. J. Schlagel, and S. H. Kahana, 1992. Cascade for relativistic nucleus collisions. *Phys. Rev. Lett.*, **68**, 2743.

[204] A. Peshier, B. Kämpfer, and G. Soff, 2000. Equation of state of deconfined matter at finite chemical potential in a quasiparticle description. *Phys. Rev.* C, **61**, 45 203.

[205] H. Pi, 1992. An event generator for interactions between hadrons and nuclei: FRITIOF version 7.0. *Comput. Phys. Commun.*, **71**, 173.

[206] R. D. Pisarski and D. H. Rischke, 1999. Superfluidity in a model of massless fermions coupled to scalar bosons. *Phys. Rev.* D, **60**, 4013.

[207] R. D. Pisarski and D. H. Rischke, 2000. Color superconductivity in weak coupling. *Phys. Rev.* D, **61**, 74 017.

[208] J. Podolanski and R. Armenteros, 1954. Analysis of V-events. *Phil. Mag.*, **45**, 13.

[209] S. Pratt, T. Csorgo, and J. Zimanyi, 1990. Detailed predictions for 2-pion correlations in ultrarelativistic heavy-ion collisions. *Phys. Rev.* C, **42**, 2646.

[210] S. D. Protopopescu *et al.*, 1973. $\pi\pi$ partial wave analysis from reactions $\pi^+ p \to \pi^+ \pi^- \Delta^{++}$ and $\pi^+ p \to K^+ K^- \Delta^{++}$ at 7.1 GeV/c. *Phys. Rev.* D, **7**, 1279.

[211] E. Quercigh, 1993. Strangeness in ultrarelativistic nucleus–nucleus interactions. In [139], page 499.

[212] C. Quintans *et al.*, NA50 collaboration, 2001. Production of the ϕ vector-meson in heavy-ion collisions. *J. Phys.* G, **27**, 405.

[213] J. Rafelski. Extreme states of nuclear matter. In R. Bock and R. Stock, editors, *Workshop on Future Relativistic Heavy Ion Experiment*, page 282. GSI-Yellow Report 81-6, Darmstadt, 1981.

[214] J. Rafelski, 1982. Extreme states of nuclear matter. *Nucl. Phys.* A, **374**, 489c.

[215] J. Rafelski, 1982. Formation and observables of the quark gluon plasma. *Phys. Rep.*, **88**, 331.

[216] J. Rafelski, 1991. Strange antibaryons from quark–gluon plasma. *Phys. Lett.* B, **262**, 333.

[217] J. Rafelski, editor. *Strangeness in Hadronic Matter: S'95, Proceedings of Tucson workshop*, volume 340. American Institute of Physics Proceedings Series, New York, 1995.

[218] J. Rafelski and M. Danos, 1980. The importance of the reaction volume in hadronic collisions. *Phys. Lett.* B, **97**, 279.

[219] J. Rafelski, L. P. Fulcher, and A. Klein, 1978. Theory of elementary particles interacting with arbitrarily strong classical fields. *Phys. Rep.*, **38**, 227.

[220] J. Rafelski and R. Hagedorn. From hadron gas to quark matter II. In H. Satz, editor, *Statistical Mechanics of Quarks and Hadrons*, page 253. North Holland, Amsterdam, 1981.

[221] J. Rafelski and J. Letessier, 1999. Expected production of strange baryons and antibaryons in baryon-poor QGP. *Phys. Lett.* B, **469**, 12.

[222] J. Rafelski and J. Letessier, 2000. Sudden hadronization in relativistic nuclear collisions. *Phys. Rev. Lett.*, **85**, 4695.

[223] J. Rafelski and J. Letessier, 2002. Importance of reaction volume in hadronic collisions: Canonical enhancement. In press in [254], hep-ph/0112151.

[224] J. Rafelski, J. Letessier, and G. Torrieri, 2001. Strange hadrons and their resonances: A diagnostic tool of quark–gluon plasma freeze-out dynamics. *Phys. Rev.* C, **64**, 54 907.

[225] J. Rafelski, J. Letessier, and A. Tounsi, 1996. Strange particles from dense hadronic matter. *Acta. Phys. Pol.* B, **27**, 1037.

[226] J. Rafelski and B. Müller, 1982. Strangeness production in the quark–gluon plasma. *Phys. Rev. Lett.*, **48**, 1066. See also *Phys. Rev. Lett.*, **56**, 2334E (1986).

[227] K. Rajagopal and F. Wilczek. The condensed matter physics of QCD. In M. Shifman, editor, *At the Frontier of Particle Physics, Handbook of QCD*, volume 2. World Scientific, 2000. Festschrift in honor of B. L. Ioffe.

[228] K. Redlich, S. Hamieh, and A. Tounsi, 2001. Statistical hadronization and strangeness enhancement from p–A to Pb–Pb collisions. *J. Phys.* G, **27**, 413.

[229] K. Redlich and L. Turko, 1980. Phase transitions in the statistical bootstrap model with an internal symmetry. *Z. Phys.* C, **5**, 201.

[230] C. Roland *et al.*, PHOBOS collaboration. First results from the PHO-BOS experiment at RHIC. In *36th Rencontres de Moriond on QCD and Hadronic Interactions, Les Arcs, France*, 2001.

[231] P. Roy, J. Alam, S. Sarkar, B. Sinha, and S. Raha, 1997. Quark–gluon plasma diagnostics in a successive equilibrium scenario. *Nucl. Phys. A*, **624**, 687.

[232] A. Rusek *et al.*, E886 collaboration, 1996. Strangelet search and light nucleus production in relativistic Si + Pt and Au + Pt collisions. *Phys. Rev. C*, **54**, 15.

[233] A. Sandoval *et al.*, 1980. Energy dependence of multi-pion production in high-energy nucleus–nucleus collisions. *Phys. Rev. Lett.*, **45**, 874.

[234] R. Santo *et al.*, WA80 collaboration, 1989. π^0 and photon spectra from central and peripheral O^{16} + Au collisions at $200A$ GeV. *Nucl. Phys. A*, **498**, 391c.

[235] R. Santo *et al.*, WA80 collaboration, 1994. Single photon and neutral meson data from WA80. *Nucl. Phys. A*, **566**, 61c.

[236] G. K. Savvidy, 1977. Infrared instability of the vacuum state of gauge theories and asymptotic freedom. *Phys. Lett. B*, **71**, 133.

[237] B. R. Schlei, D. Strottman, J. P. Sullivan, and H. W. van Hecke, 1999. Bose–Einstein correlations and the equation of state of nuclear matter. *Eur. Phys. J. C*, **10**, 483.

[238] E. Schnedermann, J. Sollfrank, and U. Heinz. Fireball spectra. In [139], page 175, 1993.

[239] M. Schroedter, R. L. Thews, and J. Rafelski, 2000. B_c meson production in nuclear collisions at RHIC. *Phys. Rev. C*, **62**, 24905.

[240] J. Schwinger, 1951. On gauge invariance and vacuum polarization. *Phys. Rev.*, **82**, 664.

[241] P. Senger and H. Stroebele, 1999. Hadronic particle production in nucleus–nucleus collisions. *J. Phys. G*, **25**, R59.

[242] M. A. Shifman. *Vacuum Structure and QCD Sum Rules*. North Holland, Amsterdam, 1992.

[243] M. A. Shifman, A. I. Vainshtein, and V. I. Zakharov, 1979. QCD and resonance physics. *Nucl. Phys. B*, **147**, 385, 448, and 519.

[244] E. V. Shuryak, 1978. Quark–gluon plasma and hadronic production of leptons, photons and psions. *Phys. Lett. B*, **78**, 150.

[245] E. V. Shuryak. *The QCD Vacuum, Hadrons and the Superdense Matter*. World Scientific, Singapore, 1988.

[246] E. V. Shuryak and J. J. M. Verbaarschot, 1992. On baryon number violation and nonperturbative weak processes at SSC energies. *Phys. Rev. Lett.*, **68**, 2576.

[247] E. V. Shuryak and L. Xiong, 1993. Dilepton and photon production in the "hot-glue" scenario. *Phys. Rev. Lett.*, **70**, 2241.

[248] F. Siklér *et al.*, NA49 collaboration, 1999. Hadron production in nuclear collisions from the NA49 experiment at $158A$ GeV. *Nucl. Phys.* A, **661**, 45c.

[249] J. Sollfrank, U. Heinz, H. Sorge, and N. Xu, 1999. Thermal analysis of hadron multiplicities from RQMD. *Phys. Rev.* C, **59**, 1637.

[250] H. Sorge, 1995. Flavor production in Pb ($160A$ GeV) on Pb collisions: Effect of color ropes and hadronic rescattering. *Phys. Rev.* C, **52**, 3291.

[251] J. Soto, 1985. Relations between quark and gluon condensates from one loop effective actions in constant background fields. *Phys. Lett.* B, **165**, 389.

[252] D. K. Srivastava, M. G. Mustafa, and B. Müller, 1997. Chemical equilibration of an expanding quark–gluon plasma. *Phys. Lett.* B, **45**, 396.

[253] D. K. Srivastava, M. G. Mustafa, and B. Müller, 1997. Expanding quark–gluon plasmas: Transverse flow, chemical equilibration and electromagnetic radiation. *Phys. Rev.* C, **56**, 1064.

[254] H. Stöcker *et al.*, editors. *Proceedings of SQM 2001. J. Phys.* G (in preparation), 2002. Recent results from SPS and RHIC.

[255] L. R. Surguladze and M. A. Samuel, 1996. Decay widths and total cross sections in perturbative QCD. *Rev. Mod. Phys.*, **68**, 259.

[256] S. Théberge, A. W. Thomas, and G. A. Miller, 1980. Pionic corrections to the MIT bag model: The (3, 3) resonance. *Phys. Rev.* D, **22**, 2838.

[257] R. L. Thews, M. Schroedter, and J. Rafelski, 2001. Enhanced J/Ψ production in deconfined quark matter. *Phys. Rev.* C, **63**, 54 905.

[258] A. W. Thomas, S. Théberge, and G. A. Miller, 1981. Cloudy bag model of the nucleon. *Phys. Rev.* D, **24**, 216.

[259] G. Torrieri and J. Rafelski, 2001. Search for QGP and thermal freeze-out of strange hadrons. *New J. Phys.*, **3**, 12.

[260] G. Torrieri and J. Rafelski, 2001. Strange hadron resonances as a signature of freeze-out dynamics. *Phys. Lett.* B, **509**, 239.

[261] B. Touschek, 1968. Covariant statistical mechanics. *Nuovo Cimento* B, **58**, 295.

[262] L. Turko, 1981. Quantum gases with internal symmetry. *Phys. Lett.* B, **104**, 153.

[263] H. van Hecke, H. Sorge, and N. Xu, 1998. Evidence of early multistrange hadron freezeout in high-energy nuclear collisions. *Phys. Rev. Lett.*, **81**, 5764.

[264] H. Vija and M. H. Thoma, 1995. Braaten–Pisarski method at finite chemical potential. *Phys. Lett.* B, **342**, 212.

[265] P. M. Vranas, 1998. Chiral symmetry restoration in the Schwinger model with domain wall fermions. *Phys. Rev.* D, **57**, 1415.

[266] X.-N. Wang, 1997. PQCD based approach to parton production and equilibration in high-energy nuclear collisions. *Phys. Rep.*, **280**, 287.

[267] S. Weinberg. *Gravitation and Cosmology*. J. Wiley, New York, 1972.

[268] S. Weinberg. *The Quantum Theory of Fields*, volume I and II. Cambridge University Press, Cambridge, 1995/96.

[269] H. A. Weldon, 1982. Covariant calculations at finite temperature: The relativistic plasma. *Phys. Rev.* D, **26**, 1394.

[270] H. A. Weldon, 1982. Effective fermion masses of order gT in high temperature gauge theories with exact chiral invariance. *Phys. Rev.* D, **26**, 2789.

[271] K. Werner, 2001. Tools for RHIC: Review of models. *J. Phys.* G, **27**, 625.

[272] U. A. Wiedemann and U. Heinz, 1999. Particle interferometry for relativistic heavy ion collisions. *Phys. Rep.*, **319**, 145.

[273] K. Wilson, 1974. Confinement of quarks. *Phys. Rev.* D, **10**, 2445.

[274] L. A. Winckelmann *et al.*, 1996. Microscopic calculations of stopping and flow from 160A MeV to 160A GeV. *Nucl. Phys.* A, **610**, 116c.

[275] S. M. H. Wong, 1996. Thermal and chemical equilibration in a gluon plasma. *Phys. Rev.* C, **54**, 2588.

[276] S. M. H. Wong, 1997. α_s dependence in the equilibration in relativistic heavy ion collisions. *Phys. Rev.* C, **56**, 1075.

[277] A. K. Wróblewski, 1985. On the strange quark suppression factor in high energy collisions. *Acta Phys. Pol.* B, **16**, 379.

[278] L. Xiong and E. V. Shuryak, 1994. Gluon multiplication in high energy heavy ion collisions. *Phys. Rev.* C, **49**, 2203.

[279] Z. Xu, STAR collaboration, 2001. Resonance studies at STAR. In [146], page 607.

[280] F. J. Ynduráin. *Quantum Chromodynamics*. Springer Verlag, Berlin, 1999.

[281] W. A. Zajc, PHENIX collaboration, 2001. Overview of PHENIX results from the first RHIC run. In [146].

[282] C. Zhai and B. Kastening, 1995. Free energy of hot gauge theories with fermions through g^5. *Phys. Rev.* D, **52**, 7232.

[283] B. Zhang, M. Gyulassy, and Y. Pang. Strangeness production via parton cascade. In *Proceedings of the Strangeness'96 Workshop, Budapest*, page 361, 1996.

Index

α_s
  bag model fit, 262
  lattice, 298, 299
  running first order, 278
  running second order, 281, 282
  at low energy, 284, 285
  scale dependence, 280, 281
  thermal, 286
    approximant, 286

antimatter
  in the Universe, 6
  matter symmetry, 19, 153
antiproton-to-proton ratio, 183
from QGP, 16
heavy ion enhancement, 19

bag model, 258
  action, 263
  bag constant, 38, 69, 261, 306
  boundary conditions, 263
  static-cavity solutions, 264
  strange-quark mass, 261
baryon, 25, 26
  barys , 1
  density
    hydrodynamic expansion,
  108
    quark–gluon liquid, 310, 312,
  313
Bessel function
  I_n, 140
  K_ν, 195

$x^2 K_2$, 197
  non-relativistic limit, 196
  relativistic limit, 196
big-bang, 1
  differences from micro-bang, 4
boson
  condensate, 193
  distribution function, 194
  entropy per particle, 207

calorimeter, 177
canonical
  conservation of strangeness, 223
  ensemble, 192
    multistrange particles, 226
    particle enhancement, 231
    particle suppression, 225, 232
  hadron yield enhancement, 233
  partition function, 224
cascade(ssq), 33
  $\Xi^*(1530)$ decay, 34
center of momentum
  accelerator energies, 72
  coordinate system, 82
  rapidity in asymmetric collisions,
  84
charged-hadron multiplicity
  energy dependence, 180, 181
  ratio, 123
charm
  canonical suppression, 234, 235
  open, 23

thermal production, 329, 333, 334
    LHC energy, 336
charmonium, 36
    enhancement, 22
    suppression, 21, 22
chemical
    entropy equilibration, 115
    equilibration
        thermal rate, 99
    equilibrium, 90
        elementary interactions, 234
        failure, 171
        strangeness in HG, 352
    gluon equilibration, 324
    nonequilibrium
        entropy, 115
        heavy quarks, 98
        SPS results, 357
    relaxation time for strangeness,
    335
chemical parameters
    Pb–Pb at $\sqrt{s_{NN}} = 130$ GeV, 367
    Pb–Pb system, 359, 360
    S–Au/W/Pb system, 358
chemical potential, 57, 212
    antiparticle, 60
    isospin asymmetry, 213
    local, 90
chiral condensate, 297
    disoriented, 269
chiral symmetry, 44
    bag model, 258
    breaking, 44
color, 7
    confinement, 38
    current, 273
    dynamic charge, 38
    flux tube, 41
    hyperfine interaction, 259
    superconductive phase, 47
confinement, 6, 38
    boundary condition, 264
    origin, 38
correlations
    Bose–Einstein, 171
cross section

geometric, 78
    momentum average, 317
    thermal average, 318
current algebra, 45

decoherence, 15
deconfinement at the AGS and SPS,
    95
degeneracy
    effective, 66
    quantum gas, 61
degrees of freedom, 66, 67
    electro-weak particles, 9
    QGP, 52
detailed balance, 316
    antibaryon production, 350
dileptons, 24
Dirac equation, 262
Doppler factor, 149

energy density, 195
    at the RHIC, 185
    hadronic gas, 209, 211
    lattice QCD, 300
    QGP, 53, 310
    quantum gas, 69
    quark–gluon liquid, 310, 312–314
ensemble
    canonical, 192
    grand-canonical, 192
    micro-canonical, 191
    statistical, 191
enthalpy, 191
entropy
    Boltzmann gas, 114
    chemical equilibration, 115
    classical gas, 205, 206
    conservation
        ideal flow, 106
        scaling solution, 128
    content
        phase-space distribution, 113
    Fermi gas, 113
    glue fireball, 116, 117
    hadronization, 126
    initial state, 129
    isolated system, 114

measurement, 121, 122
nonequilibrium, 115
particle production, 112
per baryon, 206
per particle, 114, 207
pion gas
super-saturated, 124
entropy density, 187
quark–gluon liquid, 310, 312–314
quark–gluon plasma, 315
equation of state
effect of particle mass, 199
finite-volume correction, 209
quark–gluon plasma, 303
relativistic ideal gas, 199
equilibrium
absolute chemical, 90
chemical, 90
local, 90, 95
kinetic, 96
relative chemical, 90
thermal, 90
explanation, 97
local, 96
eta function, 201
Euclidian space, 288
expansion
cooling, 112
decrease in temperature, 11
Hubble, 11
exponential mass spectrum
phase transition, 238
extensive variables, 187

Fadeev–Popov ghosts, 275
fermion
degeneracy factor, 201
distribution function, 194, 203
domain wall, 292
entropy per particle, 207
ideal-gas quark density, 204
lattice doubling, 291
fireball, 2, 95
entropy, 115
expanding, 137
thermal particle spectra, 141

expansion, 50
explosion, 51
flow of matter, 138
life span, 91
mass, 80
stages of evolution, 93
super-cooling, 51
fireball static
particle spectra, 131
first law of thermodynamics, 187
flavor
conservation, 214
symmetry, 269
free energy, 187, 188
lowest order in α_s, 304
perturbative expansion, 303
freeze-out
chemical and thermal, 158
Cooper–Frye formula, 141
surface
$1 + 1$ dimensions, 109, 110
velocity, 142
FRITIOF model, 103
fugacity, 56, 57, 213
antiparticle, 60
time dependence, 214
valence quark, 212

gauge invariance, 268
covariant derivative, 272
minimal coupling, 268
Gibbs condition
chemical potential, 50
temperature, 48
Gibbs' condition
pressure, 48
Gibbs–Duham relation, 107, 190
gluon, 6, 24, 38
current, 273
degeneracy, 53
density, 63
equilibration, 324
field, 271
field condensate, 39
field correlator, 39
spectra, 120

thermal mass, 305
yield of strangeness, 340
Goldstone boson, 44
grand-canonical ensemble, 57

hadron
    abundances, 169
        RHIC, 366
        theoretical yield error, 240
    finite-volume cluster, 247
    in modification of a medium, 345
    mass
        bag model, 259
    multiplicity of h$^-$ at SPS, 166
    ratios in A–A collisions at $14A$
    GeV, 170
    ratios in A–A collisions at $200A$
    GeV, 169
    size, 260
hadronic
    hadros , 1
    cascade, 102
    mass spectrum, 217
        exponential growth, 235–237
    resonance interactions, 243
hadronic gas, 48
    ϵ/P, 210
    asymmetry of strangeness, 215
    energy density, 210
    excluded volume, 251
        consistency, 250
        correction, 248
    finite size EoS, 209
    overheated, 51
    phase space for strangeness, 217
    pressure, 210
    properties, 65
    relativistic limit, 66
    relaxation time for strangeness,
    350
    scattering phase shifts, 243
hadronization
    deconfined matter, 350
    in a volume, 141
    statistical, 352
    sudden, 126

entropy content, 126
    surface emission, 142
Hagedorn gas, 236
    critical temperature, 240
Hagedorn temperature, 53
HBT, 171, 172
    correlations, 174, 175
    kaon, 174
    resonances, 174
    transverse mass, 176
heat function, 191
heavy ion
    baryon stopping, 74
    collision
        axis, 81
        event generators, 102
        interaction vertex, 81
        participants, 78
        spectators, 79
        systems, 72
        transport models, 102
    experimental program
        BNL–RHIC, 76
        CERN–SPS, 75
    rapidity gap, 73
hydrodynamics, 104
    equations of motion, 105
    Euler relation, 104
    flow forces, 210
    one-dimensional solution, 111
hyperon, 28
    lambda decay, 28
    lambda resonances, 28
    number, 25
    resonance, 31
    sigma–baryon, 30
    yield, 32, 33

ideal gas
    clusters in bootstrap model, 256
    energy per particle, 201, 202
    entropy, 205
    entropy per baryon, 206
    quark partition function, 203
impact parameter, 79
isospin, 25

particle counting, 214
quark current, 43

jet quenching, 23

kaon, 35

latent heat, 38, 69, 258, 261
lattice
    cell size, 294
    continuum limit, 294
    critical temperature, 301
    domain-wall fermions, 292
    dynamic quarks, 289
    energy density, 300
    infrared cutoff, 289
    mass of strange hadrons, 299
    mass of strange quarks, 299
    naive quark action, 291
    plaquette, 290
    pressure, 301
      with staggered fermions, 302
    procedure for simulations, 294, 295
    QCD action, 289
    quark mass, 298
    quenched quarks, 289, 296
    running coupling constant, 298, 299
    scaling violation, 294
    staggered quark action, 292
    ultraviolet cutoff, 289
    Wilson action, 290, 291
lepton, 6
leptos , 1
level density
    N-particle, 241
    scattering phase shift, 242
    single-particle, 61
LHC
    charge multiplicity, 181
Lorentz
    boosts, 83
    contraction, 82
    covariant gauge, 275
    invariant spectra, 132, 139

Mandelstam variables, 319

mass thermal, 305
matter–antimatter, 182
    symmetry, 4
Maxwell construction, 49
meson, 26, 27
    ϕ, 36
    mesos , 1
    strange, 25

omega, 34
    chemical equilibration, 349
    freeze-out temperature, 361
    production of decay in strangelets, 223
OSCAR, 100

pair production
    perturbative, 326
    Schwinger mechanism, 40
particle
    density, 194
      glue fireball, 119
      phase-space distribution, 317
    energy per particle, 198
    ensemble, 192
    indistinguishable, 55
    momentum, 81
    production, 95
      $\propto e^{-2m/T}$, $\propto e^{-m/T}$, 228
    secondaries, 81
    spectra, 62
      pseudorapidity, 137
      scaling solution, 128
      thermal, 135
    surface emission, 143
    temperature, 152
particle ratios
    antiproton to proton, 182
    chemical fugacities, 218
    in A–A collisions at $14A$ GeV, 170
    in A–A collisions at $200A$ GeV, 169
    in Pb–Pb collisions at $\sqrt{s_{NN}} = 130$ GeV, 367
    in Pb–Pb collisions at $158A$ GeV, 359

strange baryon–antibaryon, 160, 164
partition function
    Boltzmann, 61
    canonical, 56, 224
        generating function, 58
    grand-canonical, 57
        quantum, 61
    multicomponent system, 193
    pressure ensemble, 190
    quantum, 59
    strange particles, 217
    vacuum, 69
parton
    cascade, 102
    thermalization, 23
path integral, 288
    Fermi determinant, 293
phase
    crossover, 53
    diagram, 46, 47, 49, 50
    metastable, 51
    mixed, 49
    transition, 53
        change in g, 9
        quark-mass dependence, 53, 54
phase space
    N-particle volume element, 242
    Coulomb distortion, 215
    entropy content
        single-particle, 113
    integral, 197
    Lorentz-invariant, 139
    occupancy
        kinetic evolution, 323
    quantum particle, 193
phase transition
    early Universe, 9, 14
    finite-volume, 238
    fluctuations, 126
photon
    density, 63
    direct from QGP, 23
    production, 24
pion
    excess, 167, 168

gas, 124
    properties, 125–127
    production
        enhancement, 127
    suppression, 167
    yield
        charge asymmetry, 90
plasma
    electron–ion, 54
Podolanski–Armenteros analysis, 29, 30
pressure, 188
    critical
        early Universe, 10
    effect of particle mass, 68, 199
    hadronic gas, 209, 211
    thermal, 306
pseudorapidity, 85
    error, 87, 89
    particle emission angle, 85
    particle energy and momentum, 136
    rapidity, 86–88

QCD, 267
    K-factor
        flavor production, 330
    Λ parameter, 283
    asymptotic freedom, 278
    charge definition, 279
    color-magnetic instability, 40
    critical temperature, 301
    Feynman diagrams, 277, 326
    Lagrangian, 273
    lattice action, 289
    lattice formulation, 287
    lattice pressure, 301
    perturbative, 38, 274
    Polyakov loop, 296, 297
    renormalization, 278
        group, 280
        particle spin, 279
    running, 282
        β and γ functions, 281
    running α_s
        initial conditions, 284

sum rules, 39, 45
temperature dependence of α_s, 286
thermal Feynman diagrams, 304
transmutation of scales, 279
two-loop α_s, 286
value of Λ, 283
QED instability, 40
QGP
  energy density, 315
  entropy density, 315
  pressure, 307
quark, 6, 24, 38
  bag model, 38, 262
  cavity state, 260
  charge, 6
  chiral condensate, 297
  chiral symmetry, 44
  cold quark matter, 47
  confinement, 38
  degeneracy, 53
  energy in the bag, 266
  flavors, 6
  free, 53, 54
  ideal-gas density, 204
  Lagrangian, 273
  lattice action, 291
  mass, 7
    massless limit, 45
    phase structure, 53, 54
    running, 282, 283, 328
    sum rules, 46
  pairing, 47
  production
    running threshold, 284
  strange, 8
  sum rules, 45
  thermal mass, 305
quark density
  statistical equilibrium density, 325
quark–gluon liquid, 70, 306
quark–gluon plasma
  B_c formation, 37
  comparison of signatures, 24
  critical temperature, 10

degeneracy, 53
degrees of freedom, 52
energy density, 53, 310
equation of state, 303
equations of state, 310, 312, 313
evidence, 162
  phase-space enhancement, 363–365
formation, 153
hadronization at the RHIC, 367
in the early Universe, 5
negative pressure, 52
observability, 15, 16
partition function, 70, 304
perturbative QCD interactions, 70, 304
phase diagram, 48
strange-antibaryon signature, 351
strange-particle signature, 18
sudden hadronization, 52
super-cooling
  mechanical instability, 52
thermodynamic potential, 310
undercooled, 51
yield of strangeness, 362, 363
quasi-particle, 59, 305
quasirapidity distribution
  protons and kaons, 89

rapidity, 82
$\overline{\Lambda}$ and Λ spectra in S–S collisions, 144
asymmetric systems, 84
baryon-poor region, 184
CM reflection, 144
fragmentation region, 145
gap, 72, 73
negative-hadron distribution, 165
particle spectra, 83
  scaling solution, 128
pseudorapidity, 86–88
velocity relation, 82
reference frame
  center of momentum, 79
relaxation time, 98
  electro-weak interactions, 92

production of strangeness, 336
thermal production of charm, 337
RHIC
    charged-hadron production, 180,
    182
    first results, 178
Riemann eta and zeta functions, 200

search for strangelets, 223
spectra
    inverse transverse slope
        system size, 152
    Lorentz-invariant, 139
    pseudorapidity, 136, 138
    rapidity
        'net' baryons, 146
        massless QGP quanta, 145
        schematic representation, 147
        three-fluid model, 146
    rapidity window, 133
    strange hadron
        inverse transverse slope, 153
    thermal, 135
    thermal fit
        statistical parameters, 157
    transverse mass, 148
        Λ, $\Omega + \overline{\Omega}$, 155
        π^0 and η, 151
        strange particles, 149, 150
        strange-particle analysis, 154,
    155
spin–spin interaction, 40
statistical bootstrap, 244
    cluster formation, 256
    critical behavior, 254
    critical curve, 255
    hypothesis, 243
    idea, 244
    model, 247, 252
    physical interpretation, 257
    singularity, 246
statistical ensemble, 191
statistical hadronization, 352
    enhancement of occupancy of phase
    space, 363
    excess of omega, 360

RHIC, 366
statistical mechanics
    covariant formulation, 248
statistical significance, 358
Stefan–Boltzmann law, 52, 69
strange antibaryons
    signature of deconfinement, 160,
    164, 351
strange hadron
    inverse transverse slope, 153
strange particle
    non-statistical yield, 157
    spectra, 19
strangeness
    ϕ, 153
    abundance at the RHIC, 184
    chemical equilibrium, 98
    chemical-relaxation time, 335,
    336
    conservation, 222
        canonical, 223
        canonical QGP and HG, 229
    distillation, 222
    enhancement, 19, 160, 163, 164
    exchange reaction, 344, 345
    excitation function, 16, 17
    hyperon yield, 31, 32
    in baryons and mesons, 25
    kinetic evolution at the RHIC,
    338
    lattice quark mass, 299
    negative chemical potential, 221,
    222
    observables, 171
    Okubo–Zweig–Izuka rule, 344
    particle decays, 81
    partition function, 217
    phase-space occupancy, 339, 364,
    365
        enhancement, 363
    production, 299, 326, 327
        glue equilibration, 340
        SPS rapidity distribution, 159
    production in HG, 343, 344, 348,
    351
    QGP signature, 18

quark mass, 8, 258
relative equilibrium, 91
signature of QGP, 15, 24, 160, 164
symmetry in QGP, 214
thermal production, 332, 334
Wróblewski factor, 340, 341
yield at SPS, 159
yield at the RHIC, 369
yield in QGP, 98, 362, 363
streamer chamber, 81
$SU(3)$
adjoint representation, 270
Gell-Mann representation, 269

temperature, 188
evolution, 94
glue fireball, 118
Hagedorn, 53
inverse slope, 56
local, 90
transverse slope, 152
thermal
collision frequency, 320, 331
equilibration, 93
equilibrium, 90
Feynman diagrams, 304
particle spectra, 135
expanding fireball, 141
pressure, 306
QCD energy scale, 285
reaction rate, 318
reactivity, 318
thermal mass, 305, 308
thermalization
transport models, 104

thermodynamic potential
quark–gluon liquid, 310
three-body reactions, 349
transverse energy, 177
distribution, 177, 178
per charged particle, 179
pseudorapidity density, 180
scaling with A, 178
transverse mass, 82, 148
transverse-momentum acceptance,
134

upsilonium, 37

vacuum
energy, 41
density, 42
zero-point, 41
instability, 41
latent heat, 38, 261
partition function, 69
perturbative, 38
polarization, 277
restoration of symmetry, 46
structure, 258
true, 38
velocity
cylindrical representation, 139
relative, 319
sound, 11, 107

Wall Street, 63
Wilson action, 291

Yang–Mills fields, 271

zeta function, 200

Printed in the United States
by Baker & Taylor Publisher Services